Special Publication No. 5
of the Society for Geology
Applied to Mineral Deposits

W0017687

Base Metal Sulfide Deposits

in Sedimentary and Volcanic Environments

Proceedings of the DMG-GDMB-SGA-Meeting
Aachen, 1985

Edited by
Günther H. Friedrich and
Peter M. Herzig

With 155 Figures

Springer-Verlag Berlin Heidelberg GmbH

Prof. Dr. GÜNTHER H. FRIEDRICH
Dr. PETER M. HERZIG
Institute of Mineralogy and Geology of Ore Deposits
RWTH Aachen
Wüllnerstraße 2
D-5100 Aachen, FRG

ISBN 978-3-662-02540-6   ISBN 978-3-662-02538-3 (eBook)
DOI 10.1007/978-3-662-02538-3

© Springer-Verlag Berlin Heidelberg 1988
Originally published by Springer-Verlag Berlin Heidelberg New York 1988.

Typesetting: Overseas Typographers, Inc., Makati, Philippines; printing: Druckhaus Beltz,
Hemsbach/Bergstraße; binding: J. Schäffer GmbH & Co. KG., Grünstadt
2132/3130-543210

Preface

Research on base metal sulfide deposits is among the oldest and best-documented subjects of economic geologists worldwide, considering that copper was first mined about 3000 years ago on the island of Cyprus.

During the past 10 years, after the exciting discovery of active sulfide formation on the modern ocean floor, a considerable flow of new ideas has stimulated and influenced the discussion of ore-forming processes for copper-zinc-lead sulfides in sedimentary and volcanic environments. The development of new genetic concepts consequently led to reinterpretation of some apparently well-established formation models.

This Proceedings Volume contains a collection of carefully selected papers on current research on the geology and metallogeny of base metal sulfide deposits presented as oral or poster contributions at the DMG (Deutsche Mineralogische Gesellschaft) – GDMB (Gesellschaft Deutscher Metallhütten- und Bergleute – Fachsektion Lagerstättenforschung) – SGA (Society of Geology Applied to Mineral Deposits) Joint Meeting on Ore Deposits in Aachen, Federal Republic of Germany, September 16–19, 1985.

Base metal sulfide deposits with different ore compositions occur in a wide variety of geological and lithological settings of almost any age. This has been taken into account in organizing this volume along the lines of classical host-rock classification.

The first group of contributions focuses on sediment-hosted base metal sulfide deposits including examples of Kupferschiefer and Copperbelt-type, as well as lead-zinc mineralization in carbonate host rocks.

Volcanic- and volcanic-sediment-hosted base metal sulfides are covered by a second group of chapters describing research on hydrothermal Cyprus and Kuroko-type, Iberian Pyrite Belt and Turkish copper deposits, in addition to sulfide ore deposits in alpine volcano-sedimentary sequences.

Emphasizing lithological, structural, chemical, and temporal controls, research results and interpretations given in the presented compilation of papers will hopefully stimulate further discussion on the formation of base metal sulfide deposits – especially in the light of research work currently being done in the natural laboratories on the seafloor.

Aachen, February 1988   GÜNTHER FRIEDRICH
                                    PETER HERZIG

Contents

**Part II   Volcanic- and Volcanic-Sediment-Hosted Base
Metal Sulfide Deposits**

Contributors

You will find the addresses at the beginning of the respective contribution

Part I Sediment-Hosted Base Metal Sulfide Deposits

The Evaluation of Sedimentary Basins for Massive Sulfide Mineralization

D. LARGE[1]

Abstract

The geological setting of sediment-hosted massive sulfide deposits is reviewed in the light of new concepts on the evolution of sedimentary basins. It is shown that the sediment-hosted massive sulfide deposits are hosted by extensional basins and were formed during the post-rift phase of thermal subsidence. The mineralization event is associated with a distinct tensional pulse that is superimposed on the regional thermal subsidence of the basin and can be recognized in the stratigraphy by locally developed sedimentological indications of rapid subsidence. Basin analysis can be used to identify prospective basins and the target stratigraphy in the preliminary analysis of sedimentary sequences for sediment-hosted massive sulfide mineralization.

Introduction

Previous reviews on the geological setting of sediment-hosted massive sulfide Zn-Pb deposits have tended to identify particular geological features within a sedimentary basin, such as proximity to faults and lineaments, facies and thickness changes that may mark the transition from the shelf to basin, and the presence of "second-order" basins (Large 1980, 1983). The purpose of this paper is to reconsider the nature of those sedimentary basins, which are host to sediment-hosted massive sulfide deposits, in the light of more recent concepts on basin classification. Furthermore, an attempt will be made to show that the mineralization is associated with a predictable phase in the evolution of a particular class of sedimentary basin.

Classification of Sedimentary Basins

Bally (1982) defined a sedimentary basin as an area of subsidence that contains a sedimentary sequence of at least 1 km thickness. A simple but effective two-way classification of basins can be made on the basis of their development (Beaumont et al. 1982):

    1. basins formed as a result of crustal attenuation and extensional tectonics — these include both intracratonic rifts and Atlantic-type passive-margin basins, and

[1]Consulting Geologist, Paracelsus Strasse 40, D-3300 Braunschweig, West Germany

Base Metal Sulfide Deposits
G.H. Friedrich, P.M. Herzig (Eds.)
© Springer-Verlag Berlin Heidelberg 1988

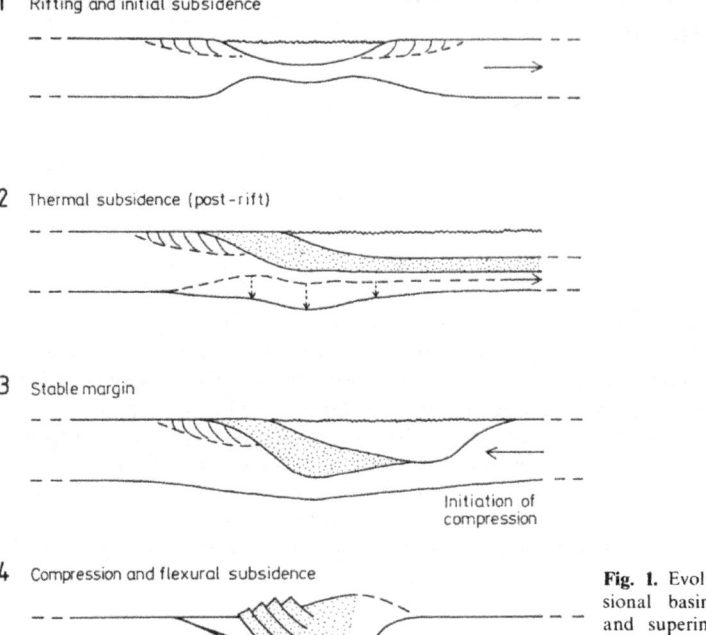

1 Rifting and initial subsidence

2 Thermal subsidence (post-rift)

3 Stable margin

                                                              Initiation of
                                                              compression

4 Compression and flexural subsidence

Fig. 1. Evolution of a tensional basin (stages 1-2) and superimposition of a foreland basin during subsequent compressional tectonism (stage 4). (After Beaumont et al. 1982)

2. basins formed as a result of tectonic and/or sediment-loading in a compressional tectonic regime – these include the foreland basins in thrust-belts.

Beaumont et al. (1982) noted that, during a tectonic cycle, a foreland-basin can be derived from and superimposed on an extensional basin (Fig. 1). Some of the geological characteristics of the foreland- and rift-type basins are summarized in Table 1.

Extensional Basins

Basins that were formed through crustal attenuation are characterised by two distinct phases in their development; a "syn-rift" and a "post-rift" event (Watts 1981).

The initial "syn-rift" phase occurs during crustal stretching and attenuation, and is marked by an enhanced geothermal gradient (McKenzie 1978; Watts et al. 1982). It is generally characterized by listric-normal faulting and rapid subsidence of graben and uplift of horst blocks. Sedimentation during the "syn-rift" phase is usually immature and consists of coarse clastics, as well as lacustrine and evaporite

Table 1. Comparison of geological features of compressional (foreland) and tensional basins

	Compressional	Tensional
Max. depth	ca. 6 km	ca. 16 km
Shape	Wedge-shaped, convex flexure	Monoclinal or graben
Setting	Bordered by fold-thrust belt	Passive cratonic margin or intracratonic
Rates of subsidence	Variable, rapid and quiescent	Rapid followed by exponential decay
Basement topography	Smooth	Rugged and faulted
Crustal structure	Uniform	Attenuated
Regional gravity	Major anomalies	Minor, isostatic equilibrium
Sediment lithology	Molasse and flysch, coal & black shale	Variable, pelites sandstones, limestones, deltaic
Thermal gradient	Normal	Enhanced

sequences. Due to the active tectonism, abrupt lateral sedimentary facies and thickness changes are common.

Rifting can proceed to crustal rupture, which results in the formation of oceanic crust and a passive continental margin (e.g., Fig. 1, Stage 2), or it may cease prior to rupture, which results in the formation of an ensialic intracratonic graben (Fig. 1, Stage 1).

The subsequent "post-rift" phase represents the period of crustal cooling and a gradual return to thermal equilibrium of the attenuated continental crust at the passive margin or intracratonic graben formed during the rifting. The rate of subsidence during the "post-rift" phase decreases exponentially with time. The sedimentation during the "post-rift" event is characterised by mature, fine- to medium-grained clastics and carbonates. In the ideal model situation (e.g., a starved basin), the lack of tectonism results in laterally persistent sedimentary facies and thicknesses. However, the facies distribution during the "post-rift" phase is often further complicated by other paleogeographic factors; for example, the input of clastics from deltaic systems and the growth of carbonate build-ups. Furthermore, regional changes in sedimentation due to any epeirogenic events may be super-imposed on the sedimentological development during this phase of subsidence.

By using the technique of "back-stripping" (Watts 1981), the amount of subsidence due to tensional tectonism can be distinguished from subsidence due to other factors such as sediment-load (including contained pore water) and height of the prevailing water column. In order to undertake the back-stripping procedure, it is essential that the following parameters must be accurately known: absolute age and thickness of the different lithologies in the package, paleobathymetry during deposition, and sediment porosity. The use of back-stripping is thus effectively restricted to the palaeontologically well-defined sequences of the Cenozoic and Mesozoic (e.g., Watts 1981), although it has been attempted for the Paleozoic (e.g. Bond and Kominz 1984; Armin and Mayer 1983).

Foreland Basins

The subsidence history of foreland basins formed in a compressive tectonic regime is generally less predictable than that for tensional basins. The subsidence in the basin generally occurs in a series of pulses, each one of which reflects the emplacement of thrust sheets and loading of the basin margin. Sediment fills the basin between each pulse and unconformities are common within the basin-fill succession (Beaumont et al. 1982). Synsedimentary faulting is not common, since subsidence occurs as a result of crustal flexure beneath the load of sediment and tectonic thrust sheets. In the central European context, the synorogenic graywacke flysch derived from the uplifted Hercynian orogen and its interdigitation with Carboniferous coal measures represents a typical foreland basin sequence formed during the compressional tectonic regime of the Hercynian orogeny (Engel et al. 1983).

Geological Setting of Sediment-Hosted Massive Sulfide Mineralization

It was previously noted by Large (1980) that sediment-hosted massive sulfide mineralization is commonly associated with host rocks that display sedimentological features indicative of penecontemporaneous vertical tectonism (e.g., intra-formational conglomerates within the host sequence, sedimentary slump and breccia structures, sedimentary facies and thickness variations). The development of the so-called second- and third-order basins was related to this tectonic event.

From detailed studies of specific deposits, it has been proposed that the mineralization occurred during a tensional tectonic event.

— Boyce et al. (1983) and Andrew (1986) have documented a period of rapid subsidence related to an extensional tectonic pulse that occurred during the formation of the Silvermines and Tynagh carbonate-hosted massive sulfide deposits in Ireland.

— Sawkins and Burke (1980) suggested that the mineralization at Meggen and Rammelsberg occurred during a Devonian extensional event. The documentation of this event with respect to these massive sulfide deposits has been refined by Large (1986, and in prep.), who proposed that the mineralization occurred during a tensional pulse superimposed on the post-rift phase of thermal subsidence that occurred subsequent to an early Devonian syn-rift event.

— For the Paleozoic sediment-hosted massive sulfide deposits in the Selwyn Basin, Canada, Lydon et al. (1985) have shown that the mineralization occurred at a time of vertical tectonism. This vertical tectonism was probably related to an extensional event (Abbott 1986), that resulted in the formation of horst and graben structures (equivalent to second-order basins in the previous studies), as well as intraformational conglomerates and chert-pebble conglomerates and abrupt facies and thickness variations (Gordey et al. 1981, 1982).

It is only by studying the complete stratigraphic development of the sequence hosting the mineralization that the evolutionary history and the type of sedimentary basin can be determined.

A summary-stratigraphy of some basins that host SMS mineralization is given in Table 2. An attempt has been made to interpret the stratigraphy in terms of the

Table 2. Evolution of selected basins that contain sediment-hosted massive sulfide mineralization

	Hercynian, Germany	Selwyn Basin, Canada	Belt Basin, Canada	Central Irish Basin
Stages in basin evolution				
Flysch Lithology	Graywacke-turbidite	Clastics	Not seen	Shales, deltaic sandstone
Age	L. Carboniferous	Permo-Triassic		Dinantian-Namurian
Post-rift Lithology	Dark grey shales grading up into calcareous red-green shales	Shales, cherts minor turbidites "starved basin" condensed sequence	Pelites and gray-wackes (deep marine) grading up into argillites and argillaceous dolomites (intertidal)	Shelf carbonates argillaceous and oolitic limestone, "Waulsortian" mud mounds
Age	M. - U. Devonian	Cambrian - Mississippian	M. Proterozoic	L. Carboniferous
Volcanism	Mafic volcanics, tuffs	Minor mafic volcs (Silurian)	Mafic sills in Sullivan area	Alkali-basalt volcanics (minor)
Mineralization	Rammelsberg Meggen	Howards Pass (Silurian) Tom, Jason (Devonian)	Sullivan	Silvermines, Navan Tynagh
Syn-rift Lithology	Sandstones and siltstones	Arkoses, arenites	Coarse clastics (Fort Steele Fm.)	Siliciclastics "Old Red Sandstone"
Age	L. Devonian	U. Proterozoic	Proterozoic	U. Devonian
Reference	Engel et al. (1983) Langenstrassen (1983)	Gordey et al. (1981, 1982) Abbott (1986)	Hoy (1982)	Andrew (1986) Boyce et al. (1983)

evolution of a tensional basin, and syn-rift and post-rift units have been identified. In some cases the flysch related to the subsequent compressional tectonic regime can also be identified.

It can be seen that the mineralization is hosted by sequences that were deposited during the post-rift phase of basin evolution. The tectonism to which the mineralization is temporally related is superimposed on the otherwise tectonically inactive, regional subsidence that is characteristic of the post-rift phase (Fig. 2). In those basins with a long history of post-rift subsidence, there can be more than one phase of extensional tectonism and associated sediment-hosted massive sulfide mineralization (e.g., the Selwyn Basin). The initial syn-rift phase of rapid subsidence and bimodal volcanism may be associated with other styles of mineralization (e.g., Pb-Zn veins, some stratabound Cu deposits, epithermal precious metal deposits).

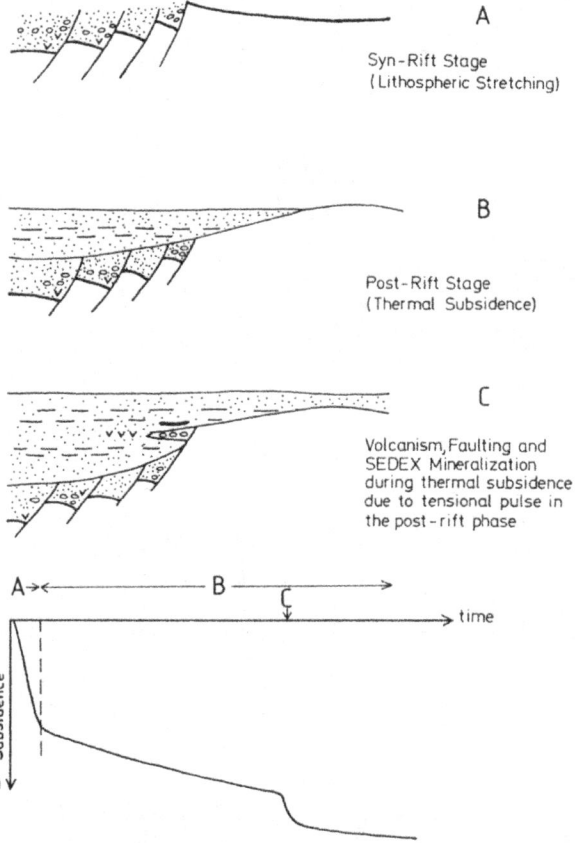

A

Syn-Rift Stage
(Lithospheric Stretching)

B

Post-Rift Stage
(Thermal Subsidence)

C

Volcanism, Faulting and
SEDEX Mineralization
during thermal subsidence
due to tensional pulse in
the post-rift phase

Evolution of a sedimentary basin during syn-rift and post-rift subsidence phases

Fig. 2. Sediment-hosted massive sulfide mineralization during the evolution of an extensional basin — a model

Discussion

The recognition of the association between sediment-hosted massive sulfide mineralization and a particular tensional tectonic event during the post-rift phase of the host sedimentary basin (Fig. 2) explains some of the features that have been previously noted as empirical parameters (Large 1983).

To date, no single sedimentary environment or lithology has been recognized as being particularly favorable for sediment-hosted massive sulfide mineralization. It is suggested that this is a reflection of the causal association of the mineralization with the extensional tectonic pulse during the post-rift subsidence phase. The mineralization will therefore be imposed on the prevailing sedimentary environment. There is no reason to expect any particular regional sedimentary environment to be developed at the time of the extensional tectonic pulse, although local facies and lithologies will characterise the synsedimentary tectonism (see next paragraph). The nature of the host lithologies and their reaction with the metal-bearing hydrothermal solutions will, however, influence the style of the mineralization (e.g., the common replacement and cavity-fill mineralization in carbonate-hosted deposits). The prevailing sedimentary environment will also largely determine if the precipitation products from the hydrothermal solution will be preserved as an orebody on the sea floor (e.g., the finely laminated synsedimentary sulfides preserved in deposits hosted by fine-grained clastics) or be dispersed into the seawater in a turbulent, oxidizing milieu.

The extensional pulse results in the sedimentological features that were previously noted. In particular the development of the so-called second- and third-order basins that were recognized as being associated with sediment-hosted massive sulfide mineralization can be related to the vertical tectonism during the extensional pulse. The observed sedimentary facies and thickness variations are also a natural consequence of this tectonic activity, as are the conglomerates and sedimentary slumps and breccias.

Extension and the concurrent crustal thinning during the post-rift extensional pulse will also be associated with an increased geothermal gradient. The volcanism that often occurs at the time of mineralization [e.g., the basic (MORB) volcanics extruded at the time of mineralization in the Hercynian, and possibly also the intrusion of hypabyssal mafic sills in the Sullivan area] is probably related to the tapping of mantle-derived magma through deeply penetrating fractures. Fractionation may result in the minor volumes of more acidic tuffs that are found associated with some sediment-hosted massive sulfide deposits.

Lydon et al. (1985) and Large (1986) suggested that the basal syn-rift clastic sequence would be an ideal reservoir for the formation of metal-bearing formation waters. The syn-rift clastic basins are confined by faults, and the overlying shale sequence of the post-rift succession acts as a cap rock for the formation waters trapped in the syn-rift clastics. The common presence of evaporites within some syn-rift sequences would provide a source for the chloride brines that leach the metals from the clastics in the reservoir (Lydon 1983).

Many sediment-hosted massive sulfide deposits are spatially associated with faults that are considered to have been active at the time of mineralization. It is possible that the faults active during the syn-rift phase of basin evolution were reactivated during the post-rift event. Such faults penetrate into the underlying syn-rift clastic sequence and would provide the ideal pathway for overpressured metal-bearing brines, contained within these permeable clastics, to be expelled to the surface.

A spatial relationship between mineralization and "hinge-lines" has also been noted — the hinge-lines being zones of abrupt sedimentary facies and thickness variations in the host succession. Watts (1981) defines a hinge-line as the zone that marks the boundary between normal continental crust and attenuated crust in an extensional basin. Such a boundary is a likely site for syn- and post-rift faulting and as such an important structure in basin analysis for sediment-hosted massive sulfide deposits. It can also be assumed that mineralization is unlikely to occur on the cratonic side (i.e., over nonattenuated continental crust) of the hinge zone. However, note that the exact location of the hinge-line may be difficult to determine due to the masking effect by the post-rift sedimentary succession.

In several basins (e.g., Selwyn Basin — Lydon et al. 1985; the Belt Basin containing Sullivan — Kanasewich 1968; Hoy 1982; the Hercynian Basin — Ziegler 1982) that contain sediment-hosted massive sulfide deposits, the mineralization is located within an area that marks the intersection of the regional trend of the passive cratonic margin with a major crosscutting structure that penetrates into the craton and, in some cases, has been interpreted as an aulacogen graben. The area of intersection between the aulacogen and the regional trend would be characterized by an anomalously high geothermal gradient (hot-spot) and crustal attenuation during both the syn-rift phase and the extensional pulse of the post-rift phase.

Conclusions

Sediment-hosted massive sulfide mineralization is associated with a predictable stage in the evolution of extensional basins. This stage is marked by sedimentological and lithological indications of rapid subsidence and volcanic activity that accompanied a tensional pulse during post-rift subsidence, subsequent to the initial syn-rift event. The syn-rift clastics, capped by impermeable post-rift shales, are considered to represent an ideal source reservoir for the hydrothermal solutions.

The recognition of the close association between the mineralizing event and a particular stage in the evolution of sedimentary basins can be used in the preliminary analysis of sedimentary basins for their potential of hosting such mineralization. Further studies should be undertaken to determine if this event can be predicted by back-stripping techniques, and if other styles of sediment-hosted mineralization are restricted to particular phases in basin evolution.

Acknowledgment. The author is grateful to the editors for making useful suggestions for improving the manuscript.

References

Abbott JG (1986) Devonian extension and wrench tectonics near Macmillan Pass, Yukon Territory, Canada. In: Turner RJW, Einaudi MT (eds) The genesis of stratiform sediment-hosted lead and zinc deposits, Conference Proceedings, Stanford University Publications. Geol Sci 20:85–89

Andrew CJ (1986) Sedimentation, tectonism, and mineralization in the Irish Orefield. In: Turner RJW, Einaudi MT (eds) The genesis of stratiform sediment-hosted lead and zinc deposits, Conference Proceedings, Stanford University Publications. Geol Sci 20:44–50

Armin RA, Mayer L (1983) Subsidence analysis of the Cordilleran miogeocline: Implications for timing of late Proterozoic rifting and amount of extension. Geology (Boulder) 11:702–705

Bally AW (1982) Musings over sedimentary basin evolution. Philos Trans R Soc Lond A Math Phys Sci 305:325–338

Beaumont C, Keen CE, Boutilier R (1982) A comparison of foreland and rift margin sedimentary basins. Philos Trans R Soc Lond A Math Phys Sci 305:295–317

Bond GC, Kominz MA (1984) Construction of tectonic subsidence curves for the early Paleozoic miogeocline, southern Canadian Rocky Mountains: Implications for subsidence mechanisms, age of breakup, and crustal thinning. Geol Soc Am Bull 95:155–173

Boyce AJ, Anderton R, Russell MJ (1983) Rapid subsidence and early Carboniferous base-metal mineralization in Ireland. Trans Inst Miner Metall (Sect B Appl earth sci) 92:55–66

Engel W, Franke W, Langenstrassen F (1983) Palaeozoic sedimentation in the northern branch of the mid-European Variscides — essay of interpretation. In: Martin H, Eder FW (ed) Intracontinental Fold Belts, Springer, Berlin Heidelberg New York, pp 9–41

Gordey SP, Wood D, Anderson RG (1981) Stratigraphic framework of southeastern Selwyn Basin, Nahanni Map area, Yukon Territory and District of Mackenzie. Curr Res Part A Geol Surv Can Pap 81-1A:395–398

Gordey SP, Abbott JG, Orchard MJ (1982) Devono-Mississippian (Earn Group) and younger strata in east-central Yukon. Curr Res Part B Geol Surv Can Pap 82-1B:93–100

Hoy T (1982) Stratigraphic and structural setting of stratabound lead-zinc deposits in southeastern B.C. Can Inst Miner Metall 75:114–134

Kanasewich ER (1968) Precambrian rift: genesis of stratabound ore deposits. Science 161:1002–1005

Langenstrassen F (1983) Neritic sedimentation of the Lower and Middle Devonian in the Rheinische Schiefergebirge East of the Rhine River. In: Martin H, Eder FW (eds) Intracontinental Fold Belts. Springer, Berlin Heidelberg New York, pp 43–76

Large D (1980) Geological parameters associated with sediment-hosted, submarine exhalative Pb-Zn deposits: an empirical model for mineral exploration. Geol Jahrb D40:59–129

Large D (1983) Sediment-hosted massive sulphide lead-zinc deposits: an empirical model. In: Sangster DF (ed) Short course in sediment-hosted stratiform lead-zinc deposits. Miner Assoc Can Short Course 8:1–29

Large D (1986) The paleotectonic setting of Rammelsberg and Meggen, Germany — a basin analysis study. In: Turner RJW, Einaudi MT (eds) The genesis of stratiform sediment-hosted lead and zinc deposits, Conference Proceedings, Stanford University Publications. Geol Sci 20:109–112

Lydon JW (1983) Chemical parameters controlling the origin and deposition of sediment-hosted stratiform lead-zinc deposits. In: Sangster DF (ed) Short Course in Sediment-hosted stratiform lead-zinc deposits. Miner Assoc Can Short Course 8:175–250

Lydon JW, Goodfellow WD, Jonasson IR (1985) A general genetic model for stratiform baritic deposits of the Selwyn Basin, Yukon Territory and District of Mackenzie. Curr Res Part A Geol Surv Can Pap 85-1A:651–660

McKenzie D (1978) Some remarks on the development of sedimentary basins. Earth Planet Sci Lett 40:25–32

Sawkins FJ, Burke K (1980) Extensional tectonics and mid-Paleozoic massive sulfide occurrences in Europe. Geol Rundsch 69:349–360

Watts AB (1981) The U.S. Atlantic continental margin: subsidence history, crustal structure and thermal evolution. In: Geology of passive continental margins: history, structure and sedimentologic record (with emphasis on the Atlantic margin), Am Assoc Pet Geol Educat Course Note Ser 19:2-1 – 2-75

Watts AB, Karner GD, Steckler MS (1982) Lithospheric flexure and the evolution of sedimentary basins. Philos Trans R Soc Lond A Math Phys Sci 305:249–281

Ziegler PA (1982) Geological atlas of western and central Europe. Elsevier, Amsterdam

Relation of Permian Base Metal Occurrences to the Variscan Paleogeothermal Field of the Fore-Sudetic Monocline (Southwestern Poland)

S. SPECZIK[1]

Abstract

Fluid inclusion studies and vitrinite reflectance measurements were performed to establish temperatures of alteration processes that affected Carboniferous, older Paleozoic, and Precambrian basement rocks of the Fore-Sudetic Monocline of southwest Poland. Surprisingly high temperatures determined for these processes (ranging from 180 to 350°C), and the locally extensive alteration indicate the importance of geothermal influence. The high-temperature geothermal field of the Fore-Sudetic Monocline shows various heat anomalies related to paleohighs and strongly tectonically disturbed basement. These anomalies are spatially related to the areas of known Zechstein base metal occurrences of the Kupferschiefer type. It is suggested that one controlling factor leading to the formation of the Kupferschiefer mineralization was the availability of thermal energy. Mobilization and gradual preconcentration of base metals during the Late Variscan was due to anomalous heat flow related to Variscan plate motions. This Variscan heat flow is considered to be mainly convective.

Introduction

Numerous methods were used to establish temperatures of diagenetic or epigenetic processes that affected sedimentary rocks. Two of the methods – fluid inclusion studies and vitrinite rank determinations – have gained wide acceptance with respect to the basement of the Central European Variscides (Wolf 1978; Speczik 1979, 1985; Speczik et al. 1986). Diagenesis and epigenesis usually represent a continuous process with gradual generation of ore solutions and conduction of endogene heat. Therefore, the problem of distinguishing between diagenetic and epigenetic processes is principally of formal concern and has no real meaning for paleogeothermal interpretations.

Paleogeothermal observations are commonly used for tectono-magmatic discussions, since tectonophysical phenomena are believed to be related to the geothermal gradient and its variability at the contact of the lithosphere and the upper mantle. During most of the Carboniferous the Southern European continent was probably a rather thin, hot plate with a rigid crust, about 20 km thick. This area was

---

[1]Institute of Geology, The University of Warsaw, 02-089 Warsaw; al Zwirkii Wigury 93 Poland

---

Base Metal Sulfide Deposits
G.H. Friedrich, P.M. Herzig (Eds.)
© Springer-Verlag Berlin Heidelberg 1988

probably thermally active during Lower Permian time, as indicated by extensive Autunian volcanic activity. Therefore, the geothermal paleogradients of the Central European Variscides are generally thought to be elevated with the upper crustal rocks subjected to high temperature-low pressure metamorphism. (Zwart 1967; Lorenz 1976). The Rhenohercynian zone is distinguished by having the highest paleogeothermal gradient among the European Variscides.

It has already been established that all known Zechstein base metal occurrences of economic importance rest exclusively above the tectonically and thermally active transition zone between the Rhenohercynian and Saxothuringian of the European Variscides (Rentzsch 1981; Speczik et al. 1986). The availability of thermal energy is an important factor in the development of most ore-generating systems. Heat can stimulate and promote base metal mobilization and preconcentration processes, e.g., leaching of sedimentary and basement rocks by subsurface brines rich in chlorides that can occur as intrastratal, connate, and metamorphic waters (Rose 1976; Carpenter 1978; Bischoff et al. 1981). High heat flow was also determined to be effective in producing highly reactive gas mixtures rich in base metals from any type of black shale (Walker and Buchanan 1969). Similar processes can also form red-bed-type base metal deposits (Brown 1978; Berendsen and Speczik 1984).

The geothermal anomaly of southwestern Poland is interpreted by Majorowicz (1982) as being a deep Variscan thermal disturbance of the upper mantle. The Dolsk deep fracture zone (Fig. 1) is still believed to be active and produces remnant heat. Jowett and Jarvis (1984) argue that the present thermal anomaly of southwestern Poland is probably not a remnant from Permian times, but rather of Tertiary age and related to the Alpine orogeny. Nevertheless, the time span of virtual Variscan heat flow is the same as the time of base metal mobilization and preconcentration. The hydrothermal transport of leached metals to the surface by mostly chloride complexes seems to be a rational source for metals deposited in stratiform Zechstein occurrences.

This paper attempts to reevaluate the energetic factor of the above concept, and to correlate the position of known base metal occurrences to the Variscan paleogeothermal field of the Central European Variscides.

Geologic Overview

The Fore-Sudetic Region is located in southwestern Poland (Fig. 1) and is geologically and spatially equivalent to the Rhenohercynian and to some extent the Saxothuringian Zones of Central European Variscides. The southern part of the Fore-Sudetic Region — the Fore-Sudetic Block — is separated from the Sudetes by the Marginal Sudetic Fault, and from its northern part by the Middle Odra step- like fault structure. The geological structure of the northern part of the Fore-Sudetic Region can be conveniently discussed in terms of the folded and faulted Precambrian and Paleozoic basement, and gently dipping (from 2° to 5°C) Permian, Mesozoic, and Cenozoic sedimentary cover — the Fore-Sudetic Monocline.

The Fore-Sudetic Monocline is divided into southern and northern parts by the deep seated Dolsk fracture (Guterch et al. 1975). The border zone between the Fore-Sudetic Monocline and the Fore-Sudetic Block, called the Middle Odra

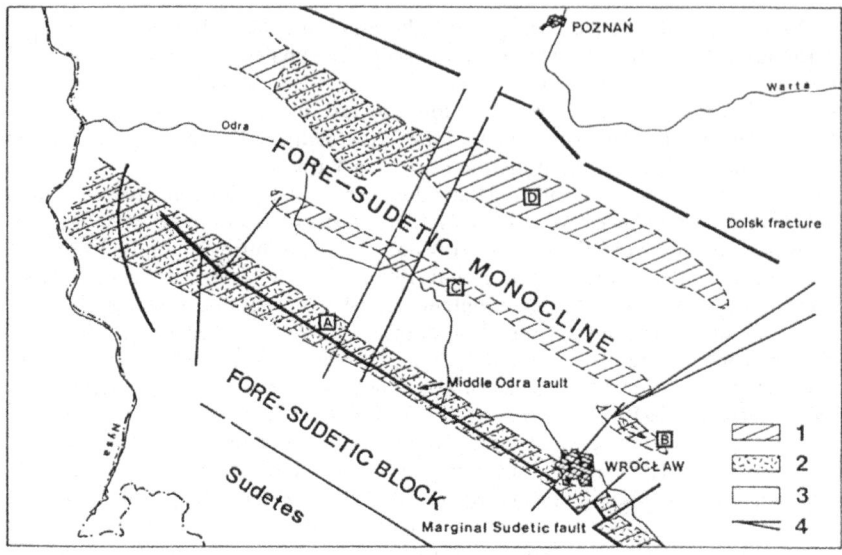

Fig. 1. Geologic-structural sketch map of the Fore-Sudetic Region. *1* Anticlinal structures in the basement of the Fore-Sudetic Monocline; *2* Precambrian and Early Paleozoic rocks; *3* Lower and Upper Carboniferous (only in the area of the Fore-Sudetic Monocline); *4* major faults. *A* Middle Odra Crystalline Zone; *B* Dobrzeń Unit; *C* Trzebnica-Bielawy Antycline; *D* Krotoszyn-Wolsztyn Uplift

Crystalline Zone, represents a transition belt between Rhenohercynian and Sax-othuringian Zones of the Central European Variscides (phyllite-schist zone – Germany). Precambrian and Early Paleozoic rocks occur in the inner parts of four anticlinal zones trending parallel (SE-NW) to the Middle Odra fault structure. At least two of those zones, the Middle Odra Crystalline Zone and Krotoszyn-Wolsztyn Uplift, are thought to represent paleohighs during the Zechstein time. Among the Precambrian and Early Paleozoic rocks numerous rock types are recognized: granites, gneisses, schists, hornfelses, and variously altered phyllites.

The basement rocks of the Fore-Sudetic Monocline are composed mainly of Lower and Upper Carboniferous rocks. These basement rocks consist of sandstones, graywackes, wackes, arenites and quartz arenites interbedded or intermixed with siltstones and shales with minor intercalations of conglomerates. In addition, convergent geotectonic processes were recorded by a long chain of Upper Car-boniferous granitoid and subvolcanic intrusions that trend parallel to the Middle Odra fracture zone. The Carboniferous rocks are variously tectonically disturbed, by folding and faulting. The intensity of folding decreases in a northerly direction. Carboniferous and old Paleozoic rocks of the Fore-Sudetic Monocline basement have been subjected to various types and intensities of alteration processes. Detailed petrographic and mineralogic studies of those rocks revealed that the intensity of the alteration processes increases toward the paleohighs and in the strongly tectonically disturbed regions. In these areas albitization, chloritization, silicification, hemati-tization, and carbonatization were observed. Intensity of alteration often increases with depth of examined strata. However, in many drilling profiles, variable patterns of alteration were observed. These patterns suggest that vertical, or in some cases horizontal flow of heat and fluids occurred that were parallel to abundant micro-tectonic structures. The direction of this flow is usually marked by a dense network of epigenetic veinlets that intersect Carboniferous strata.

Rotliegendes sedimentary and volcanic rocks were deposited in depressions of the basement and are overlain discordantly by Zechstein and Mesozoic strata, that are succeeded by Tertiary and Quaternary rocks. The huge Kupferschiefer type deposits of the Lubin-Sieroszowice area overlay highly metamorphosed rocks of the Middle Odra Crystalline Zone.

Fluid Inclusion Studies

Quartz, anhydrite, and minor sphalerite were used for both homogenization and decrepitation temperature determinations. Mineral-bearing fluid inclusions were selected during routine petrographic investigations of core material from more than 40 drilling profiles. These borings pierced mostly Carboniferous rocks, but in some regions (Krotoszyn-Wolsztyn Elevation, Middle Odra Crystalline Zone), early Paleozoic and Precambrian rocks, were also penetrated.

Fluid inclusions occur abundantly in epigenetic veinlets that cut basement rocks as well as in the adjacent host rocks (Fig. 2). Basement rocks in areas related to paleohighs and tectonically disturbed zones were pervasively penetrated by an extensive hydrothermal system that caused various alteration processes. The direction of hydrothermal metasomatism is marked and emphasized by a network system of secondary fluid inclusions which are arranged perpendicular or vertical to the bedded rock structure. Epigenetic veinlets cut basement rocks of the Fore-Sudetic Monocline at various angles from 20° to 90°. These veinlets are mainly 2 to 5 mm thick, although thicker veins (up to 5 cm) were observed. The texture of these veinlets is mainly crustification, crustification-banded, and collomorphic type, with crustification as well as and segment-type of filling. At the contact of these veinlets with the wall rock, signs of corrosion and metasomatic alteration were observed. Some of the veinlets were formed in one continuous cycle, while others were healed by a multistage process.

Most of the veinlets are filled by gangue minerals. Siderite, adularia, albite, chlorite, ankerite, dolomitic ankerite, quartz, dolomite, anhydrite, calcite, chal-cedony, kaolinite, barite, fluorite, and gypsum were recognized in order of prevailing paragenetic sequence. Ore minerals of hydrothermal veinlets are chalcopyrite, hematite, pyrite, pyrrhotite, and marcasite with minor bornite and tetrahedrite, and rarely other minerals (Fig. 3). Covellite, cuprite, azurite, and limonite were formed by supergene alteration of hydrothermal minerals. A number of veinlets had features that proved to be either of diagenetic or epigenetic formation, but the most common alteration was ambiguous or intermediate between diagenetic and epigenetic. These observations might be interpreted as confirmation of a continuous process of veinlet formation starting with diagenesis. The veinlets and signs of alteration ceased abruptly at the contact between Carboniferous and Lower Per-mian (Rotliegendes) strata. As the veinlets cut continually both Carboniferous and older formations, their Variscan (Sudetic or Asturian) age seems correct.

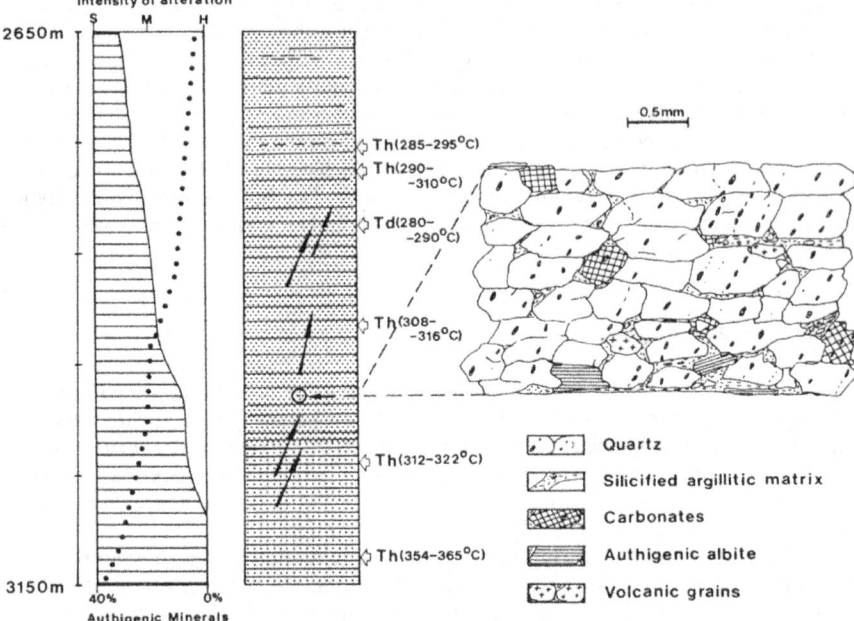

Fig. 2. Schematic diagram showing a network of secondary fluid inclusion associated with extensive alteration. (Profile Wycislowo IG-1). *Arrows* mark the direction of mineralizing fluids. Relative intensity of alteration: *S* Semi-mature stage; *M* mature stage; *H* hypermature stage. *Th* temperature of homogenization; *Td* temperature of decrepitation

Homogenization temperatures (Th) were measured with the microscope heating stage to a temperature accuracy equal to $+2°C$. Pt/Pt 10% Rh thermocouples were used. Homogenization temperature was measured at least twice for each inclusion, and only those results were accepted that were reproducible within the accuracy limit. The decrepitation method (visual observation in the microscope heating stage) was used only for the auxiliary determinations. The grain size used for decrepitation temperature (Td) determinations was 0.1 to 0.25 mm. The indicated Td values are temperatures for the beginning of abundant decrepitation. Sample preparation for fluid inclusion studies was very difficult, because of the small size of veinlets and the minute size of crystals and inclusions. Because of these size problems, chemical analysis of remnant liquid contained in the inclusions was impossible. However, the presence of liquid carbon dioxide, halides, common daughter and trapped minerals, and in several cases bituminous material, were stated by optical means. Homogenized inclusions from epigenetic veinlets had features that are exclusively of primary origin, while those found in detrital material were mostly of secondary (trapped) origin.

The temperatures of homogenization and decrepitation are very high, attaining values up to 400°C in some samples. Data of homogenization temperatures were already presented by Speczik (1979, 1985). All those results, as well as new data were

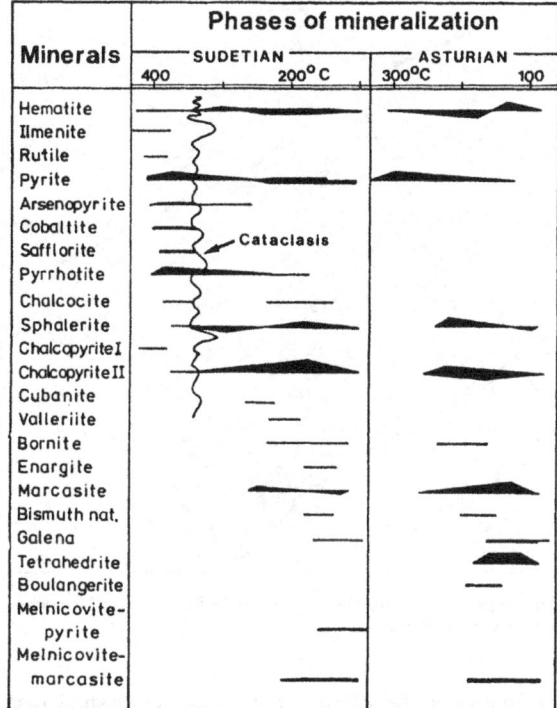

Fig. 3. Ore mineral succession and phases of mineralization

used to construct a map of paleoisotherms of the uppermost basement of the Fore-Sudetic Monocline (Fig. 4). The results do not relate to actual depth of the questioned rocks, and there is no clear correlation between the depth of sample and temperature of fluid inclusion homogenization or decrepitation in a particular drillhole. The latter problem caused difficulty in estimation of the paleogeothermal gradient in those drillholes where veinlet material was examined. In vertical profiles that were extensively and pervasively affected by hydrothermal metasomatic processes, determination of the paleogeothermal gradient was easy, e.g., in profiles Wycislowo IG-1 and Czeszow-4, where positive correlation exists between sample depth and temperature (cf. Fig. 2).

Vitrinite Rank Determination

The degree of metamorphism of organic matter changes during subsidence because of the increase in temperature. This metamorphism factor can be determined by different methods with examination by optical means — vitrinite reflectivity measurements — being the most common. Numerous research efforts are known to link rank of vitrinite with the geothermal gradient, temperatures of alteration processes, and burial history of sediments (Kontorowicz et al. 1967; Castano and Sparks 1974;

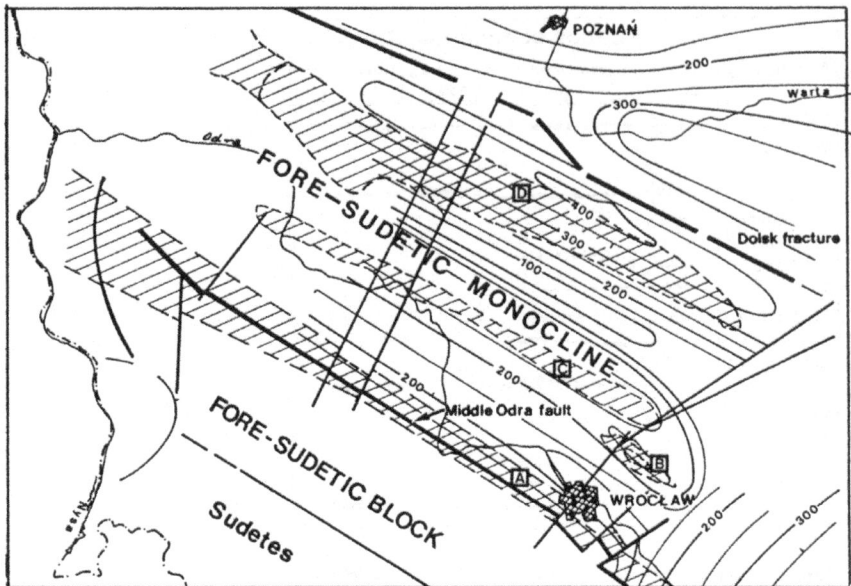

Fig. 4. Paleoisothermal map (°C) of the uppermost Carboniferous strata in the Fore-Sudetic Monocline based on fluid inclusion studies. Abbreviations as in Fig. 1

Ammosov and Utkina 1975; Bostic 1979). Results of these works clearly show that the temperature and sedimentary history (erosion and elevation processes) seem to be critical to interpretations of the change in vitrinite reflectance. Pore fluid chemistry does not appear to have any effect on coal rank, and the time factor can be neglected in sedimentary basins with rather long burial time such as the Paleozoic basin of the Fore-Sudetic Monocline. The burial-erosional history of the Fore-Sudetic Monocline was reconstructed based on the work by Majorowicz et al. (1984), and this study. Thus, the maximum burial depth was achieved in the Late Cretaceous. A deep burial depth also existed for the Late Carboniferous, especially in the southern part of the Fore-Sudetic Monocline. The maximum temperature for the paleodepth of about 2000 m was estimated according to various previous work, particularly that of Hood and Castano (1974).

Polished sections used in mineralogic studies were sorted to select those sections that contain vitrinite. The samples selected were mostly from Lower and Upper Carboniferous rocks with a few samples from Devonian rocks (e.g. Klepinka drillhole, Table 1). These rocks ranged in metamorphic alteration up to greenschist facies. Material suitable for investigations was found in only 18 of the 43 drillholes that were examined. The random (mean) reflectivity coefficient of vitrinite particles was measured on those samples. In some areas related to paleohighs or tectonically disturbed zones, e.g., the Middle Odra Crystalline Zone and the Trzebnica-Bielawy Anticline, the organic material was entirely expelled from the rocks due to extensive processes of alteration (especially hematitization). On the other hand, lack of

Table 1. Representative results of vitrinite reflectance determinations in the basement rocks of the Fore-Sudetic Monocline

Drilling profile	Stratigraphy	Depth (m)	Mean reflectance (%)	Drilling profile	Stratigraphy	Depth (m)	Mean reflectance (%)
Szymonkow IG-1	Lower Carboniferous	1096.0	1.7-1.8	Klepinka IG-1	Devon	623.1	2.8-3.0
		1141.5	1.6-1.7			652.1	2.4-2.6
		1161.4	2.8-3.0				
Czernczyce IG-1	Visen-Namur	1154.2	1.1-1.3	Brzostowo-1	Lower Carboniferous	1678.2	3.0-3.2
		1165.0	1.0-1.1				
Pogorzela-7	Upper Carboniferous	2206.8	2.2-2.5	Wrzesnia IG-1	Upper Carboniferous	5170.3	1.8-2.0
		2211.2	3.2-3.4			5205.3	2.8-3.0
						5313.1	3.2-3.4
						5497.4	2.6-2.8
Wolczyn IG-1	Lower and Upper Carboniferous	922.9	2.1-2.2	Siciny IG-1	Lower Carboniferous	2181.9	1.8-2.1
		1305.0	2.0-2.2			2265.2	1.3-1.4
		1514.1	2.8-3.0			2313.4	2.2-2.4
		2090.0	3.6-3.8				
		2148.4	3.6-3.8				
		2148.6	4.0-4.2	Dabcze-2	Westfal A	2202.1	2.6-2.8
Sieciejow IG-1	Turney	1114.6	3.2-3.4	Czeszow-4	Namur B,C	1818.2	3.0-3.2

vitrinite particles in some regions (Krotoszyn-Wolsztyn Elevation), contrasts the abundance of high rank inertinite grains. This might suggest an oxidizing environment of observed epimetamorphic processes. Published results of vitrinite reflectivity measurements of the Fore-Sudetic Monocline (Wilczek 1982; Speczik 1985) as well as those presented here (Tab. 1) clearly show that rank of vitrinite in the basement rocks of the Fore-Sudetic Monocline is generally higher than 2% R_{oil}. In some polished sections the values are as high as 4–5% R_{oil}.

All of these results do not correlate with the present geothermal conditions or in some cases the maximum burial depth of the examined rocks, but this might point to a variable pattern of Late Variscan heat flow. A better correlation between the depth of the sample and the measured vitrinite reflectivity was established when the vitrinite reflectance was compared with the results of fluid inclusion studies. The areas showing vitrinite reflectivity < 1.5% R_{oil}, within a range from 1.5% to 2% R_{oil}, and >2% R_{oil} were plotted on a geologic-structural map of the Fore-Sudetic Monocline (Fig. 5). There are no indications (except for Middle Odra Crystalline Zone), that the analyzed vitrinite was in direct contact with the magmatic bodies. The presented data are approximate in view of the uncertainties in the data on burial-erosional development.

Discussion

Results of fluid inclusion studies and vitrinite rank determinations plotted together on a geologic-structural map of the Fore-Sudetic Monocline (Fig. 6) show a distinct spatial correlation pattern between the rank of vitrinite and the temperature of fluid inclusion homogenization a range of 1.5 to 2.5% R_{oil}. Therefore a rough comparison of homogenization temperatures and vitrinite rank determinations is made only for three roughly corresponding temperature and vitrinite rank classes (Fig. 6). These classes include below 200°C, (< 1.5% R_{oil}), between 200 and 250°C (1.5 to 2.0% R_{oil}), and above 250°C (>2.5% R_{oil}), respectively.

The results of vitrinite rank determinations and fluid inclusions studies are quite similar, as shown on Figs. 4, 5, and 6. The paleogeothermal field parameters of the Fore-Sudetic Monocline are found to be generally high. This corresponds to reported earlier findings for the Rhenohercynian belt of the North-Central Europe (Teichmüller and Teichmüller 1979). Inside the paleogeothermal field of the Fore-Sudetic Monocline, characterized by high geothermal gradient values additional positive anomalies in close proximity to paleohighs and highly tectonically disturbed regions were recognized. Some tens of kilometers north from the Dolsk deep fracture, the high geothermal paleogradients of the Fore-Sudetic Monocline are adjacent to areas of lower paleogeothermal gradients. Similar low geothermal field parameters were reported in the North-Sudetic Syncline. The results yielded in both those areas virtually correspond to a Variscan geothermal gradient (Teichmüller et al. 1979), and to the present depth of the investigated rocks. The irregular spatial distribution pattern of Th and Td temperatures in individual drillholes, and the abundance of secondary fluid inclusions in extensively altered portions of the Variscan basement allows the explanation that the heat flow that caused and governed those alterations was mostly of hydrogenic character. The idea

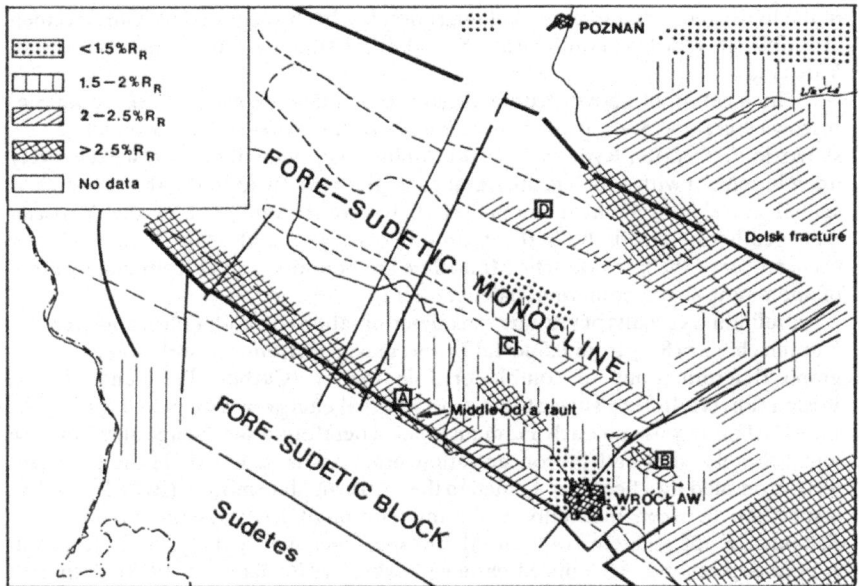

Fig. 5. Map of equal reflectance fields. The isorank fields are superimposed on a geological-structural map of the Fore-Sudetic Monocline. Explanations as in Fig. 1

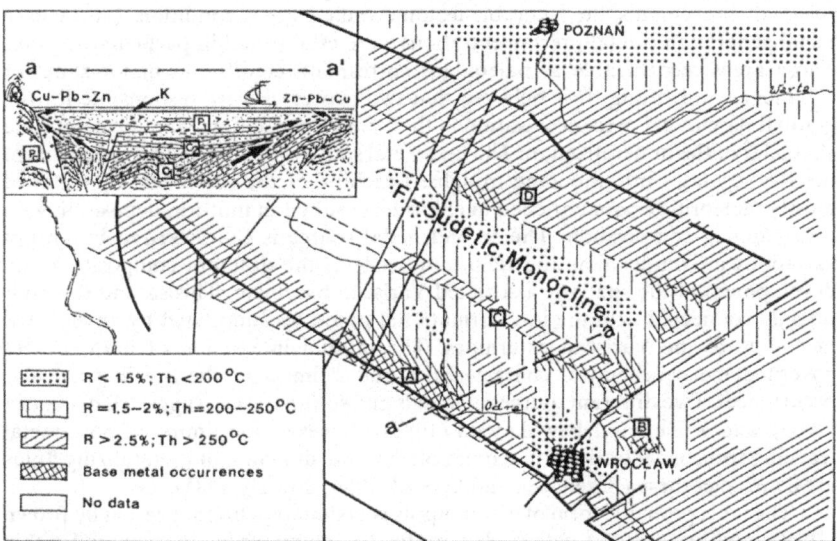

Fig. 6. Paleogeothermal basement fields of the Fore-Sudetic Monocline as revealed by fluid inclusion studies and vitrinite reflectance determinations, and profile (a-a') showing known base metal occurrences. P_c Precambrian; C_1 Lower Carboniferous; C_2 Upper Carboniferous; P_1 Lower Permian (Rotliegendes); K Kupferschiefer. *Arrows* mark the direction of mineralizing fluids. Other explanations as in Fig. 1

of the hydrogenic character of heat migration has been suggested by Van Breemen et al. (1982) for the Moldanubian and Saxothuringian zones of the Central European Variscides.

The location of known Kupferschiefer base metal deposits of the Lubin-Sieroszowice area, as well as other occurrences of the Fore-Sudetic Monocline, are shown in Fig. 6 (after Rydzewski 1978). All the base metal deposits rest exclusively upon basement with a high geothermal paleogradient. In addition, they show clear spatial correlation to heat anomalies that were encountered in the Variscan paleogeothermal field. Both the Lubin-Sieroszowice and other Kupferschiefer occurrences of the Fore-Sudetic Monocline are bounded and surrounded by areas of anomalous high geothermal paleogradients.

Preliminary results of vitrinite rank determinations on both Polish and German Kupferschiefer (R_{oil} from 0.5 to 1.3%) and mass spectrometer and gas-chromatograph determinations of soluble organic matter (Carbon Preference Index, Pristan/Phytan Index), suggest that most thermal energy was delivered during the Late Carboniferous and Early Permian times. Therefore, a similar age of major base metal mobilization and preconcentration processes is assumed. Those processes were promoted by a heat flow related to the Variscan plate motion (Behr 1978). The association between thermally active intracontinental rift systems (White Pine, Katanga, Kupferschiefer of Central Europe), hot spot activity, and stratiform copper deposits was recognized earlier (Sawkins 1976; Bauman 1978). Although situated at the base of marine Zechstein formation, Kupferschiefer mineralization seems to be related to a late phase of crystalline basement development.

During the Late Variscan, probably as a consequence of deviations in particular sub-plate movements, the favorable tectonic and energetic conditions for the base metal generation (metal preconcentration) were established in particular regions. Those areas recognizable as paleohighs correspond credibly to major zones of subfluence and rifting. It is believed that two processes were responsible for the gradual base metal preconcentration in the Late Variscan time, and the composite character of the mineralization. The first metal source is from red-bed sediments of the Rotliegendes intramontane troughs. Leaching of the sediments by water of mostly meteoritic origin enriched in chlorides resulted in uniform deposition with wide-ranging composition. The second metal source is characterized by abrupt variations in composition and tenor, and reflects the different composition and development of the spatially associated Variscan basement. This second source is directly connected with the intraformational processes stimulated by an elevated heat flow of the Variscan externides. The mutual interaction of both sources governed the geochemical characteristics and richness of the ore in particular occurrences. Two different sources of metals and sulfur are also suggested by Kucha (1985), and Kucha and Pawlikowski (1986). However, the timing of base metal leaching (preconcentration), precipitation, and the direction of mineralizing fluids are contradict these (Berendsen and Speczik 1984; Speczik 1985).

The most probable form of metal supply was similar to that suggested by Brown (1978) and Lury'e (1986), with cooled ore-bearing solutions having emerged within the epigenetic Rote Fäule in areas closely associated with the Variscan paleohighs. The discharge was restricted to those areas where hydraulic communications between subsurface water and Kupferschiefer horizon existed, with the Kupfers-chiefer acting as a geochemical hydrogen-sulfide barrier. These conditions can explain the observed metal and mineral zonation patterns of the Kupferschiefer (Fig. 6).

Conclusions

A positive correlation in spatial distribution of the Zechstein base metal occurrences and areas of anomalously high paleoheat flow of the Fore-Sudetic Monocline was recognized. It is suggested that the occurrence of Kupferschiefer mineralization is related to the paleotectonical development of intimately associated basement rocks. Availability of thermal energy at the convergent and divergent boundaries of the South European plate during the Late Variscan was a critical factor in mobilization and preconcentration of base metals from associated Rotliegendes as well as from basement formations. The leaching of base metals mainly from crustal sources by water enriched in chlorides, coupled with strong input of thermal energy, provided the basic environment in which Kupferschiefer ore-generating systems could operate. The Variscan heat flow was mainly of hydrogenic character.

Acknowledgments. Sincere thanks are expressed to Prof. Dr. L. Brady (Kansas Geological Survey) and two anonymous referees who kindly read an early draft of this paper, and offered many helpful comments. Special thanks are extended to Prof. Dr. G. Friedrich (RWTH Aachen) for aid and personal encouragement to write this paper. The Alexander von Humboldt Foundation generously supported this research at the Institut für Mineralogie und Lagerstättenlehre, RWTH Aachen, FRG.

References

Ammosow II, Utkina AJ (1975) Paleotemperatures, lithification and oil and gas occurrences in Neogene deposits of northern Sakhalin. In: Veremin IV (ed) Paleotemperatury zon nieftoobrazowania. Nauka, Moscow, pp 70–93
Bauman L (1978) Zur Bedeutung der Plattentektonik für die Metallogenie — Mineralogie. Z Geol Wiss 6:1357–1377
Behr HJ (1978) Subfluenz-Prozesse im Grundgebirgs-Stockwerk Mitteleuropas. Z Dtsch Geol Ges 129:283–318
Berendsen P, Speczik S (1984) A comparison of Polish and U.S. Midcontinent stratiform copper occurrences. Arch Miner 40:1–23
Bischoff JL, Radtke AS, Rosenbauer RJ (1981) Hydrothermal alteration of graywacke by brine and seawater: roles of alteration and chloride complexing on metal solubilization at 200°C and 300°C. Econ Geol 78:659–676
Bostic AC (1979) Microscopic measurements of the level of catagenesis of solid organic matter in sedimentary rocks to aid exploration. SEPM Spec Publ 26:141–158
Brown AC (1978) Stratiform copper deposits — evidence for their post-sedimentary origin. Miner Sci Eng 10:172–171
Carpenter AB (1978) Origin and chemical evolution of brines in sedimentary basins. Okla Geol Surv Circ 79:60–77
Castano JR, Sparks DM (1974) Interpretation of vitrinite reflectance measurements in sedimentary rocks and determination of burial history using vitrinite reflectance and authigenic minerals. Geol Soc Am Spec Pap 153:31–52

Guterch A, Materzok R, Pajchel J, Perchuc E (1975) Crustal structure from deep seismic sounding along International VII Profile. Przegl Geol 4:153–163

Hood A, Castano J (1974) Organic metamorphism: its relation to petroleum generation and application to studies of authigenic minerals. Coord Comm Offshore Prosp Tech Bull 8:85–118

Jowett CE, Jarvis GT (1984) Formation of foreland rifts. Sediment Geol 40:51–72

Kontorowicz AE, Parparova GM, Trusnikov PA (1967) Metamorphism of organic matter and several questions of oil content (by example from Mesozoic deposits of the West-Siberian Lowland). Akad SSSR Sibirsk Otdeleniye Geologiye i Geofizika 2:16–29

Kucha H (1985) Feldspar, clay, organic and carbonate receptors of heavy metals in Zechstein deposits (Kupferschiefer type) Poland. Trans Inst Miner Metall Sec B:133–146

Kucha H, Pawlikowski M (1986) Two-brine model of the genesis of strata-bound Zechstein deposits (Kupferschiefer type) Poland. Miner Dep 21:70–80

Lorenz V (1976) Formation of Hercynian subplates, possible causes and consequences. Nature 262:374–377

Lury'e AM (1986) Formation conditions of copper sandstone and shale type deposits. In: Friedrich GH, Genkin AD, Naldrett AJ, Ridge JD, Sillitoe RH, Vokes FM (eds) Geology and Metallogeny of Copper deposits. SGA Spec Publ 4:477–491

Majorowicz J (1982) The ambiguities in tectonic interpretation of geothermal field patterns in the platformic areas in Poland. Przegl Geol 2:86–94

Majorowicz J, Marek S, Znosko J (1984) Paleogeothermal gradients by vitrinite reflectance data and their relation to the present geothermal gradient patterns of the Polish Lowlands. Tectonophysics 103:141–156

Rentzsch J (1981) Mineralogical-geochemical prospection methods in the Central European Copper Belt. Erzmetall 34:492–495

Rose AW (1976) The effect of cuprous chloride complex in the origin of red-bed copper and related deposits. Econ Geol 71:1036–1049

Rydzewski A (1978) Facja utleniona cechsztynskiego lupku miedzionosnego na obszarze monokliny przedsudeckiej. Przegl Geol 2:107–107

Sawkins FJ (1976) Metal deposits related to intracontinental hotspot and rifting environments. J Geol 84:653–671

Speczik S (1979) Ore mineralization in the basement Carboniferous rocks of the Fore-Sudetic Monocline (SW Poland). Geol Sudetica (Warsaw) 14:77–122

Speczik S (1985) Metallogeny of pre-Zechstein basement of the Fore-Sudetic Monocline. Geol Sudetica (Warsaw) 20:35–105

Speczik S, Skowronek C, Friedrich G, Diedel R, Schumacher C, Schmidt FP (1986). The environment of generation of some base metal occurrences in Central Europe. Acta Geol Pol 36:1–34

Teichmüller M, Teichmüller R (1979) Diagenesis of coal (Coalification). In: Larsen G, Chilingar G (eds) Diagenesis in sediments and sedimentary rocks. Elsevier, Amsterdam, pp 207–246

Teichmüller M, Teichmüller R, Weber K (1979) Inkohlung und Illit-Kristallinität, vergleichende Untersuchungen im Mesozoikum und Paläozoikum von Westfalen. Fortschr Geol Rheinl Westfalen 27:201–276

Van Breemen O, Aftalion M, Bowes DR, Dudek A, Misar Z, Povondra P, Vrana S (1982) Geochronological studies of the Bohemian massif, Czechoslovakia, and their significance in the evolution of Central Europe. Trans R Soc Edinb, Earth Sci 73:89–108

Walker AL, Buchanan AS (1969) The production of hydrothermal fluids from sedimentary sequences, Part 1. Econ Geol 64:919–922

Wilczek T (1982) Zastosowanie bádan kategenezy SO do oceny mozliwosci powstawania weglowodorow w osadach karbonu i aalenu gornego. International Conference Geonafta, Serock, pp 239–250

Wolf M (1978) Inkohlungsuntersuchungen im Hunsrück (Rheinisches Schiefergebirge). Z Dtsch Geol Ges 129:217–227

Zwart HJ (1967) The duality of orogenic belts. Geol Mijnbouw 46:283–309

Geologic Setting and Genesis of Kupferschiefer Mineralization in West Germany

F.-P. SCHMIDT[1] and G. FRIEDRICH[2]

Abstract

Three principle types of Kupferschiefer mineralization can be distinguished, characterized by different geologic setting, lithological features, geochemical pattern, and paragenesis.

Syngenetic formation seems likely for the low grade mineralization, which hosts either in Kupferschiefer over thick Rotliegend or directly over Paleozoic basement. Characteristic minerals are pyrite, marcasite, chalcopyrite, galena, and sphalerite, coupled with high $Zn/Cu + Pb$ and $Zn + Pb/Cu$ ratios respectively. With regard to the great extension of low grade mineralization, a preconcentration of metals within the Rotliegend red-beds, succeeded by reworking due to transgression of Zechstein paleosea, is presumed.

High grade mineralization is restricted to marginal parts of Rotliegend basins, characterized by the association of hematite, Fe-hydroxides, chalcocite, bornite, digenite, and covellite as well as by a high $Cu/Pb + Zn$ ratio. There exists evidence for diagenetic formation of high grade mineralization and associated Rote Fäule by alkaline basinal brines, enriched in Na and sulfate S, which generated due to progressing subsidence of the sedimentary basin.

Structure bound mineralization occurs mainly adjacent to areas, where post-Variscan (Saxonian) tectonism caused strong basin and range formation in connection with abundant fault structures. Ideal sites therefore have been found around paleohighs, uplifts, and graben structures. This type of mineralization is characterized by tennantite, enargite, loellingite, arsenopyrite, millerite, safflorite, rammelsbergite, barite, and several other Co-Ni-As-Ag-bearing minerals. Structure bound mineralization can be referred to an epigenetic formation by hydrothermal fluids, probably due to progressing mantle diapirism.

Taking into account that all known Kupferschiefer ore districts in Europe are situated along the Rhenohercynian/Saxothuringian boundary, there seems to be evidence that the formation of Kupferschiefer type deposits can ultimately be traced back to the evolution of the sedimentary basin in relation to the crustal setting of the basement rocks and their regional geotectonic history.

[1]Gerling Institut Pro Schadenforschung und Schadenverhütung, Friesenwall 89, 5000 Köln 1, FRG
[2]Institut für Mineralogie und Lagerstättenlehre der RWTH, Wüllnerstraße 2, 5100 Aachen, FRG

Base Metal Sulfide Deposits
G.H. Friedrich, P.M. Herzig (Eds.)
© Springer-Verlag Berlin Heidelberg 1988

Introduction

The European Permian system contains in its basal Zechstein unit one of the world's most famous sediment-hosted base metal enrichments, referred to as "Kupferschiefer type" mineralization.

Due to the typical development from continental red-beds (Lower Permian Rotliegend) into a shallow marine environment (Upper Permian Zechstein) mineralization occurs in the first reduced rock sequence above oxidized red-beds within the Weissliegend (S 1), Kupferschiefer (T 1), and Zechsteinkalk (Ca 1). Typical Weissliegend rocks are conglomerates, often alternating with cross-laminated sandstones, which form fining upward sequences. According to Schumacher (1985b) depositional environment of the Weissliegend can be generally referred to as braided-river formation.

A mixture of clay and organic matter with minor amounts of quartz debris characterizes the most frequent Kupferschiefer type. Intercalations of organic and anorganic components cause fine lamination, which is disturbed by slumping structures in some places. Close to Weissliegend sandbars, which formed shoals in the paleosea, Kupferschiefer bears significant amounts of quartz debris.

The Zechsteinkalk contains limestones and dolostones; adjacent to major coastlines marlstone is abundant.

The discussions concerning the genesis of Kupferschiefer mineralization have been often controversial, due to investigations of more or less local occurrences by different workers.

The presumed genetic processes ran the whole gamut from pure synsedimentary deposition as a result of metal supply by surface waters to epigenetic formation by hydrothermal fluids. Many workers have regarded the Kupferschiefer as the prototype of a stratiform, syngenetic base metal enrichment in an euxinic environment (e.g. Wedepohl 1964). For example Rentzsch (1974) has shown that, viewed on a large enough scale, the mineralization is discordant to the bedding plane and further, that it surrounds areas of post-depositional alteration, referred to as "Rote Fäule". From the Polish deposits, Rydzewski (1978) described the replacement of framboidal and euhydral pyrite by hematite within the Rote Fäule and suggested an epi(dia)genetic formation of the mineralization.

The local influence of hydrothermal solutions, which ascended along faults, have been reported from the Mansfeld-Sangerhausen district in East Germany as well as from the Richelsdorf district in West Germany (Gunzert 1953).

Apart from genetical viewpoints, two major systems for classification of Kupferschiefer mineralization had been developed. Knitzschke (1966) used the relation between copper, lead, and zinc to distinguish several metal facies types, whereas Rentzsch and Knitzschke (1968) pointed out ten distinct mineral associations for a classification based on microscopical studies.

Thorough research as result of an exploration campaign targeted on Kupferschiefer-type deposits in the Federal Republic of Germany, which comprises to date more than sixty drill holes on behalf St. Joe Explorations GmbH, Hannover, yielded a tremendous amount of new data from the Spessart, Rhön, and Richelsdorf area. Additionally data from other areas, where investigations had recently been carried out by different workers, have been taken into consideration, e.g. from the

Niederrheinische Bucht (Diedel 1986), SW-Harz (Paul 1982), Korbacher Bucht (Kulick et al. 1984), and Norddeutsches Becken (Wedepohl 1964) (Fig. 1).

As a consequence, a strict relation between geologic setting, metalfacies and mineralfacies, and formation conditions could be established (Table 1). This paper deals with the ascertained facts with the aim of establishing an improved model for the Kupferschiefer mineralization in West Germany and Europe.

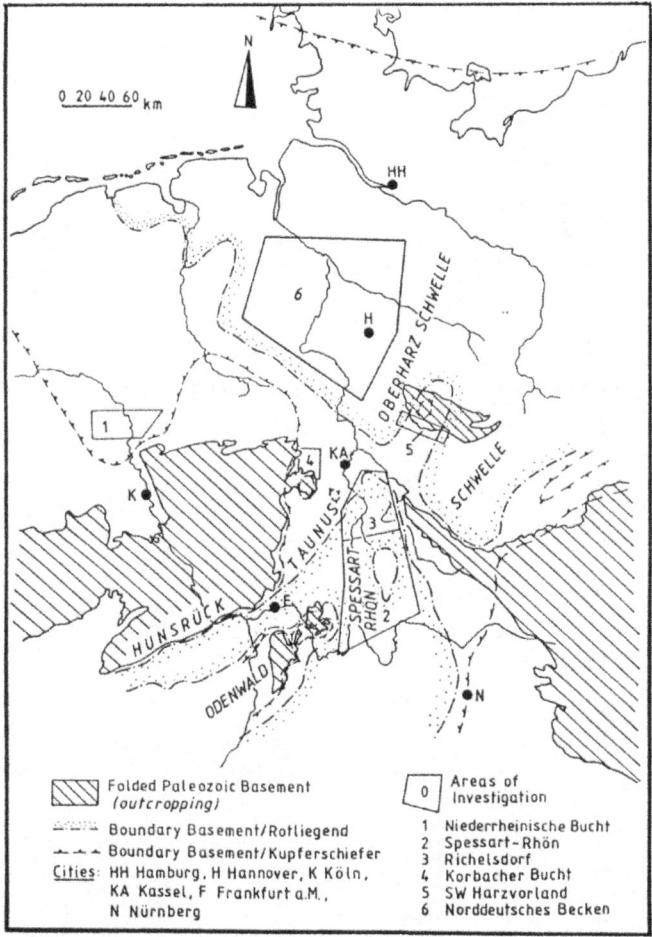

Fig. 1. Sketch map showing the paleogeographical situation of Central Germany at the Rotliegend/ Zechstein boundary

Table 1. Geological, mineralogical, and geochemical features of outlined three principle types of Kupferschiefer mineralization

Classification	Metallogeny	Average Base Metal Content	Zonation Pattern	Paragenesis	Alteration Pattern	Formation Conditions	Geologic Setting	Geotectonic Event
Low Grade Miner. schwellen-type basin-type	$Zn+Pb>Cu$ $Zn>Pb+Cu$	2000 ppm	$Cu \rightarrow Pb, Zn$	py1, mc, sph1, cp1, gn1, tn1, ll1	–	SYNGENETIC slight alkaline, reducing $T \triangleq 25°C$.	basal Zechstein sequence over paleozoic basement. basal Zechstein sequence over thick (>200 m) Rotliegend	Transgression of Zechstein sea due to breakdown of Ringkøbing Fyn High
High Grade Miner.	$Cu \gg Pb+Zn$	3%	$Fe^{3+} \rightarrow Cu \rightarrow Pb \rightarrow Zn$	hm, Fe-hyd, cc, bn, dg, cv, cp2, gn2, sph2	oxidation [Rote Fäule, anhydrite]	DIAGENETIC $T<103°C$, metal influx by basinal brines from Rotl.	margin of Rotliegend basin	major subsidence of sedimentary basin in lower Zechstein
Structure Bound Miner.	$Cu>Pb+Zn$	7000 ppm	$Cu–Pb–Zn$ $Cu–Zn–Pb$	tn2, tn3, en, ll2, ap, py2, sc, mi, ba, ra, pc, str, ag	kaolinization	EPIGENETIC acidic hydrothermal fluids. $T>103°C$	adjacent to paleohighs and uplifts	basin and range formation due to Saxonian tectonism

Low Grade Mineralization

The great extension of the Kupferschiefer stratum in particular and its mineralization over an area of about 100,000 km^2 gave, very early on, rise to the assumption of a syngenetic formation of the contained metals.

According to Wedepohl (1964) the zinc-lead type governs nearly the entire mineralized area. As a rule this type of mineralization consists of pyrite, marcasite, sphalerite, galena, and chalcopyrite and occurs either within the basal Zechstein sequence over paleohighs (schwellen-type) or over paleobasins (basin-type), Fig. 2.

Over paleohighs, which are for example portrayed by the Niederrheinische Bucht, SW-Harz, Werra-Grauwacken-Gebirge and Spessart-Rhön Schwelle, Weissliegend and Kupferschiefer transgressed directly over Paleozoic basement. In those places Kupferschiefer is characterized by high carbonate contents resulting in a great thickness (Paul 1982). In numerous places Kupferschiefer is underlain by a limestone unit, referred to as "Productuskalk" (Kulick 1968). According to Wedepohl (1964) and Paul (1982), schwellen-type Kupferschiefer as well as the Zechsteinkalk bear high dolomite/calcite ratios in contrast to sediments which were deposited within basins.

Basin-type Kupferschiefer appears as a fine laminated black shale, containing abundant organic matter grading up to 8% C_{org}. Below the basal Zechstein sequence up to several hundred meters of Rotliegend molasse sediments occur. Such areas are found in the Saar-Werra trough, Kraichgau-Saale trough and Norddeutsches Becken.

Lower Rotliegend was governed by igneous activity, whereas the Upper Rotliegend rock sequence bears large amounts of clastics which generated under arid red-bed conditions. In some places classic sedimentary basins had not developed. The Schneverdinger Graben, situated within the Norddeutsches Becken, is one

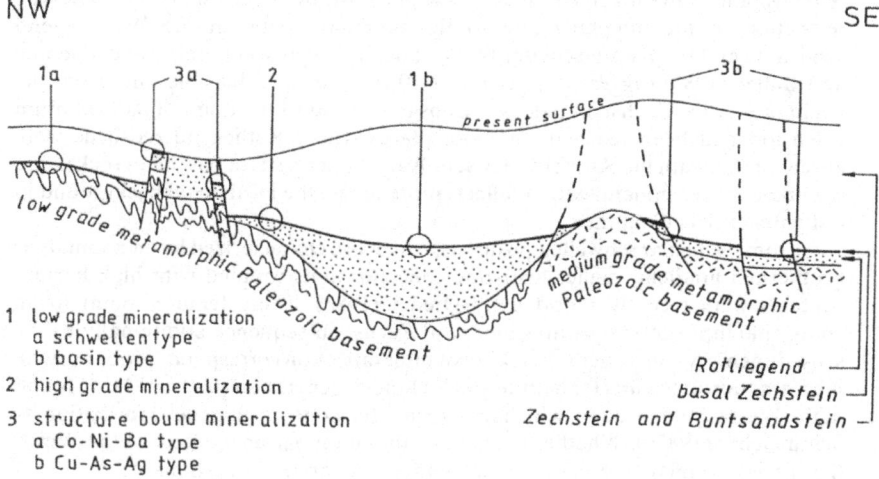

NW   SE

1 low grade mineralization
   a schwellen type
   b basin type
2 high grade mineralization
3 structure bound mineralization
   a Co-Ni-Ba type
   b Cu-As-Ag type

Rotliegend
basal Zechstein
Zechstein and Buntsandstein

Fig. 2. Geologic setting of low grade, high grade, and structure-bound mineralization types

example where the huge pile of Rotliegend volcanoclastics and sediments accumulated within a narrow graben structure (Drong et al. 1982).

Investigations by Wedepohl (1964) have shown that a high Zn + Pb/Cu ratio, a vertical zonation from the bottom to the top into Cu → Pb + Zn as well as an average base metal content of 2000 ppm seems typical for the low grade mineralization type.

Sulfur-, carbon- and oxygen isotope studies by Marowsky (1969) point to a sulfide formation within the sediment due to seawater-sulfate reduction by bacteria. The large variation of sulfide δS-values (−4% to −44%) with an average δ^{34}S-maximum of −34%, as well as the lack of any systematical fractionation pattern for sulfur within the stratigraphic sequence, support the theory of a penecontemporaneous formation of sulfides and host sediment.

In spite of these facts, the investigations are not suitable to give any direct hint to the provenance of metals and the mechanism of transportation. So far Wedepohl et al. (1978) have shown that lead isotopes from the Kupferschiefer mineralization are generally in accordance with some lead isotope compositions from veins which are exposed in the Rhenish Massif, whose age is suggested as Variscan. With reference to the observed geochemical pattern, a dissolution of the contained metals by weathering, followed by transportation in surface waters and redeposition within the Kupferschiefer in an euxinic, reducing environment has been presumed by some authors. With regard to the great extension of the mineralization, even in areas several hundred kilometers away from known ancient coastlines, this model seems questionable.

The conclusion drawn by Schmidt (1985), that most of the metals which are concentrated within the Kupferschiefer were leached from the upper parts of Rotliegend sediments due to the transgression of Zechstein paleosea, is a modification of Lur'ye's formation model (1977) and could explain the moderate mineralization over such a large area.

Taking into account that overflooding of the Rotliegend, reworking, and redeposition of the involved metals took place within a very short time interval, restriction of the mineralization to the boundary between the Weissliegend sandstone and the Kupferschiefer black shale no longer seems enigmatic. Glennie and Buller (1983) suggested a period of 10 (!) years as a feasible time from the beginning of the Zechstein paleosea transgression until the point of its maximum extension had been reached. In those places where Rotliegend red-beds were absent underneath the Kupferschiefer, only the upper parts of basement rocks were reworked, hence mineralization reflects more or less the local metal background in a slightly enriched manner.

From particular schwellen positions, which are characterized by an anomalous geothermal gradient, higher lead-zinc concentrations coupled with high barium contents were recently reported by Diedel (1986). These features point to an endogenic input of metals into the basal Zechstein sequence additionally to the synsedimentary component. This kind of mineralization corresponds probably with the formation of the first tennantite and loellingite generation (tn-1 and 11-1) found in the Werra-Fulda trough, which had been referred to an epigenetic formation by Schumacher (1985a). Whether the always abundant barium content of the Weissliegend is also related to the suggested endogenic input, is questionable.

High Grade Mineralization

In the 1960's and 1970's, new data were published by East German and Polish authors, which focused on the geology of East German Mansfeld-Sangerhausen, Spremberg-Weißwasser, and the giant Polish Lubin deposit. In contrast, only scarce information exists with regard to the ore control of the West German Richelsdorf district, where mining ceased in the mid 1950's.

Recent investigation of Kupferschiefer type deposits in West Germany (Schumacher and Schmidt 1985, Schmidt et al. 1986b), yielded a tremendous amount of new data. In particular, in the Richelsdorf area, most of the parameters display remarkable similarities to the ore-bearing areas of Mansfeld, Spremberg, and Lubin.

Principally less than 5% of Kupferschiefer mineralization is governed by a copper predominance, counted on the total amount of base metals trapped in the Kupferschiefer. These areas may contain ore deposits, which are mined chiefly for copper and silver. Precious metals like gold and platinum group elements, whose occurrence has recently been reported by Kucha (1982), gain only economic importance due to the high mining output (20,000 t/d per mine) in Poland. As a rule, ore deposits or, if subeconomic, high grade mineralization, occur over marginal positions of Rotliegend troughs, not in basin or in schwellen positions (cf. Fig. 2).

Occurrence of ore deposits in such transitional zones adjacent to basement highs is a very common feature for sediment-hosted base metal deposits and has been described by Bjørlykke and Sangster (1981).

A metal zonation into Rote Fäule, copper facies, and lead/zinc facies has been reported by Rentzsch (1974) for the Mansfeld-Sangerhausen district in East Germany (Fig. 3). Rote Fäule represents the oxidized side of a redox boundary, which had been formed after deposition of the basal Zechstein sequence. Paleomagnetic dating of Rote Fäule occurrences from Poland by Jowett (1986) indicated a Middle Triassic age of 210 m.y. and thus a late diagenetic formation. Depending on host rock lithology, Rote Fäule occurrences display distinct shapes, e.g., shreds, dots, spots, schlieren, and clouds (Table 2). Based on our own observations from Polish mines,

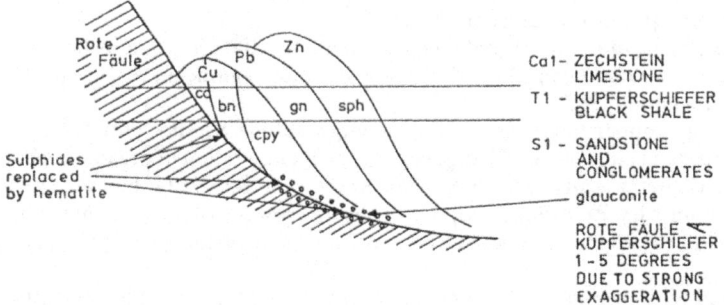

Fig. 3. High grade mineralization related metal and mineral zonation within the basal Zechstein sequence. (Simplified after Rentzsch 1974)

the following structures are typical:

Table 2. Relation between occurrence of Rote Fäule and host rock lithology

Stratigraphy	Host Rock lithology	Structural development of Rote Fäule
Weissliegend	Sandstone	Dots, spots, clods, Schlieren
Kupferschiefer	Black shale	Stratiform layers
Zechsteinkalk	Limestone	Shreds, layers, dots

Research work focused on the ore control, with special reference to the Rote Fäule-related high grade mineralization (Schmidt 1985), resulted in the detection of numerous geochemical and mineralogical zonation patterns, which improved the existing knowledge gained from deposits outside West Germany.

Comparable to the situation in Spremberg-Weiβwasser, within the Richelsdorf district, Rote Fäule corresponds apparently with sandbars of the Weissliegend, which had been shoals in the Kupferschiefer paleosea (Fig. 4). The depositional system was interpreted by Schumacher (1985b) as fluviatile braided-river, characterized by alternating sandbar-conglomeratic channel sequences.

Approaching the sandbar, Kupferschiefer wedges out or changes into a sandy facies equivalent with minor amounts of bituminous schlieren (Fig. 5).

Related to distribution of Rote Fäule a metal zonation into Rote Fäule → copper facies → lead/zinc facies had been observed, which lies discordant to the bedding plane (Fig. 6). Principally Rote Fäule is characterized by a low metal tenor due to the instability of most of the base metal sulfides in an oxidized environment (Fig. 7). Caused by lack of organic matter, sulfate reduction could not take place, so sulfate S is most abundant (Fig. 8). Typical minerals within Rote Fäule are Fe-hydroxides and hematite. Whereas the former occur in an admixture with the carbonaceous-sulfaceous matrix, the latter crystallized within the intergranular spaces or as rim cements around quartz and feldspar debris.

Trace element studies displayed an enrichment in sulfate S in connection with an increase of Na_2O/K_2O and Fe^{3+}/Fe^{2+} ratios as most significant feature for the Rote Fäule (Schmidt 1987).

Succeeding copper facies shows a 2:1 copper/sulfide S ratio, indicating the formation of chalcocite (Cu_2S) and digenite (Cu_9S_5). The content of organic carbon is fairly high (2.5 wt.%) and relatively constant apart from Rote Fäule-bearing areas.

At the boundary Rote Fäule/copper facies, the lowest total iron content has been found, due to solubility of both ferrous and ferric iron in an E_h and pH neutral environment.

Proximal to Rote Fäule, copper concentration took place in the lower part of Zechsteinkalk as well as within the Kupferschiefer horizon, whereas the lead-zinc content is low and chiefly concentrated within the higher parts of the Zechsteinkalk.

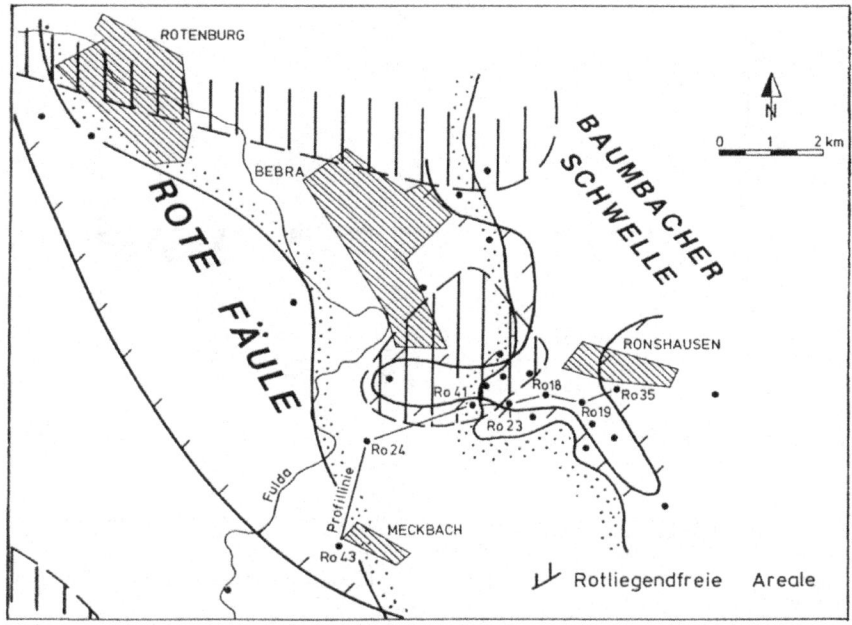

Fig. 4. Distribution of basement (Rotliegendfreie Areale), Weissliegend sandbar (*dotted*) and Rote Faüle in the southern part of the Richelsdorf district. *Ro* = drill holes

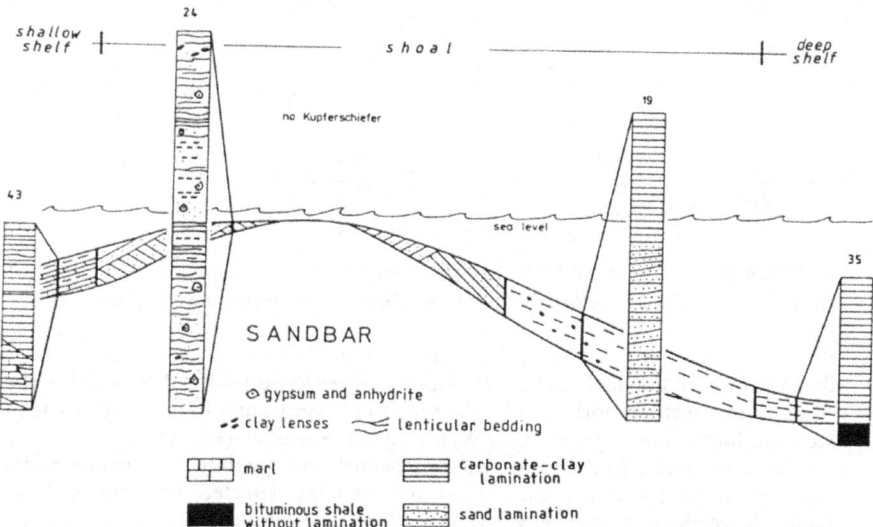

Fig. 5. Facies types of Kupferschiefer in relation to Weissliegend sandbar. (After Schumacher 1985a)

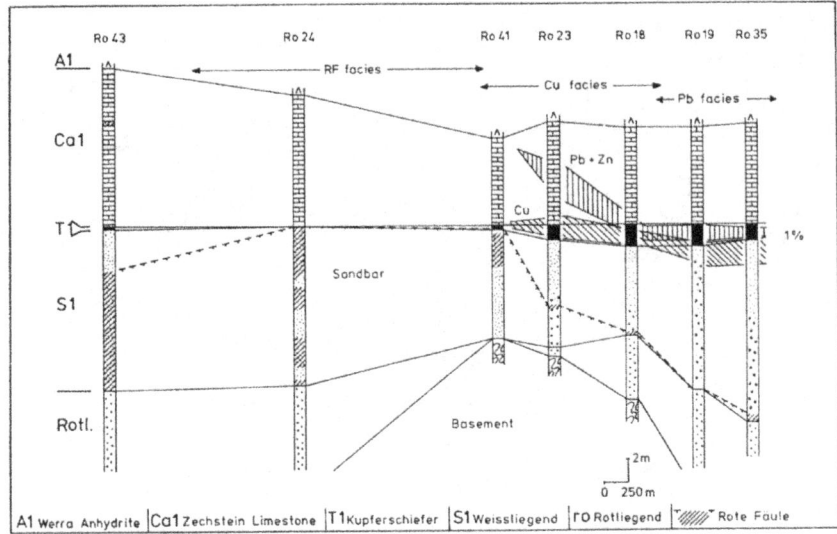

RELATION   BETWEEN   PALEOGEOGRAPHY   AND   METAL   DISTRIBUTION

Fig. 6. Cross-section through the southern part of the Richelsdorf district (for location of drill holes cf. Fig. 4)

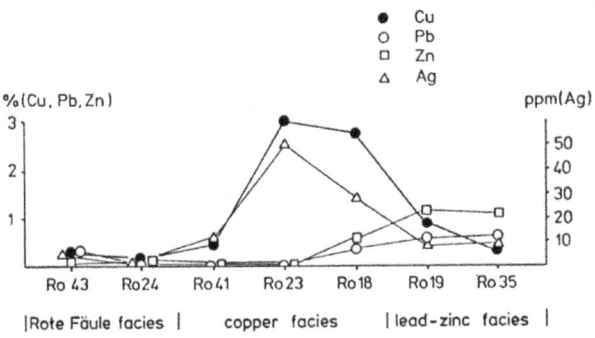

Fig. 7. Distribution of Cu, Pb, Zn, and Ag in the Kupferschiefer in relation to metal facies

By approaching the lead-zinc facies, copper concentration shifts downward in the stratigraphic section, portrayed by the highest concentration in the lower parts of Kupferschiefer and upper parts of Weissliegend sandstone (Fig. 9).

With reference to different positions of drill holes within the distinct metal zones, an apparent and significant feature has been detected. Proximal to Rote Fäule, the vertical distance between copper, lead, and zinc peaks is larger than distal to Rote Fäule, due to the listric shape of the Rote Fäule boundary (cf. Fig. 3). The term "spacing" has been applied to this feature (Schmidt 1985). As illustrated in

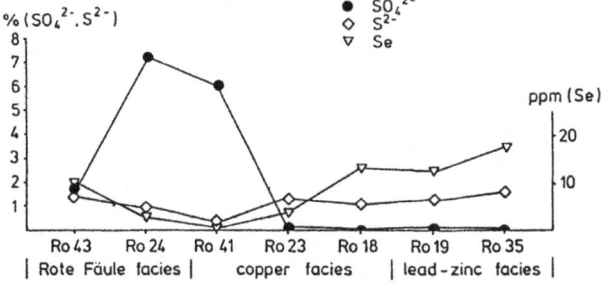

Fig. 8. Distribution of SO_4^{2-}, S^{2-}, and Se in the Kupferschiefer in relation to metal facies

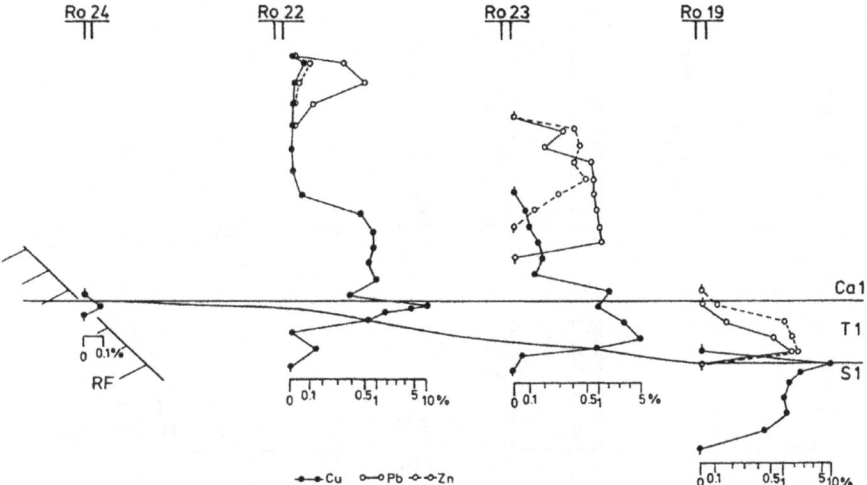

Fig. 9. Distribution of Cu, Pb and Zn within the basal Zechstein sequence. *Ca 1* Zechsteinkalk; *T 1* Kupferschiefer; *S 1* Weissliegend in relation to Rote Fäule (*RF*)

Fig. 10, spacing (measured in meters) is characteristic for certain metal facies types and thus represents a useful tool for exploration.

In general, distribution of mineralization is disseminated and as stratiform layers, the latter tending to be more massive. In some places parts of the Zechsteinkalk are interlaced by cracks and fissures, which are filled with chalcocite and gypsum. Within the stratigraphic sequence a vertical mineral zonation exists similar to the observed metal zonation, also depending on its position to Rote Fäule alteration (Friedrich et al. 1982, 1983a,b) (Fig. 11).

Detailed investigations focused on the apparently sharp contact between copper-rich sulfide mineralization and galena and sphalerite respectively, e.g., portrayed by drill hole Ro 18, did not show any significant change in redox conditions, indicated by Fe^{3+}/Fe^{2+} and S^{2-}/Se ratios (Schmidt 1985).

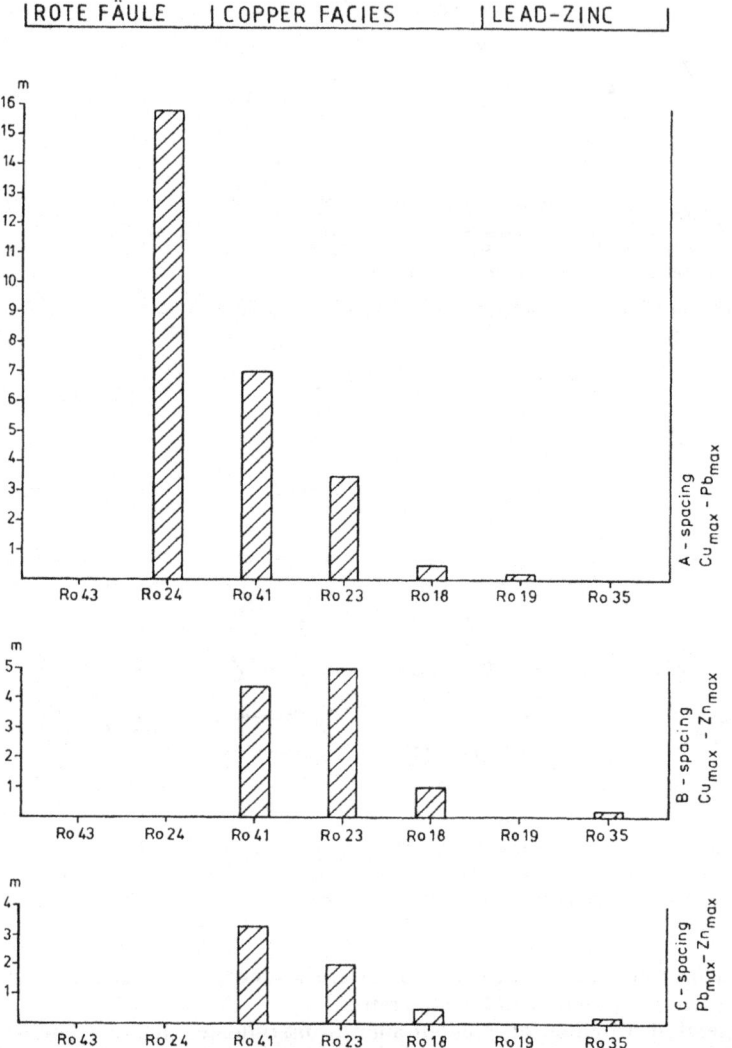

Fig. 10. Spacing for some typical drill hole intersections in the Rote Fäule, copper facies and lead-zinc facies. *A-spacing* vertical distance between copper and lead peak; *B-spacing* vertical distance between copper and zinc peak; *C-spacing* vertical distance between lead and zinc peak, all measured in meters

---

Fig. 11a-c. Spatial distribution of ore minerals in the vertical section of three typical drill holes. **a** Drill hole Ro 24 (Rote Fäule facies); **b** Drill hole Ro 18 (copper facies); **c** Drill hole Ro 21 (lead facies). *hm* hematite; *cc* chalcocite; *bn* bornite; *dg* digenite; *cv* covellite; *cp* chalcopyrite; *gn* galena; *sph* sphalerite; *py* pyrite; *mc* marcasite; *tn* tennantite

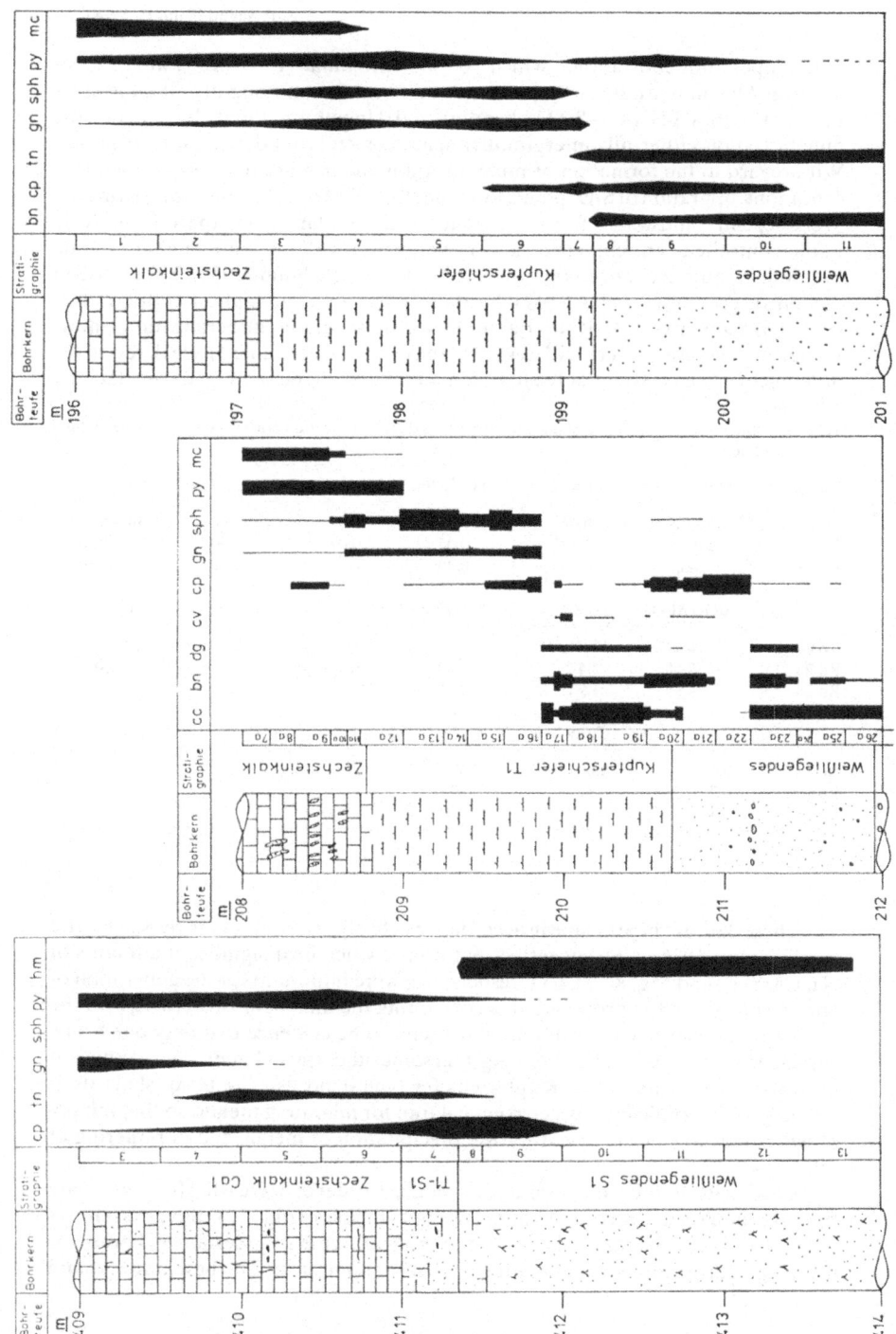

Copper mineralization, which occurs proximal to Rote Fäule, hosts hematite/digenite intergrowth, pointing to a formation in a slightly alkaline environment with a pH of 7–8 (Tischendorf and Ungethüm 1965). In general, the mineralization either fills intergranular spaces or replaced detrital quartz grains. With regard to the formation of mineralization under presumed low temperature conditions, migration of SiO_2 points to an influx of alkaline solutions. Comparing the Weissliegend sandstones from ore zones with similar rocks apart from Rote Fäule-controlled enrichments, lower potassium/sodium and higher sulfate S/sulfide S ratios have been reported for the former (Schmidt 1987). Taking a 2-m section of economic interest (usually those parts with the highest copper/silver contents) into account, Rote Fäule-related high grade mineralization shows numerous zonation pattern with regard to the metal facies (Table 3) which reveal the difference between Rote Fäule alteration an related copper and lead/zinc facies.

Table 3. Characteristic geochemical features of high grade mineralization portrayed by example of three typical drill holes

Drill hole	Metal facies	Cu %/2m	Pb %/2m	Zn %/2m	Ag gt⁻¹/2m	Cu:Pb:Zn
Ro 24	Rote Fäule	0.04	0.03	0.01	− 2	36:42:22
Ro 23	copper	2.81	0.05	0.03	50	68:22:10
Ro 28	zinc	0.08	0.26	0.86	− 2	13:28:59

	SiO_2/Al_2O_3	Sr/Rb	K_2O/Na_2O	Cr/V	SO_4^{2-}/S^{2-}	Fe^{3+}/Fe^{2+}
Ro 24	10.44	12.44	2.22	4.09	12.0	0.71
Ro 23	7.65	2.42	3.43	2.32	1.5	0.40
Ro 28	4.94	1.93	3.59	0.84	1.2	0.26

	C_{org}	Cu/S	S^{2-}/Se
Ro 24	0.2	0.003	3093
Ro 23	0.7	0.125	2500
Ro 28	0.8	0.047	2110

These data are in principle in accordance with observations made by Kucha and Pawlikowski (1986), who suggested that brines which bear significant amounts of Na, Ca, SO_4^{2-} and Mg, K, Cl, CO_3 respectively were important for the generation of ore. In contrast to the proposed idea concerning the mixing of two brines with the previously described composition, there seems to be evidence that only one brine, enriched in sulfate S, Na, Fe^{3+}, Cu, Ag, and some other trace elements, was sufficient to explain the formation of Kupferschiefer-type deposits. The black shale itself represented a significant physicochemical trap for migrating metals, so that mixing of two brines was not required for the precipitation of metals due to buffering of solutions.

Chalcocite, bornite, digenite, and covellite, in order of decreasing frequency, are the prominent ore minerals within the copper facies (Fig. 12). Silver is always connected to copper minerals; microprobe measurements ascertained bornite as main Ag-bearing mineral, containing up to 3500 ppm silver. The best grades found

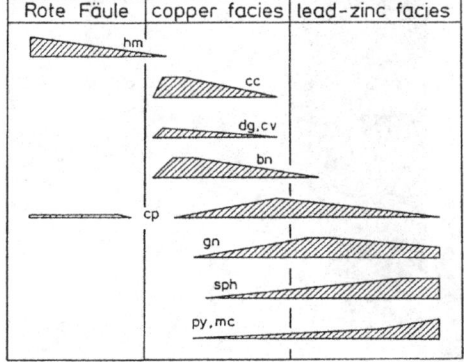

Fig. 12. Metal facies related distribution of ore minerals in the Kupferschiefer bed. *hm* hematite; *cc* chalcocite; *dg* digenite; *cv* covellite; *bn* bornite; *cp* chalcopyrite; *gn* galena; *sph* sphalerite; *py* pyrite; *mc* marcasite

were 10.5% Cu plus 160 g/t Ag within the Kupferschiefer and 2.8% Cu plus 50 g/t Ag counted over 2 m mining heigh.

The succeeding lead-zinc facies contains only minor amounts of bornite, accompanied by traces of digenite, covellite, and chalcocite. Chalcopyrite is typical, but galena and sphalerite are most abundant. Detailed ore petrographical studies of more than 800 polished sections exhibited manifold replacement structures (Fig. 13), which led to the definition of several replacement sequences. Notwithstanding the description following on p. 48, some of the features must be mentioned briefly in advance, due to their importance as evidence of the diagenetic nature of high grade mineralization.

Summarizing the observed geological, mineralogical, and geochemical features, the idea of a diagenetic formation is supported by the following facts:

1. the discordant course of Rote Fäule and related mineralization to the stratigraphic boundaries;
2. the replacement of grey-bed hosted, synsedimentarily formed pyrite, galena, sphalerite, and chalcopyrite by copper-rich sulfides and/or hematite and Fe-hydroxides;
3. positive element correlations between Fe^{3+}/S^{2-} and Cu^+/SO_4^{2-} in the euxinic Kupferschiefer black shale;
4. deformation structures within host rocks, e.g., bending of Kupferschiefer lamination around large mineral grains (hieken), which point to a mineral growth within the still unconsolidated sediment;
5. early fissure fillings with chalcocite and gypsum;
6. absence of feeder zones or feeder structures and of alteration pattern, which are typical for SEDEX deposits.

With regard to the suggestion that oxidizing solutions enriched in metals, emerged from the Rotliegend molasse red-beds (Schmidt 1985; Schumacher 1985a; Schmidt et al. 1986b), according to Tourtelot and Vine (1976) the term "diagenetic" can be applied, explained as: "post-depositional formation of new minerals by equilibrium

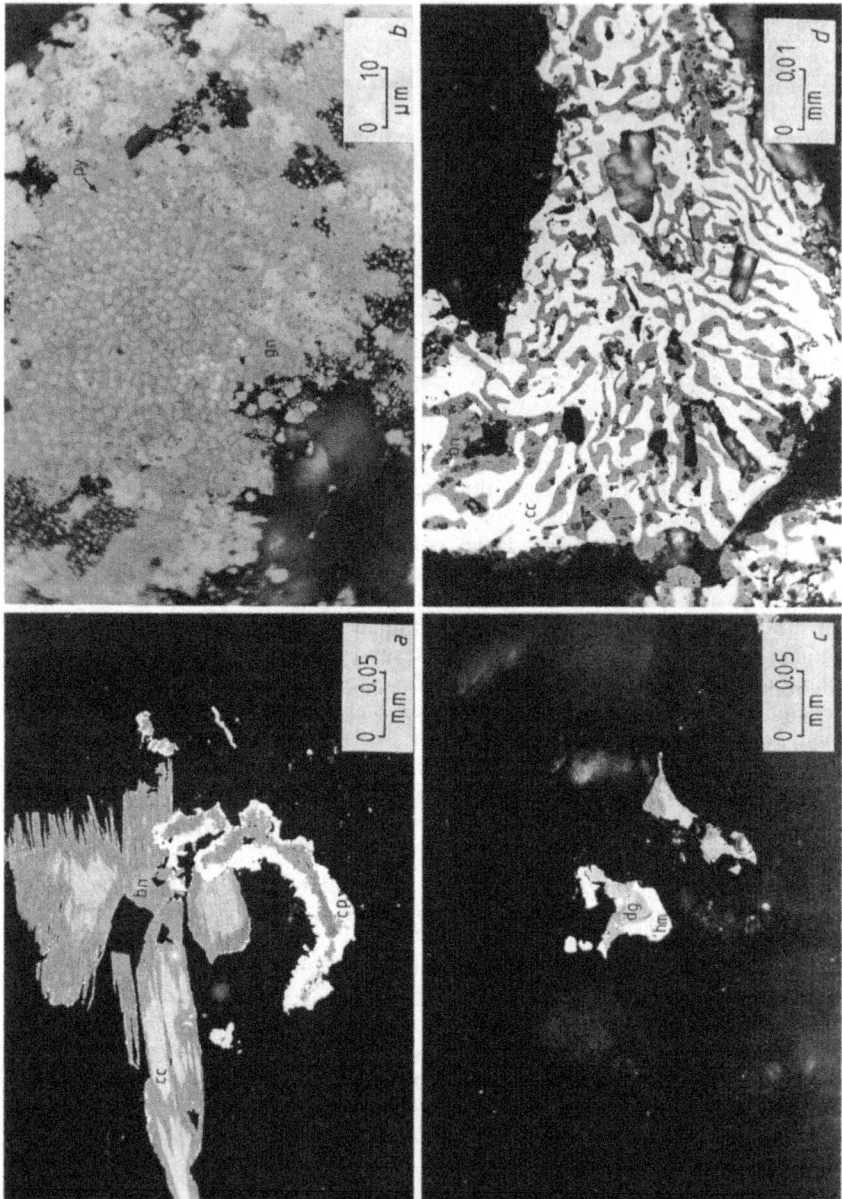

Fig. 13a-d. Paragenetic features of diagenetic high grade mineralization: **a** chalcocite replaced by bornite (*bn*), which is further replaced by chalcopyrite (*cp*), drill hole Ro 26, AS 20658. **b** Framboidal and euhydral pyrite (*py*), cemented by galena (*gn*), drill hole Ro 21, AS 20002. **c** Intergrowth of hematite (*hm*) and chalcocite (*cc*), drill hole Ro 22, AS 20082. **d** Chalcocite/bornite (*cc/bn*) myrmekite, drill hole Ro 19, AS 19379

reactions between original rock constituents, both detrital and chemical, and interstitial fluids and gases within the sequence", by extending the term sequence to the Rotliegend and basal Zechstein as an integral part of the depositional system.

According to Gustafson and Williams (1981), the so-called "source-transport-trap chain of coincidence" represents the prerequisite for the formation of an ore deposit. In our special case not only the input of metals by basinal brines which generated within the Rotliegend was of significance, but also the host rocks, which were actually the physicochemical trap for the migrating solutions, must have had adequate physical and chemical composition in case of an encounter. This means, for example, good porosity, good permeability, as well as abundant reductants to fix the metals as sulfides.

It has been already mentioned, that particularly in the vicinity of sandbars, the Kupferschiefer contains high amounts of quartz debris. Ultimately the content of organic carbon and so the H_2S, HS^- and S^{2-} tenor can be decisive for or against the formation of an ore deposit. As shown in Fig. 14 by the example of two drillholes which were sunk in a similar position to Rote Fäule, bore hole Ro 41 bears lower metal concentrations due to lower C_{org} and hence lower S^{2-} contents in comparison to bore hole Ro 22. At drill site Ro 41 most of the metals were percolating through Kupferschiefer into the Zechsteinkalk and farther on. In these places, a dilution over a larger distance took place in contrast to the strong concentration in the Kupferschiefer horizon of drill hole Ro 22.

Perelman (1972) applied the term "contrast" for the capacity to concentrate various amounts of metals over a certain distance. As a rule a sharp contrast means a strong metal concentration, whereas a weak contrast characterizes a low concentration of metals. Following Perelman's idea, that in the case of lacking H_2S, HS^-, or S^{2-}, a mineralization is principally redox-controlled, whereas in case of abundant H_2S, HS^- or S^{2-} solubility products of metals are the primary control, the Kupferschiefer mineralization seems no longer principally redox-controlled as was emphasized by several authors. High grade mineralization can not be easily compared to "roll-front type" uranium deposits, especially since the occurrence of selenides along the supposed redox boundary has not been noticed. According to Jung et al. (1973), berzecianite (Cu_2Se) and/or umangite (Cu_3Se_2) should have formed with respect to high selenium contents due to a better stability of those selenides in a more oxic environment of formation compared to the stability of sulfides. Referring to Fig. 8, where the selenium content has been plotted with regard to the metal facies types, an increasing selenium content was reported toward the lead/zinc facies. As a consequence, a relation between organic carbon and selenium seems very likely, which has been established by positive C_{org}/Se correlations and had been already suggested by Rentzsch (1981) for the East German deposits.

Summarizing the observations, high grade mineralization seems more controlled by solubility of involved metals than by redox conditions. This can probably be the explanation for the sharp contact between the chalcocite-bornite-digenite-covellite mineralization on one hand and the galena-sphalerite-chalcopyrite mineralization on the other, as it has been previously demonstrated by the example of drill hole Ro 18 (cf. Fig. 11b).

Taking into account the configuration and position of the mineralized zone, there appears to be strong evidence for a direct provenance of the metaliferous solutions from the Rotliegend basins, as was assumed by several authors (Rentzsch

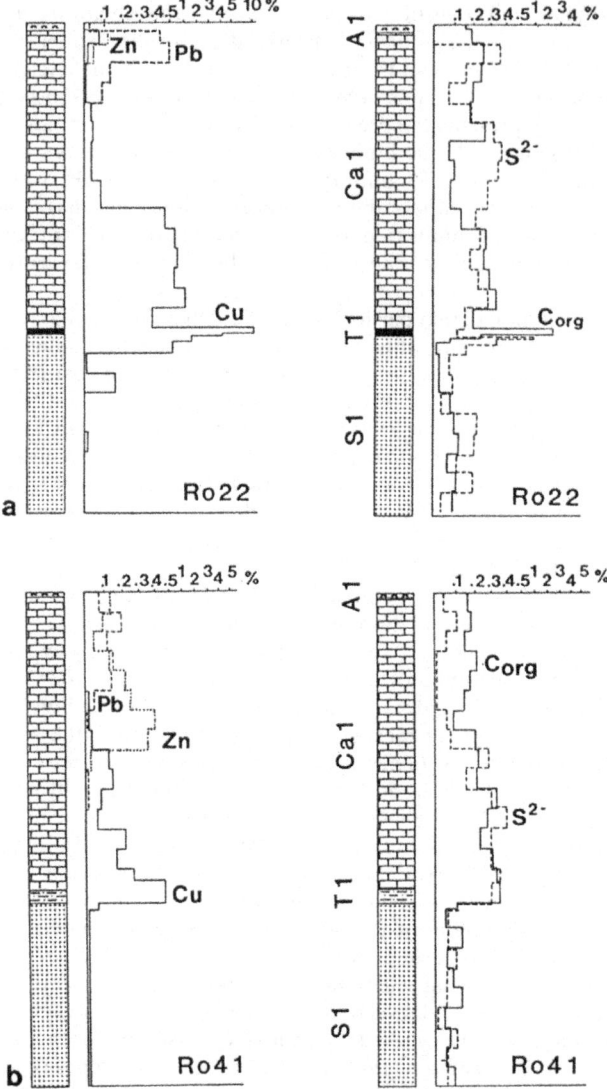

Fig. 14. Relation between base metal content, tenor of organic carbon and S^{2-}, illustrated by example of drill holes Ro 22 (**a**) and Ro 41 (**b**). *A 1* Werra-Anhydrite; *Ca 1* Zechsteinkalk; *T 1* Kupferschiefer; *S 1* Weissliegend

1974; Schmidt et al. 1986b; Speczik et al. 1986). The beds overlying the cupriferous zone are pyritic, whereas the underlying beds are red, hematitic sediments. The location of the cupriferous zone in the basal reduced beds of an originally iron sulfide-rich stratigraphic unit lying directly above oxidized sediments would be consistent with a concept whereby copper and silver was chemically screened from upward moving ore solutions.

The restriction of ore deposits to marginal basin positions, as well as the arrangement of the metal zonation toward the paleoblock, strongly support this assumption.

Structure Bound Mineralization

Fault structures, which intersect the basal Zechstein sequence, are known from numerous sites. The most spectacular and well-investigated fault-related mineralization is referred to as "Rücken", reported from Mansfeld (G.D.R.) and from Richelsdorf (F.R.G.). The term Rücken designates morphological ridges as a consequence of barite veins, which are associated with that type of mineralization and are very resistant against weathering. Adjacent to those places, where barite veins intersect the Kupferschiefer, black shale is ennobled in cobalt (up to some percent) and other elements, e.g., nickel. Several Kupferschiefer locations in the old Richelsdorf mining district display alteration pattern, which had been explained as the result of the influence of so-called "empty hydrotherms" (Gunzert 1953), lacking supply of additional metals. Recently completed investigations could show that some of these alteration features must be referred to the diagenetic formation process. Remarkably, Fe-hydroxides and hematite did not always generate due to a low total iron content, but nevertheless, the process is similar to the formation of Rote Fäule and reveals all other characteristic geochemical patterns, i.e., high sulfate-S content, lower potassium/sodium ratio, etc. Nevertheless, the Rücken-type mineralization generated apart from Rote Fäule formation and its "invisible" affiliate and represents a separate type of mineralization.

Significant and characteristic ore minerals associated with that type of epigenetic mineralization are scutterudite, millerite, nicelite, rammelsbergite, bravoite, safflorite and minor amounts of tennantite, tetraedrite, and loellingite (Messer 1955).

Within the Spessart/Rhön area another type of structure-bound mineralization has been detected, which reveals evidence for a hydrothermal affiliation (Friedrich et al. 1984). In contrast to the apparent connection of Co-Ni-Ba mineralization with exposed vein systems, in that case the features are less obvious.

Mineralization of the Spessart/Rhön area is characterized by a high content of arsenic, so Cu-As sulfides (tennantite, enargite) and arsenides (loellingite, arsenopyrite) were formed.

Delineation of arsenic-related mineralization yielded a remarkable correspondence of high Cu-As-Ag concentrations with fault structures of approximately mesozoic to tertiary age. According to Diedel (1984), three types of tennantite could be distinguished by microprobe analyses, which differ by varying amounts of arsenic, antimony, and silver (Table 4). With respect to these observations, tn-1 refers to the previously mentioned low grade mineralization, which exists only in

remnants. Tn-2 and tn-3 represents the subsequently formed copper-arsenic mineralization and thus replaces all types of sulfides, which generated prior to the
epigenetic process. In this coherence the occurrence of chalcocite-bornite replacements by tennantite 2 and 3 within the Spessart area must be mentioned,
pointing to an alteration of a probably Rote Fäule-related, high grade mineralization into Cu-As sulfides (Fig. 15). To date no Rote Fäule has been reported from

Table 4. Major chemical characteristics of different tennantite types. (After Diedel 1984)

Tennantite type	Ag wt.%	Sb wt.%	As wt.%
tn-1	−1	−2	−20
tn-2	−1	−2	−17
tn-3	−1	−2	−15

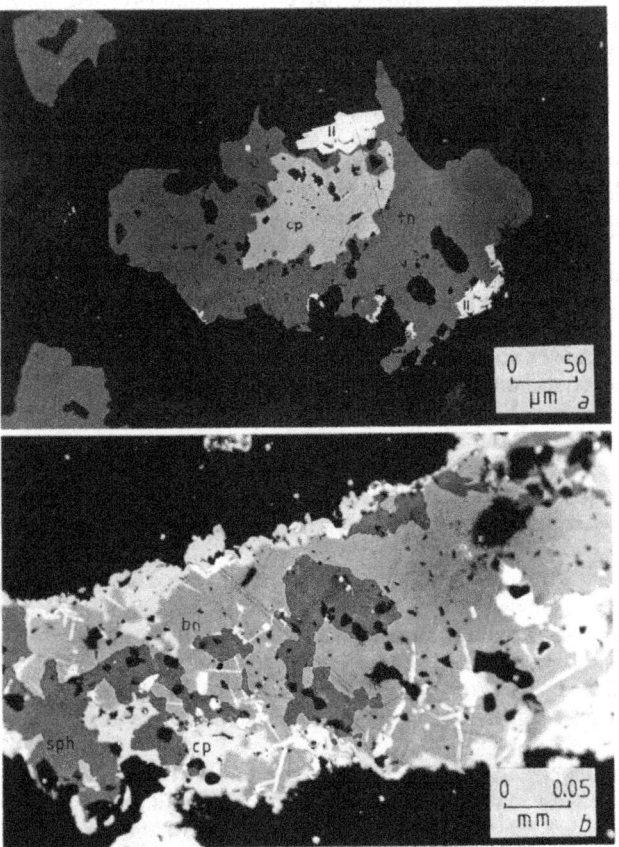

Fig. 15. Paragenetic features of epigenetic structure bound mineralization type: **a** chalcopyrite (*cp*)
replaced by tennantite (*tn*) replaced by loellingite (*ll*), drill hole SR 10, AS 19700. **b** Sphalerite (*sph*)
replaced by bornite (*bn*), replaced by chalcopyrite (*cp*), drill hole SR 9, AS 19554

the Spessart/Rhön area; red layers, which are restricted to the Zechsteinkalk and were thought to represent Rote Fäule (Kulick et al. 1984; Schumacher et al. 1984) had to be reinterpreted due to geochemical-mineralogical investigations and could be distinguished from ore controlling Rote Fäule (Schmidt 1987). The absence of probably to the chalcocite-bornite mineralization related Rote Fäule could be explained by the remobilization of Fe-hydroxides and hematite due to the influx of the hydrothermal fluids. It is worth noting that within the Spessart/Rhön area postvolcanic activities still exist, which refer to the mafic volcanism of Tertiary age. As a consequence, carbon dioxide gas deposits result, mainly outgasing in the vicinity of major fault structures. Associated wall rock alteration, e.g., kaolinization of Weissliegend sandstones and siderite ribbons within the Zechsteinkalk had been observed, whereas a slight gold enrichment (300 ppb) had been found within the Kupferschiefer black shale.

Particularly in the Rhön area, a third generation of tennantite (tn-3) has formed, characterized by significant amounts of silver. Discrimination between background and anomalous populations estimated a local background of 6 ppm (!) for the Rhön area (Schmidt 1985), reflecting the very untypical composition of the country rocks. Stromeyerite, which has been found in several drill cores from the Spessart/Rhön area, generated eventually as decomposition from tn-3. Mineralization is further accompanied by pyrite, indicating its hydrothermal affiliation due to a $Co/Ni > 1$ ratio.

In general, the base metal distribution shows a zonation into $Cu \rightarrow Zn \rightarrow Pb$ and $Cu \rightarrow Pb \rightarrow Zn$ respectively. Stratiform mineral layers are less frequent than in the other types, whereas lenses, fissures, and cavity fillings appear to be more abundant. Average grades are up to 7000 ppm copper, but nevertheless grades range up to 9% Cu plus 350 g/t Ag within the Kupferschiefer and 1.5% Cu plus 56 g/t or 1.1% Cu plus 73 g/t Ag over 2 m mining heigh for the two best holes in the Rhön area.

Genetic Models

Preconcentration and Synsedimentary Stage

In accordance with most of the recent authors who have carried out investigations on sandstone- and shale-hosted copper deposits (e.g., Moine et al. 1986; Lur'ye 1986; Rose et al. 1986), the underlying red-beds are considered as source for the metals entrapped in the sandstone, shale, or limestone units.

With special reference to the West German Kupferschiefer mineralization, these red-beds refer to the Rotliegend molasse troughs or basins, which generated from conjungated shear systems due to the relaxation of the highly compressed Variscan orogene (Drong et al. 1982). Igneous activities, as well as weathering of basement rocks, filled the basins with volcanics, pyroclastics, fanglomerates, conglomerates, and sandstones. Deposition of clastics in the Upper Rotliegend took place chiefly under arid conditions and led to preconcentration of various metals. As a result of lacking organic matter, a redox-controlled metal segregation developed, governed by decreasing E_h-values toward the center of the sedimentary basin, leading to a zonation into $Cu \rightarrow Pb \rightarrow Zn$ (Fig. 16a). For example, Lützner and Rentzsch (1975) described such a situation from the Goldlauter Schichten for an intramontanous molasse trough in the G.D.R. Due to an increasing grade of evaporation, sabkha sediments (e.g., caliche horizons and playa lake sediments) have formed in several places. As a consequence of the better solubility of lead and zinc in contrast to copper, the majority of preconcentrated zinc and lead in combination with only minor amounts of copper had been transported upward in the stratigraphic sequence and was reprecipitated at the top of the Rotliegend (Fig. 16b).

According to Rose (1976), solutions rich in chlorine can transport large amounts of metals as chloride-complexes. These chloride-brines preferably generated in red-bed environments. Controlled by the process of preconcentration, two major metal zones could have developed, revealing a copper predominance within the lower portion of Rotliegend and at the edges of the basin, and a zinc-lead predominance within higher parts of the Rotliegend and in the more basinward position. At least that zonation reflects the situation which has been encountered later on in the low grade mineralization of the Kupferschiefer after its deposition.

During transgression of Zechstein paleosea the upper parts of Rotliegend were penetrated by circulation of seawater and had been reworked (Fig. 16c). Precon-

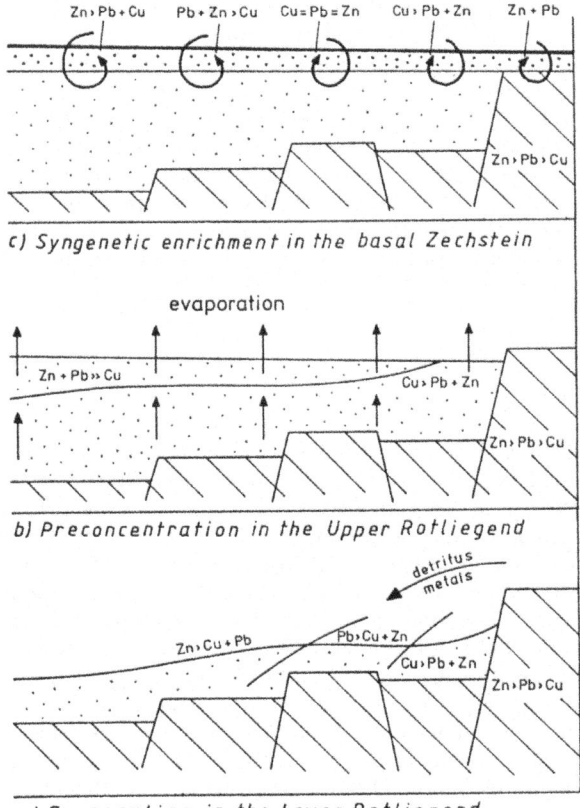

c) Syngenetic enrichment in the basal Zechstein

b) Preconcentration in the Upper Rotliegend

a) Segregation in the Lower Rotliegend

Fig. 16a-c. Model of preconcentration and synsedimentary stages of Kupferschiefer mineralization: **a** Metal segregation and preconcentration due to weathering of Variscan rocks in the Lower Rotliegend. **b** Preconcentration and further segregation due to sabkha processes in the Upper Rotliegend. **c** Reworking of Rotliegend sediments due to transgression of Zechstein paleosea and synsedimentary deposition of metals simultaneous to deposition of basal Zechstein host rock sequence

centrated metals were leached and had been redeposited penecontemporaneously with the Kupferschiefer black shale in an euxinic, reducing environment. Concerning involved mineral association, pyrite 1, marcasite 1, sphalerite 1, galena 1, chalcopyrite 1, tennantite 1 and loellingite 1 were formed. The formation of tn-1 and 11-1 could have been the result of weak hydrothermal activities, probably due to the waning stage of the Lower Permian volcanism, especially since abundant arsenides had recently been reported from Permian volcanics of the Saar-Nahe area (Dreyer 1981). Over paleohighs, where no preconcentration took place due to lacking underlying Rotliegend red-beds, the metals trapped in the basal Zechstein sequence originated directly from the reworked basement rocks. As a consequence the metal tenor is lower than within paleobasins. With reference to the proposed endogenic input, according to Diedel (1986) deep-sitting igneous bodies or incipient mantle diapirism may have been responsible for these mineralizations to a certain extend.

Notwithstanding that in some places early diagenetic features had been observed, this kind of mineralization is referred to as syngenetic and is so far characteristic for the low grade mineralization type.

Diagenetic Stage

It has already been discussed that formation of high grade mineralization and, if further accumulations have gained economic significance, of ore deposits, refers almost to the diagenetic stage due to an input of copper and silver from the underlying Rotliegend sediments into the Weissliegend-Kupferschiefer-Zechsteinkalk host rock sequence.

Transportation of metals by basinal brines could have been the result of a progressing subsidence of the sedimentary basin. Load of overburden rocks caused differences in the paleohydrological system; as a consequence migration of initially deep-seated aquifers was triggered, which moved solutions into the direction of lowest overload, which had been the marginal positions of the basins (Fig. 17). While percolating through the Lower Rotliegende sediments, which were relatively enriched in copper due to its prior depletion in lead and zinc, the basinal brines were mainly enriched in copper and silver. The sandbars, which so far had been "gaps" within the impermeable and reducing Kupferschiefer black shale, were favorable escape structures for the metaliferous solutions. Consequently the brines vented into the overlying, still soft sediments, where a precipitation of metals could take place along the flanks, governed by changing E_h and pH conditions. The slight crosscutting of mineralization could be interpreted as the result of the movement of solutions through the interface between higher and lower permeable rocks. According to v.Engelhardt (1973), the flow of fluids, which encounters a sedimentary interface at a steep angle, can change completely into an intrastratal flow when changing from a high permeable layer into a layer of lower permeability. Caused by the distinct host rock lithology and thus differences in permeability for the basal Zechstein sequence ($S\ l_{perm} > T\ l_{perm} > Ca\ l_{perm}$), the fluids could gain higher migration velocity within the Weissliegend than within the Zechsteinkalk. These features, in connection with further inhomogenities in host rock composition, could have caused the irregular shape of the Rote Fäule boundary. A model, how the Rote Fäule front and

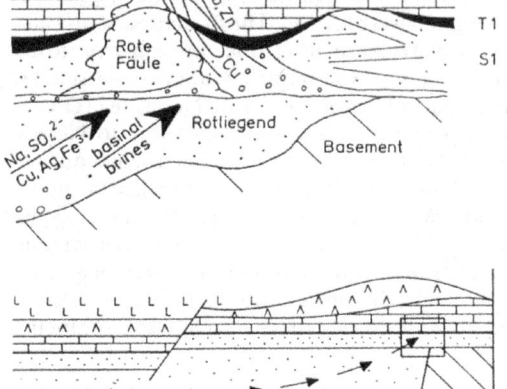

Fig. 17. Diagenetic formation of high grade mineralization due to flow of basinal brines to site of deposition as a consequence of progressing basin subsidence during Lower Zechstein. *A 1* Werra-Anhydrite; *Ca 1* Zechsteinkalk; *T 1* Kupferschiefer; *S 1* Weissliegend

associated mineralization could develop due to these circumstances, is illustrated below (Fig. 18).

Some amounts of the dissolved iron precipitated as ferrous iron together with carbonate and sulfate within the Rote Fäule. An E_h positive and pH neutral environment is presumed for the generation of Fe-hydroxides, whereas hematite was formed at the Rote Fäule margin in a slightly alkaline and more or less E_h-neutral environment. At the flanks of sandbars, where organic matter is scarce, a relatively H_2S deficit evolved in relation to Kupferschiefer beds, which are located in depressions. As a consequence, precipitation of metals was controlled by solubility, displaying decreasing Cu/S ratios with increasing distance to Rote Fäule and associated sandbars. Mineral deposition took place in the following sequence: $FeOOH \rightarrow Fe_2O_3 \rightarrow Cu_2S \rightarrow CuS \rightarrow Cu_5FeS_4 \rightarrow PbS \rightarrow ZnS \rightarrow FeS_2$.

Whereas hematite, chalcocite, covellite, digenite, and bornite appear for the first time, chalcopyrite 2, galena 2, and sphalerite 2 represent the remobilized and reprecipitated metals which resulted from the mineralization formed prior by the preconcentration and synsedimentary stages.

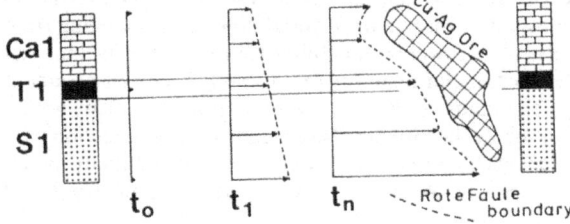

Fig. 18. Formation of Rote Fäule boundary and associated ore body as a function of time and host rock lithology

Epigenetic Stage

Formation of both the "Rücken" type Co-Ni-Ba mineralization in Richelsdorf and the Cu-As-Ag mineralization in Spessart/Rhön must be considered in relation to so-called Saxonian (post-Variscan) tectonism, which comprises a time span from Mesozoic to Tertiary and characterizes a tectonic style which is predominantly governed by basin and range formation with strong vertical and minor amounts of horizontal movement.

Probably some of the fault structures were arranged in the Permian or are even older, and had been reactivated, principally around paleohighs. Along these structures, which could act as channelways, fluids enriched in Co-Ni-Ba and Cu-As-Ag respectively ascended and altered the pre-existing Kupferschiefer mineralization, which was formed due to syngenetic and/or diagenetic processes (Fig. 19).

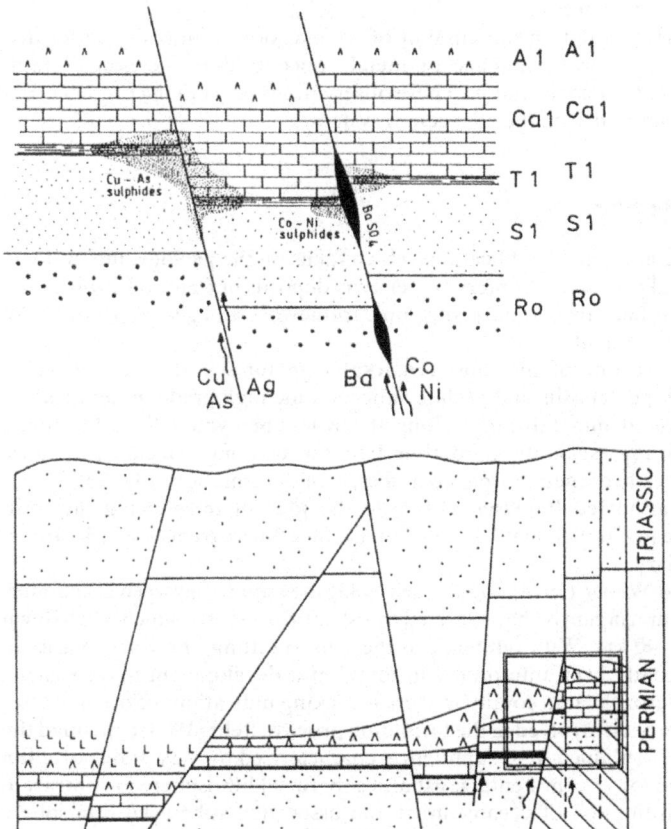

Fig. 19. Model of post-Permian epigenetic mineralization of structure bound mineralization in relation to basin and range formation

The distinct chemical composition of the solutions which formed the Co-Ni-Ba subtype and the Cu-As-Ag subtype respectively reflects probably regional variations in the composition of basement rocks. The basement of the Richelsdorf area consists of graywackes and phyllites of Carboniferous or Devonian age, which underwent only slight metamorphism, whereas the basement of the Spessart/Rhön area belongs to the Mid German Crystalline Rise, where rocks of Late Proterozoic to Ordovician age are exposed. Most common rock types are para- and orthogneisses, mica schists, amphibolites, diorites, quartzites, and other metasediments, which had undergone metamorphism at the Lower Carboniferous/Upper Carboniferous boundary (Schmidt et al. 1986a). The grade of metamorphism ranges from greenschist to amphibolite facies.

Presuming the leaching of metals from basement rocks by circulating, hot aqueous solutions, followed by their transportation to the site of deposition, the introduction of heat into the system coupled with the opening of fault structures are primary factors of ore control.

Studies which focused on the amount of coalification of organic matter displayed an apparent correspondence of metal concentration, location of fault structures, and grade of coalification. These observations point to the presence of a heat source which has driven the convection system.

Geotectonic Framework

Despite the large amount of publications which focus on the geology, mineralogy, geochemistry, and genesis of Kupferschiefer-type deposits in Central Europe, only little attention has been paid to the geotectonic framework and geodynamics of the surrounding environment.

In this context, one of the most remarkable features is the occurrence of Kupferschiefer-type deposits and to date subeconomic high grade mineralization and structure-bound mineralization along the Rhenohercynian/Saxothuringian boundary, which represents the borderline between two major tectonic regimes, both distinct in grade of metamorphism and style of deformation (Fig. 20).

This observation led Rentzsch (1974) to the idea of introducing the term "Central European Copper Belt" for those base metal occurrences which stretch along the boundary.

According to Weber (1978) and Giese (1983) the contact between Rhenohercynian and Saxothuringian is characterized by listric thrust faults, which reach down to a depth of 30–40 km. With reference to the crustal setting, the whole Variscan complex has recently been interpreted in terms of a development in an ensialic terrain over Proterozoic, thin continental crust, lacking indications of oceanic crust which could have been involved in the formation process. Behr (1978) explained the geodynamics as A-subduction or subfluence, characterized by mass transportation within the heated lower continental crust and subcrustal lithosphere in contrast to sinking and consumption of oceanic crust and associated subcrustal lithosphere which is typical for B-subduction zones.

Whether A-subduction complexes represent the modification of B-subduction zones with reference to the ensialic position and are so far comparable to the latter,

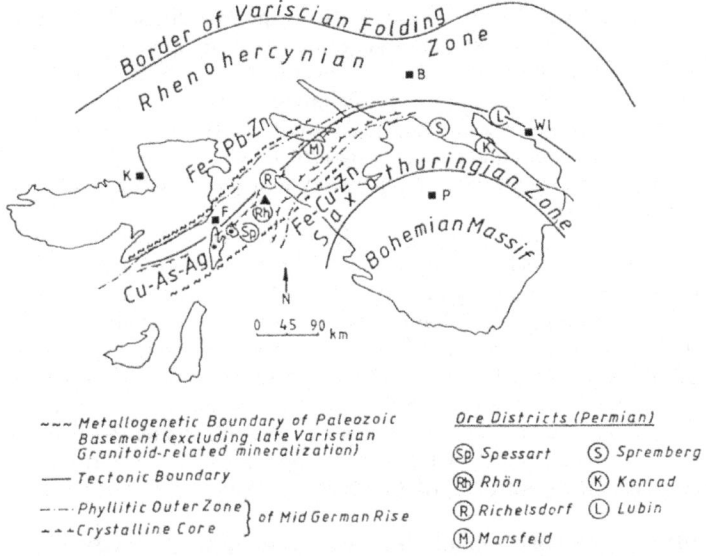

Legend:

~~~ *Metallogenetic Boundary of Paleozoic*
Basement (excluding late Variscian
Granitoid-related mineralization)

—— *Tectonic Boundary*

— · — *Phyllitic Outer Zone* ⎫ *of Mid German Rise*
- - - *Crystalline Core* ⎭

Ore Districts (Permian)

(Sp) *Spessart*   (S) *Spremberg*

(Rh) *Rhön*   (K) *Konrad*

(R) *Richelsdorf*   (L) *Lubin*

(M) *Mansfeld*

Fig. 20. Sketch map showing the tecto-stratigraphic situation of Central European Variscan Basement. Cities: *F* Frankfurt/M; *B* Berlin; *P* Prague; *Wl* Wrocław; *K* Köln; *triangle* sample sites of felsic metavolcanics analyzed; *points* sample sites of mafic metavolcanics analyzed

or lie in any spatial/temporal relation to convergent plate boundaries, seems unclear.

Based on the geochemistry of mafic metavolcanics (Okrusch et al. 1985), and from the metallogenetic point of view there seems to be evidence that parts of the Mid German Crystalline Rise, which represents the northernmost part of the Saxothuringian zone, could be referred to an ensialic island arc of late Proterozoic to early Paleozoic age. Compared to modern ensialic island arcs, which are favorable sites for iron, copper, and precious metal occurrences, similar characteristics have been presumed for ancient ensialic island arc settings (Schmidt in prep.). The already suggested copper-silver-arsenic significance of the Spessart/Rhön basement could probably be traced back to such a setting. Taking into account that in the Mid German Crystalline Rise most of the igneous rocks are of late Proterozoic to early Paleozoic age, in locations further southeast Zn-Cu-Fe massive sulfide deposits developed, which could be referred to the "Kieslager" type. According to Hutchinson (1980) these kind of deposits are typical for a fore-arc trough or trench setting. Investigations of the mafic country rocks ascertained a Fe-rich tholeiitic basaltic composition with mixed MORB/IAT affiliation, which could be related to an immature island arc setting. (Schmidt in prep).

The Rotliegend troughs and basins developed mainly north of the Mid German Crystalline Rise and were filled with debris which had been derived chiefly from the proposed ensialic arc. Hence the sediments evolving could have been initially enriched in copper, silver, and other base metals. Initial stage for the development of the Rotliegend basins was the opening of SW-NE and NW-SE striking shear systems due to relaxation of the Variscan stress regime. This development continued into the Zechstein and must be seen in direct connection to the previous orogenic activities. Consequently, the subsidence of sedimentary basins and thus the diagenetic formation of ore deposits and high grade mineralization does not refer to the early stage of an intracontinental rifting, as described by Sawkins (1984), but to the waning stage of the Variscan orogeny, probably in a sense such as has been described by Mitchell (1985).

In contrast to the interpretation of the Rotliegend volcanics as bimodal suits (e.g., Lorenz and Nicholls 1984), investigations by Eckhardt (1979) could show that these volcanics which occur within the Norddeutsches Becken display differentiation into basalt-andesite-rhyolite sequences and bear calc-alkalic composition comparable to rocks from island arcs like the Andes or Anatolie.

Several authors (e.g., Russel and Smythe 1978; Segalstad 1978) suggested subduction of an oceanic North Sea plate underneath the European continent at the Upper Carboniferous/Lower Permian boundary. Evidence for such an oceanic plate as a part of the ancient north Atlantic ocean appears in the occurrence of Permian tholeiites from the Central North Sea, which display "ocean floor basalt" affinity.

With regard to the presumedly small size of oceanic crust which formed between the Caledonian Orogene in the north and the Variscan Orogene in the south, consumption must have taken place of a very short duration. In general, young oceanic crust is less dense than old one, so the subduction zone dipped down gently.

In contrast to the calc-alkalic volcanics of Northern Germany, which so far represent the principle magmatic arc, the Permian volcanics further south, e.g., within the Saar-Nahe trough, show increasing K_2O/SiO_2 ratios (Jung and Vinx 1973), but still display calc-alkalic character.

According to observations made along the South American active continental margin, this type of volcanism can be referred to an inner-arc setting.

In relation to the abundance of rhyolite and melaphyre, andesite seems to be subordinated, hence a generation due to partial melting of oceanic crust in the vicinity of a Benioff zone seems less obvious. On the other hand the Saar-Nahe trough and its further northeast heading extension, for example, are situated over an area where thrust faults and subfluence planes appear within the continental crust, as described by Weber (1978) and Behr (1978). According to Giese (1983), intracrustal discontinuities have been discovered in a similar position, which were interpreted as lower crust and parts of the crust/mantle transition. With reference to the occurrence of the Gießen greywacke nappe complex north of the Mid German Crystalline Rise, the present lower crust beneath the northern part of the Saxothuringian can be regarded as subduced continental segments.

If subduction of the oceanic North Sea plate took place at the presumed shallow angle, a southward-directed mass transport within the upper mantle/lower lithosphere could have resulted. Due to compressional forces within the inner arc, deformation and probably upwelling of the mantle/crust boundary could be triggered below the complex of the Rhenish Massif. This incipient stage of mantle diapirism could have caused further tensional forces in marginal positions, where deep-reaching thrust faults existed and had been reactivated. These areas were

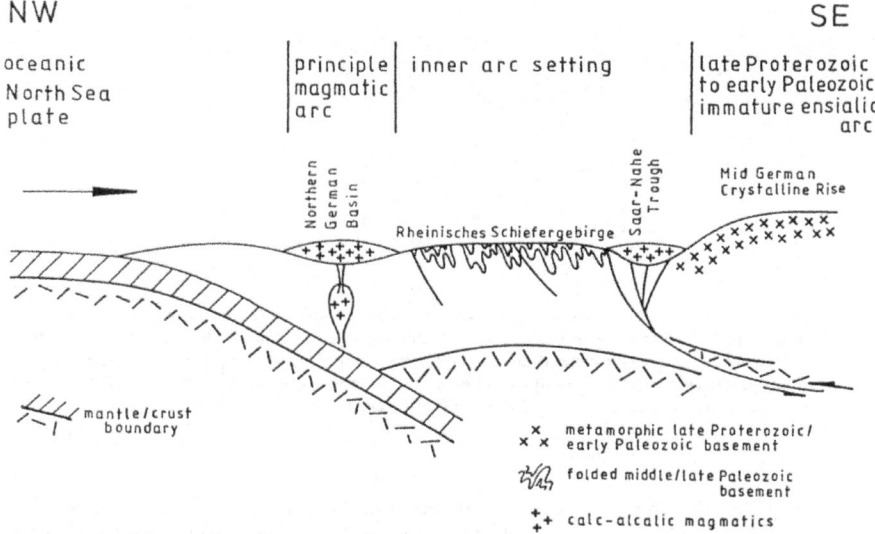

Fig. 21. Geotectonic model for the Northern German and Central German region at the Upper Carboniferous/Lower Permian time boundary

favorable sites for the development of Rotliegend troughs and basins in combination with the rise of magma which led ultimately to the generation of the basalt-andesite-rhyolite sequence (Fig. 21).

Whether the Lower Permian volcanics itself could represent source rocks for the contained metals of the Rotliegend, or if pene-exhalative activities were coupled with volcanism, as suggested by Brown and Chartrand (.1986) for some deposits in the USA, seems questionable.

Referring to the interpretation of geodynamics at the Upper Carboniferous/Lower Permian boundary, two models for the introduction of metals into the Rotliegend troughs, which lie adjacent to ore deposits, are discussed:

1. weathering of Cu-Ag enriched basement rocks which generated within an ensialic island arc setting;
2. emplacement of Permian volcanics which contain significant amounts of Cu and Ag and/or an additional input of pene-exhalative solutions enriched in Cu and Ag forming a reservoir within the Rotliegend red-beds (Fig. 22).

With regard to the occurrence of Kupferschiefer-type deposits along the Rhenohercynian/Saxothuringian boundary, model (1) is favored by the authors, especially since the Rotliegend trough between the Richelsdorf and Spessart-Rhön area bears only minor amounts of Autunien volcanics.

Incipient development of the previously mentioned mantle diapir could have been responsible for the endogenetic input of metals into the Kupferschiefer in particular places; progressing evolution could communicate with the uplift of the

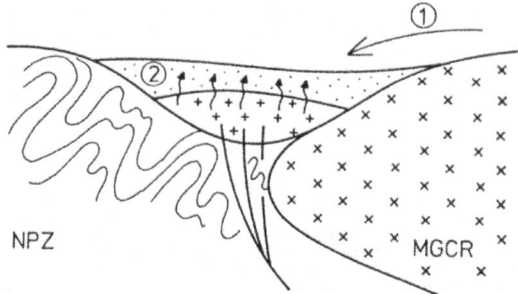

Fig. 22. Model for the provenance of Rotliegend red-bed hosted base metals, particular along the Rhenohercynian/Saxothuringian boundary. *1* Weathering of rocks from the Mid German Crystalline Rise which are enriched in base metals due to their ensialic island arc setting. *2* Input of pene-exhalative solutions enriched in Cu-Ag due to emplacement of Lower Permian calc-alkalic volcanics. *NPZ* Northern phyllite zone of Rhenohercynian; *MGCR* Mid German crystalline rise of Saxothuringian

Rhenish Massif, the early stage of Rheintalgraben rifting, and development of post-Variscian mineralization cycle. In accordance with the hydrothermal affiliation of the post-Variscan vein deposits, the genetic model, which involved leaching of metals from surrounding country rocks, has been accepted by most authors. In contrast to the conventional formation model by a heat-driven convection cell due to mantle diapirism, seismic pumping due to waning tectonic activities of the Variscan orogenesis was presumed by Behr and Horn (1984) as driving force for the formation of the epigenetic ore deposits.

Concluding Remarks

In general, the paragenesis of the Kupferschiefer-type mineralization can be described as a replacement sequence of iron sulfides → copper sulfides → arsenides with respect to the syngenetic, diagenetic, and epigenetic formation. In more detail, the following paragenetic features have been established (Fig. 23). Data from several locations of the Central European Copper Belt showed that the complete genetic sequence exists more or less within all known mineralization districts (Table 5). It has to be emphasized that economically important ore deposits are always coupled with the formation of Rote Fäule, which represents so far the visible alteration zone around the deposits.

Indeed, monster ore bodies like the Lubin district, where a minimum of 1500 million tons of ore had been reported, are unique and due to the coincidence of several important geological factors, i.e., pre-enriched source rocks, favorable host rock lithology and at least the correct timing of the formation process.

Comparing the Lubin district in Poland with Richelsdorf in West Germany, despite many similarities, some remarkable differences in the geologic setting were found, which might explain the difference in size and homogenity of the contained ore bodies (cf. Speczik et al. 1986).

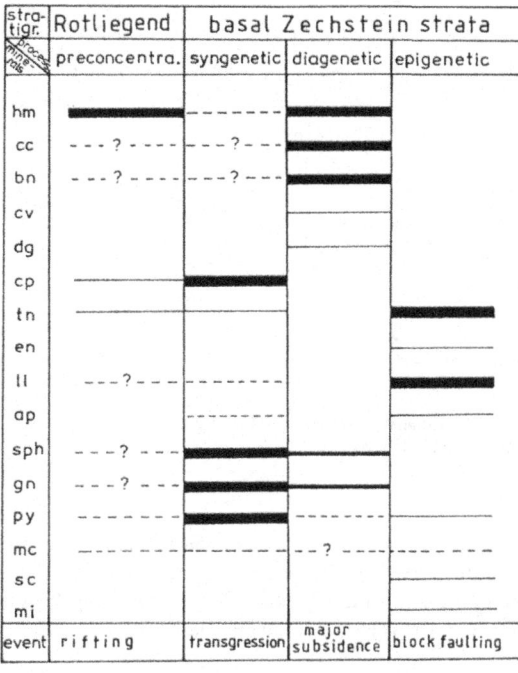

Fig. 23. Occurrence of ore minerals in relation to host strata, genetic process and geologic event. *hm* hematite; *cc* chalcocite; *bn* bornite; *cv* covellite; *dg* digenite; *cp* chalcopyrite; *tn* tennantite; *en* enargite; *ll loellingite;* *ap* arsenopyrite; *sph* sphalerite; *gn* galena; *py* pyrite; *mc* marcasite; *sc* scutterudite; *mi* millerite

Table 5. Processes of mineralization in the Central European Kupferschiefer districts. *Brackets* minor occurrence

"Kupferschiefer"-districts	Processes of mineralization
Lubin-Polkowice-Rudna (Poland)	Syngenetic-diagenetic-(epigenetic)
Mansfeld-Sangerhausen (G.D.R.)	Syngenetic-diagenetic-epigenetic
Spremberg-Weißwasser (G.D.R.)	Syngenetic-diagenetic-(epigenetic)
Richelsdorf (F.R.G.)	Syngenetic-diagenetic-epigenetic
Spessart/Rhön (F.R.G.)	Syngenetic-(diagenetic)-epigenetic
Harzvorland (F.R.G.)	Syngenetic-(diagenetic)-(epigenetic)
Korbacher Bucht (F.R.G.)	Syngenetic-(diagenetic)-epigenetic
Niederrheinische Bucht (F.R.G.)	Syngenetic-(diagenetic)
Norddeutsches Becken (F.R.G.)	Syngenetic-(diagenetic)

Abrupt facies changes within the Weissliegend, which are typical for the Richelsdorf and Spessart/Rhön area, have not been reported so frequently from Lubin. Further, in contrast to Richelsdorf, where the main tectonic lines crosscut the paleohigh trends, in Lubin the major fault structures run parallel in strike to the basement-basin configuration (Fig. 24).

As a consequence, in Poland the paleohydrological features probably corresponded better with the pregiven paleogeographic scenario, so that the emerged fluids encountered and mineralized the Kupferschiefer in a broader scale.

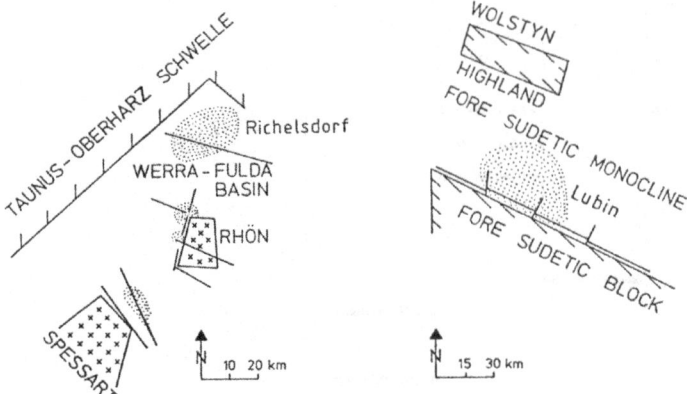

Fig. 24. Major tectonic lines and basement-basin strike directions of the West German Richelsdorf-Spessart-Rhön area and the Polish Lubin district in comparison

Taking all ascertained and described parameters into account it can be finally suggested, that formation of Kupferschiefer mineralization and associated ore deposits appears as a multi-stage process, ultimately controlled by the evolution of the sedimentary basins as a function of the regional geotectonic setting and history.

Acknowledgments. The authors wish to thank the St. Joe Minerals Corp., USA, for permission to publish results from the Kupferschiefer Project, which was financially supported by the Bundesministerium für Wirtschaft, Bonn. Further thanks are due to C. Schumacher (Braunschweig), R. Diedel (Aachen), M. Kotnik (Aachen), S. Speczik (Warzawa), A. Rydzewski (Warzawa), J. Rentzsch (Berlin) for numerous discussions, to N. Janz (Hanover), who carried out most of the draftings and to Mrs. S. Maßmann (Hanover) for improvement of the English manuscript.

References

Behr H (1978) Subfluenz-Prozesse im Grundgebirgsstockwerk Mitteleuropas. Z Dtsch Geol Ges 129/1:283–318
Behr HJ, Horn E-E (1984) Quarzbildung und Verkieselungsprozesse in den Karbonatkomplexen des Rheinischen Schiefergebirges. In: Postvaristische Gangmineralisation in Mitteleuropa, GDMB Schriftenreihe 41:27–45
Bjørlykke A, Sangster DF (1981) An overview of sandstone lead deposits and their relation to red-bed copper and carbonate hosted lead-zinc deposits. In: Skinner BJ (ed) Econ Geol 75th Anniv Vol:179–213
Brown AC, Chartrand FM (1986) Diagenetic features at Wh᠁ Pine (Michigan), Redstone (NW Territories, Canada) and Kamoto (Zaire). Sequence of mir ization in sediment-hosted copper deposits (Part 1). In: Friedrich GH, Genkin AD, Naldrett AJ ge JD, Sillitoe RH, Vokes FM (eds) Geology and metallogeny of copper deposits. SGA Spec P 1:390–397
Diedel R (1984) Mikrosondenuntersuchungen an Sulfiden des Kupferschiefers der Gebiete Spessart, Rhön und Richelsdorf. Diplom-Thesis, Rheinisch Westfälische Technische Hochschule Aachen

Diedel R (1986) Die Metallogenese des Kupferschiefers in der Niederrheinischen Bucht. PhD-Thesis, Rheinisch Westfälische Technische Hochschule Aachen

Dreyer G (1981) Erzmineralisation im Permokarbon des Saar-Nahe Gebiets. In: Proceedings of the International Symposium Central European Permian in Warsaw 1978, pp 646–647

Drong HJ, Plein E, Sannemann D, Schuepbach MA, Zimdars J (1982) Der Schneverdinger-Sandstein des Rotliegenden – eine äolische Sedimentfüllung alter Grabenstrukturen. Z Dtsch Geol Ges 133:699–725

Eckhardt FJ (1979) Der permische Vulkanismus Mitteleuropas. Geol Jahrb D 35:3–84

Engelhardt W v (1973) Die Bildung von Sedimenten und Sedimentgesteinen. Sediment-Petrologie, Teil 3. Schweizerbarth'sche Verlagsbuchhandlung, Stuttgart

Friedrich G, Diedel R, Hürter H, Kotnik K (1982) Petrographische und erzmikroskopische Untersuchung von Dünn-und Anschliffen der Bohrung Ro 18, Untersuchungsbericht Nr. 19, Kupferschieferprojekt. Rheinisch Westfälische Technische Hochschule Aachen

Friedrich G, Diedel R, Hürter H, Kotnik K (1983a) Petrographische und erzmikroskopische Untersuchung von Dünn- und Anschliffen der Bohrung Ro 21, Untersuchungsbericht Nr. 25, Kupferschieferprojekt. Rheinisch Westfälische Technische Hochschule Aachen

Friedrich G, Diedel R, Hürter H, Kotnik K (1983b) Petrographische und erzmikroskopische Untersuchung von Dünn- und Anschliffen der Bohrung Ro 24, Untersuchungsbericht Nr. 28, Kupferschiefer-projekt. Rheinisch Westfälische Technische Hochschule Aachen

Friedrich G, Diedel R, Schmidt F-P, Schumacher C (1984) Untersuchungen an Cu-As Sulfiden und Arseniden des basalen Zechsteins der Gebiete Spessart/Rhön und Richelsdorf. Fortschr Miner 62(1):63–65

Giese P (1983) The evolution of the Hercynian crust – some implications to the uplift problem of the Rhenish Massif. In: Fuchs K, Gehlen K von, Mälzer H, Murawski H, Semmel A (eds) Plateau uplift Springer, Berlin Heidelberg New York pp 303–314

Glennie AW, Buller AT (1983) The Permian Weissliegend of NW Europe: the partial deformation of eolian dune sands caused by the Zechstein transgression. Sediment Geol 35:43–81

Gunzert G (1953) Über die Bedeutung nachträglicher Erzverschiebungen in der Kupferschieferlagerstätte des Richelsdorfer Gebirges. Notizbl Hess Landesamtes Bodenforsch Wiesb 81:258–283

Gustafson LB, Williams N (1981) Sediment-hosted stratiform deposits of copper, lead and zinc. In: Skinner BJ (ed) Econ Geol 75th Anniv Vol:139–178

Hutchinson RW (1980) Massive base metal sulfide deposits as guides to tectonic evolution. In: Strangway DW (ed) The continental crust and its mineral deposits. Geol Assoc Can Spec Pap 20:659–684

Jowett EC (1986) Late diagenetic origin of Kupferschiefer Cu-Ag deposits in Poland by convective fluid flow of Rotliegendes brines, vol 11. Abst GAC MAC CGU, p 87

Jung D, Vinx R (1973) Einige Bemerkungen zur Geochemie der Magmatite des Saar-Nahe-Pfalz Gebietes. Ann Sci Univ Besancon Geol 3 ser 18:197–202

Jung G, Knitzschke G, Gerlach R (1973) Zur Selenführung des Kupferschiefers im SE-Harzvorland. Z Angew Geol 19/2:57–67

Knitzschke G (1966) Zur Erzmineralisation, Petrographie, Hauptmetall und Spurenelementführung des Kupferschiefers im SE Harzvorland. Freib Forschungsh C 207:1–147

Kucha H (1982) Platinum group metals in the Zechstein copper deposits, Poland. Econ Geol 77 (6):1578–1591

Kucha H. Pawlikowski M (1986) Two-brine model of the genesis of strata-bound Zechstein deposits (Kupferschiefertype), Poland. Miner Dep 21:70–80

Kulick J (1968) Erläuterungen zur geologischen Karte von Hessen: Blatt 4719 Korbach, Wiesbaden

Kulick J, Leifeld D, Meisl S et al. (1984) Petrofazielle und chemische Erkundung des Kupferschiefers der Hessischen Senke und des Harz-Westrandes. Geol Jahrb D 68:3–223

Lorenz V, Nicholls JA (1984) Plate and intraplate processes of Hercynian Europe during the Late Paleozoic. Tectonophysics 107:25–56

Lützner H, Rentzsch J (1975) Sedimentation und Metallogenie in einem intramontanen Becken der variszischen Molasse. Z Geol Wiss 3 (11):1473–1490

Lur'ye AM (1977) Zur Herkunft des Kupfers in den Basisschichten des Zechsteins der Kasanstufe. Z Angew Geol 23/6

Lur'ye AM (1986) Formation conditions of copper sandstone and copper shale deposits. In: Friedrich GH, Genkin AD, Naldrett AJ, Ridge JD, Sillitoe RH, Vokes FM (eds) Geology and Metallogeny of Copper Deposits. SGA Spec Publ 4:477–491

Marowsky G (1969) Schwefel-, Kohlenstoff- und Sauerstoff-Isotopenuntersuchungen am Kupfers-
    chiefer als Bietrag zur genetischen Deutung. Contrib Miner Pet 22:290-334
Messer E (1955) Kupferschiefer, Sanderz und Kobaltrücken im Richelsdorfer Gebirge. Hess Lager-
    stättenar 3
Mitchell AHG (1985) Mineral deposits related to tectonic events accompanying arc-continent collision.
    Trans Inst Miner Metall 94:B115-B125
Moine B, Guilloux L, Audeoud D (1986) Major element geochemistry of the host rocks in some
    sediment-hosted copper deposits. In: Friedrich GH, Genkin AD, Naldrett AJ, Ridge JD, Sillitoe
    RH, Vokes FM (eds) Geology and metallogeny of copper deposits. SGA Spec Publ 4:443-460
Okrusch M, Müller R, El Shazly S (1985) Die Amphibolite, Kalksilikatgesteine und Hornblendegneise
    der Alzenauer Gneis-Serie am Nordwest-Spessart. Geol Bavarica 87:5-37
Paul J (1982) Zur Rand- und Schwellenfazies des Kupferschiefers. Z Dtsch Geol Ges 133/4:571-605
Perelman A (1972) Geochemie epigenetischer Prozeße (Die hypergene Zone Teil 1 und 2). Akademie,
    Berlin
Rentzsch J (1974) The Kupferschiefer in comparison with the deposits of the Zambian Copper Belt. Cent
    Soc Geol Belg Gisem Strat Prov Cup:295-418
Rentzsch J (1981) Mineralogical-geochemical prospection methods in the Central European Copper Belt.
    Erzmet 34 (9):492-495
Rentzsch J, Knitzschke G (1968) Die Erzmineralparagenesen des Kupferschiefers und ihre regionale
    Verbreitung. Freib Forschungsh C 231:189-211
Rose AW (1976) The effect of cuprous chloride complexes in the origin of red-bed and related deposits.
    Econ Geol 71:1036-1048
Rose AW, Smith AT, Lustwerk RL, Ohmoto H, Hoy LD (1986) Geochemical aspects of stratiform and
    red-bed copper deposits in the Catskill Formation (Pennsylvania, USA) and Redstone Area
    (Canada). Sequence of mineralization in sediment-hosted copper deposits, part 3. In: Friedrich GH,
    Genkin AD, Naldrett AJ, Ridge JD, Sillitoe RH, Vokes FM (eds) Geology and metallogeny of
    copper deposits. SGA Spec Publ 4:412-421
Russel MJ, Smythe DK (1978) Evidence for an early Permian oceanic rift in the Northern Atlantic. In:
    Neuman ER and Ramberd IB (eds) Petrology and Geochemistry of Continental Rifts. pp 173-180
Rydzewski A (1978) Oxidated facies of the copper bearing Zechstein shales in the Fore-Sudetic
    Monocline. Przege Geol 26:102-108
Sawkins FJ (1984) Metal deposits in relation to plate tectonics. Springer, Berlin Heidelberg New York
Schmidt F-P (1985) Erzkontrolle im Kupferschiefer Osthessens, Bundesrepublik Deutschland. PhD-
    Thesis Rheinisch Westfälische Technische Hochschule Aachen
Schmidt F-P (1987) Alteration zones around Kupferschiefer-type base metal mineralization in West
    Germany. Miner Dep 22:172-177
Schmidt F-P (in prep) The Mid German Crystalline Rise — example of a late Proterozoic ensialic arc.
    Earth Planet Sci Lett
Schmidt F-P, Gebreyohannes Y, Schliestedt M (1986a) Das Grundgebirge der Rhön. Z Dtsch Geol Ges
    137 (1):287-300
Schmidt F-P, Schumacher C, Spieth V, Friedrich G (1986b) Results of recent exploration for copper-
    silver deposits in the Kupferschiefer of West Germany. In: Friedrich GH, Genkin AD, Naldrett AJ,
    Ridge JD, Sillitoe RH, Vokes FM (eds) Geology and metallogeny of copper deposits. SGA Spec
    Publ 4:572-582
Schumacher C (1985a) Die Kupfervererzungen des basalen Zechsteins im Rahmen der sedimentären
    Entwicklung des Werra-Fulda Beckens. PhD-Thesis, Freie Universität Berlin
Schumacher C (1985b) Die Grenze Rotliegendes/Zechstein im Werra-Fulda Becken. Z Dtsch Geol Ges
    136 (1):121-128
Schumacher C, Schmidt F-P (1985) Kupferschieferexploration in Osthessen und Nordbayern. Erzmet
    38 (9):428-432
Schumacher C, Kaidies E, Schmidt F-P (1984) Der basale Zechstein der Spessart-Rhön Schwelle. Z Dtsch
    Geol Ges 135 (2):563-571
Segalstad TV (1978) Petrology of the Skien basaltic rocks and the early basaltic (B1) volcanism of the
    permian Oslo rift. In: Neuman ER, Ramberd IB (eds) Petrology and Geochemistry of Continental
    Rifts. pp 209-216
Speczik S, Skowronek C, Friedrich G, Diedel R, Schumacher C, Schmidt F-P (1986) The environment
    of generation of some Zechstein base metal occurrences of Central Europe. Acta Geol Pol 35:1-35

Tischendorf G, Ungethüm H (1965) Zur Anwendung von E_h-pH Beziehungen in der geologischen Praxis. Z Angew Geol 11 H 2

Tourtelot EB, Vine JD (1976) Copper deposits in volcanogenic and sedimentary rocks. US Geol Surv Prof Pap 907C:1–34

Weber K (1978) Das Bewegungsbild des Rhenohercynikums — Abbild einer varistischen Subfluenz. Z Dtsch Geol Ges 129 (1):249–281

Wedepohl KH (1964) Untersuchungen am Kupferschiefer in NW-Deutschland; ein Beitrag zur Deutung der Genese bituminöser Sedimente. Geochem Cosmochem Acta 28:305–364

Wedepohl KH, Delevaux MH, Doe BR (1978) The potential source of lead in the Permian Kupferschiefer bed of Europe and some selected Paleozoic mineral deposits in the Federal Republic of Germany. Contrib Miner Pet 65:273–281

Base Metal Mineralization and Maturation of Organic Material in the Kupferschiefer of the Lower Rhine Basin

R. DIEDEL[1] and W. PÜTTMANN[2]

Abstract

Inorganic and organic geochemical investigations have been carried out on the Kupferschiefer of 30 drillings of the Lower Rhine Basin. The target area covers a region of 600 km². It represents the most southern part of the former Zechstein sea in this area and is situated north of the Rhenohercynian zone in the sub-Variscan foreland of the Variscan Mountains. The folded Carboniferous rocks of the Variscan basement are discordantly overlain by the Permian Kupferschiefer. Rotliegend sediments are totally lacking. The area is characterized by Variscan striking synclines and anticlines, which are perpendicularly cut by deep, northwest-southeast striking faults. In the west and the north two areas with an increased Zn-Pb-mineralization occur. They are both linked to the main tectonic structures, and, in the western part, additionally to an anomalous maturation of the Kupferschiefer. The base metal mineralization of the Kupferschiefer reflects the metal distribution of the basement and is suggested to represent the final stage of intraformational processes. A positive correlation between mineralization and organic carbon has not been observed.

Geology

The Lower Rhine Basin constitutes the southern, NNE to SSE-striking part of a vast geological structure situated near the border of Westphalian (sub-Variscan foreland) and Rhenohercynian zones of the Variscan externides. This area of virtual Variscan subsidence that stretches along the southern shorebelt of the Zechstein sea is currently continually subsiding. The extension of the Lower Rhine Basin is limited on the east by the Münsterländer Cretaceous Plateau, and on the south and southeast by the Rheinisches Schiefergebirge.

The geological structure of the Lower Rhine Basin is composed of two structural stages, the folded and faulted Carboniferous basement, and the gently (from 2° to 5°) north-dipping, Permian-Mesozoic and Caenozoic sedimentary cover. The base-

---

[1]Institut für Mineralogie und Lagerstättenlehre, RWTH Aachen, Wüllnerstr. 2, 5100 Aachen, FRG
[2]Lehrstuhl für Geologie, Geochemie und Lagerstätten des Erdöls und der Kohle, RWTH Aachen, Lochnerstr. 4–20, 5100 Aachen, FRG

Base Metal Sulfide Deposits
G.H. Friedrich, P.M. Herzig (Eds.)
© Springer-Verlag Berlin Heidelberg 1988

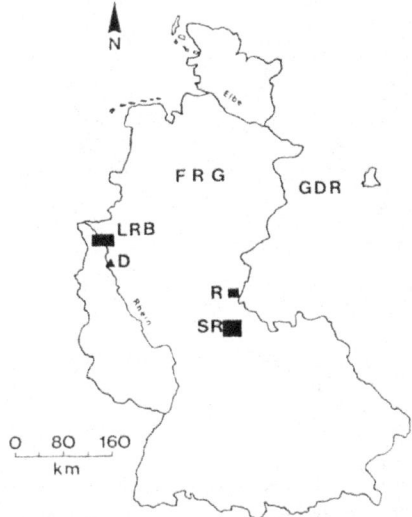

Fig. 1. German Kupferschiefer occurrences; *LRB* Lower Rhine Basin: *R* Richelsdorf: *SR* Spessart/Rhön; *D* Düsseldorf

ment crops out southeast, at the northern rim of the Rheinisches Schiefergebirge. Older formations of Devonian and Ordovician age, similar to that of the Stavelot-Venn Anticline of the Rheinisches Schiefergebirge are thought to compose the deeper parts of the Lower Rhine Basin basement. The latter assumption has not been confirmed by consequent drillings. The basic pluton of the Permo-Carboniferous 'Krefelder Gewölbe' domes up to a depth of 3.5 km beneath the top of the Upper Carboniferous rocks. The basic character of the pluton is suggested both by geophysical work (Buntebarth et al. 1982) and by basalt-olivine veins (Niemöller et al. 1973) that abundantly cut Carboniferous formations. Up to several thousand meters thick Lower and Upper Carboniferous formations consist of: limestone, quartzite, sandstone, clay-schist, and shale with countless minable coal seams, more in the upper part of the section (Westfal A and B). After Westfal C, the Carboniferous strata has been folded, with decreasing intensity from the south to the north. Simultaneously, the distance between the axis of northeast to southwest-trending anticlines and synclines increases. The folding system is accompanied by perpendicular system of northwest-southeast striking faults, that shifted or uplifted some elements of the structure, up to 400 m. The movement along these faults, in the post-Variscan time, resulted in actual horst and graben morphology (Fig. 2).

Rotliegend sediments were thinly deposited over the entire area of the Lower Rhine Basin and eroded immediately. A thin (about 25 cm, max. 1 m) bed of conglomerate or fine to coarse-grained sandstone represents the first sediment of the transgressing sea. The main components of pebbles and cobbles are very poorly rounded and sorted fragments of Carboniferous limestones, sandstones, and schists, rarely Upper to Middle Devonian limestones. Limestones similar to the latter compose the main body of the Velberter Anticline, situated about 20 km southeast

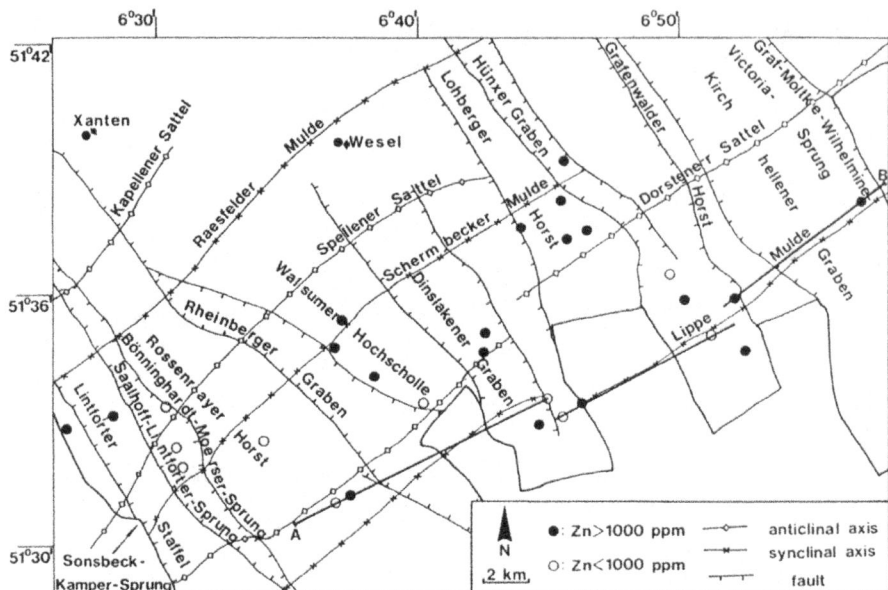

Fig. 2. Main tectonic structures in the Lower Rhine Basin (from Geologische Karte des Ruhrkarbons, 1982) and average zinc content of investigated drillings

of Lower Rhine Basin. The cements consist mainly of calcite, dolomite, ankerite, anhydrite, gypsum, and barite. At places it has an elevated content of argillaceous material. The barite also occurs in a form of veinlets and nodules up to 3 cm thick.

A thin (2 to 5 cm) irregular patchy bed of basal dolomitic limestone sometimes separate the Kupferschiefer from underlying basal conglomerates. The Kupferschiefer is developed here as a dark grey, bituminous and laminated marly limestone varying in thickness from 50 to 380 cm. At places it is intercalated with thin clayey laminae, and at flanks of some sea paleohighs it pinches out. Overlying Zechstein sediments of Z1 to Z4 cyclotherms show facies thickness development and distribution typical of marginal basins. They are overlain by about 300 m Triassic Buntsandstein succeeded by Paleocene, Oligocene, and Miocene (about 300 m).

The Kupferschiefer reveals remarkable distribution pattern of base metals (see next Sect. 2), barium, and strontium. When overlying Westfal C strata the Kupferschiefer shows relatively low barium content, about 200 ppm; above the Westfal A and B strata the barium content increases to an average of about several thousand ppm and reaches a maximum of 5% (Diedel and Friedrich 1986). Such high values are also well known from the Carboniferous basement lead-zinc-barium hydrothermal vein- and impregnation-type mineralization of the 'Erzprovinz Ruhrgebiet' in the southeast and east of the Lower Rhine Basin (Buschendorf et al. 1957; Pilger 1961).

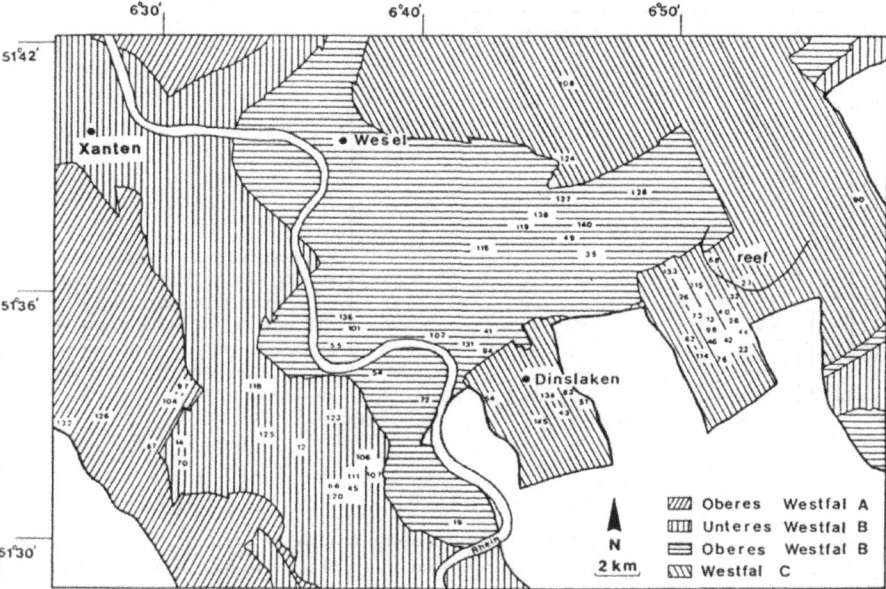

Fig. 3. Top of the Upper Carboniferous, Zechstein cover removed. *Numbers* indicate locations of drillings (from Geologische Karte des Ruhrkarbons, 1982)

Base Metal Mineralization

The lithostratigraphic unit of the Kupferschiefer has been investigated for copper, lead, and zinc. Base metal contents of "normally" composed black shales are given by Knitzschke (1966): 20–300 ppm for copper, 20–400 ppm for lead, and 100–1000 ppm for zinc. The distribution pattern of these metals in the Kupferschiefer of the Lower Rhine Basin is as follows:

The *copper* content is always very low, ranging from a few ppm to 280 ppm. These values correspond to those above for "normal" black shales.

The *lead* content normally varies between 20 and 350 ppm. Anomalous high values up to 1900 ppm are observed in ten drillings.

With the exception of drillings 43, 72, 98, and 133 anomalous high *zinc* contents occur in all drillings. The highest values are 1.25% Zn in drilling 90 and 0.55% over a vertical section of 75 cm in drilling 132.

The distribution of both lead and zinc are characterized by two populations (Diedel 1986), representing background (population 1) and anomalies (population 2). Population 1 ranges from 0 to 350 ppm for lead (Fig. 4) and from 0 to 750 ppm for zinc. These values correspond significantly with those after Knitzschke (s.a.) and are defined as syngenetic. The second population covers the wide field from 350 to 1900 ppm (lead) and from 750 to 12.500 ppm (zinc), representing an anomalous

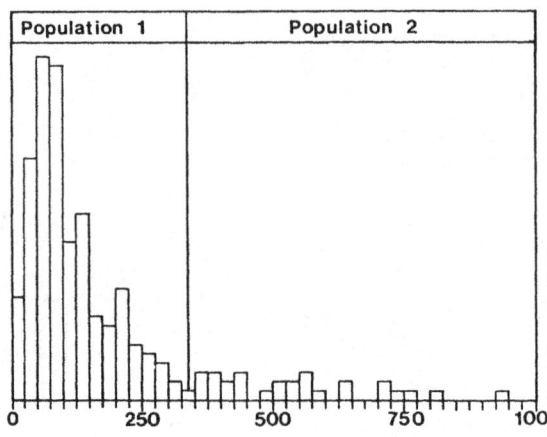

Fig. 4. Histogram for lead showing the relative frequency (y-axis) and the concentration range from 0 to 1000 ppm (x-axis); divided into 40 classes. Population 1: background values; Population 2: anomalies (after Diedel, 1986)

enrichment of both elements in a diagenetic/epigenetic stage.

Within a vertical section the anomalous high lead and zinc contents are concentrated without exception in the basal part (1 m) of the Kupferschiefer (Table 1). Maxima of zinc mostly occur 30–50 cm above the border of Kupferschiefer to underlying clastic sediments (Zechstein-Konglomerat). The maxima of lead are always situated above the maxima of zinc. Therefore a vertical zoning (bottom-top) Zn-Pb is evident, in contrast to the common Cu-Pb-Zn-zoning in the well known German (Richelsdorf) and Polish deposits (Lubin, Polkowice, Lena, Konrad, Speczik et al. 1986). The inverse zoning in the Lower Rhine Basin can be explained by an input of lead and zinc transported as chloride complexes in ascending oxidized solutions. Because of the lower stability of zinc-chloride-complexes compared with lead-chloride-complexes (Helgeson 1964), zinc sulfides have been the first precipitates. Another indication for ascending solutions is the fact that increased base metal contents are displaced to the basal part of the overlying Zechstein limestone (Zechsteinkalk), if the thickness of Kupferschiefer is lower than 50 cm.

A comparison of the average lead-zinc-contents of the Kupferschiefer (Fig. 5) with the underlying strata of Carboniferous basement (Fig. 3) shows a relationship between strong base metal mineralization and the strata of Oberes Westfal A and Oberes Westfal B and a relationship of low base metal contents and the strata of Unteres Westfal B (Table 2). Comparable results are reported by Diedel (1986) for the distribution of barium in the Kupferschiefer.

Using the diagram Cu-Pb-Zn after Knitzschke (1966) it is possible to characterize the distribution of base metals in the Kupferschiefer. Figure 6 shows that nearly all values are situated in the field of the zinc type. Only the Kupferschiefer of three drillings (43, 98, 133), representing a calcareous facies and characterizing the synsedimentary mineralization, belongs to the mixed lead-zinc-subtype.

Table 1. Average values of copper, lead, and zinc (ppm) in the lowermost part of the Kupferschiefer horizon (1 m; four specimens)

Drilling	Cu	Pb	Zn	BM	Basement
104	5[a]	85	288	378	
126	176	173	1032	1381	Oberes
132	70	675	4177	4922	Westfal A
14	53	129	1503	1658	
20	4[a]	178	982	1164	
45	40	333	2062	2435	Unteres
70	57	119	615	791	Westfal B
125	11[a]	116	564	691	
41	45	423	2024	2492	
49	47	828	2456	3331	
54	48	113	1221	1382	
55	52	125	1018	1195	
72	60	71	60	191	Oberes
94a	60	643	2980	3683	Westfal B
119	18[a]	606	1551	2175	
127	47	319	2987	3353	
136	60	207	1540	1807	
140	50	245	2386	2681	
22	50	124	1705	1879	
26	13	151	1235	1386	
32	55	511	3528	4094	
43	34	127	141	302	
51	11[a]	378	1347	1725	
90	59	155	3425	3639	Westfal C
98	38	233	241	512	
105	59	199	1302	1560	
124	53	397	3089	3539	
133	34	67	210	311	
134	48	163	768	979	

BM: sum of base metals
[a]values below detection limit.

Maturation of Organic Material

The Kupferschiefer of the Lower Rhine Basin is situated in depths between 300 m and 1000 m. It is a sediment with a low stage of maturation, and the vitrinite reflectance (R_{oil}) only reaches up to a maximum of 0.5%. Therefore, the method of vitrinite reflectance determination is not suitable to recognize low local differences of maturation.

A more sensitive parameter to indicate the maturation of sediments and petroleum has been found in the relative stereochemistry of pristane (Patience et al. 1978). This compound is a saturated isoprenoid hydrocarbon ($C_{19}H_{40}$). It is derived from the phytyl side chain of chlorophyll and is present in most sediments and crude oils. Because of the possible stereochemical situation in the position 6 and 10 of the

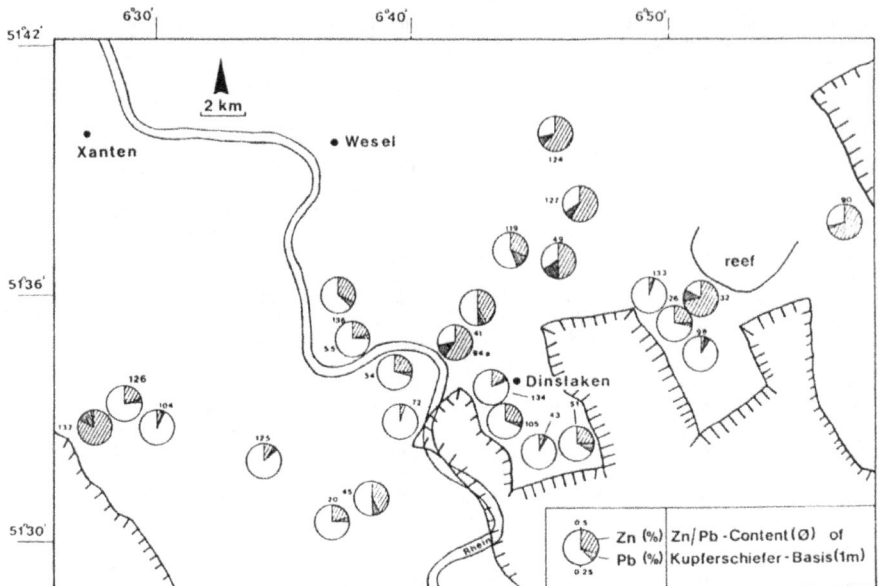

Fig. 5. Regional distribution of lead and zinc in the Lower Rhine Basin, represented by 25 drillings. Data are given for the lowermost part (1 m) of the Kupferschiefer horizon

Table 2. Statistic parameters of the copper-, lead-, and zinc-values of 30 drillings (values of each single drilling as in Table 1)

Basement	n		Cu	Pb	Zn	BM
		x_{min}	5[a]	85	288	378
Oberes	3	x_{max}	176	675	4177	4922
Westfal A		$\bar{x}$	84	311	1832	2227
		s	70	260	1686	1949
		x_{min}	4[a]	116	564	691
Unteres	5	x_{max}	57	333	2062	2435
Westfal B		$\bar{x}$	33	175	1145	1348
		s	22	82	561	641
		x_{min}	18	71	60	191
Oberes	10	x_{max}	60	828	2987	2435
Westfal B		$\bar{x}$	49	358	1822	2229
		s	12	245	879	1048
		x_{min}	11[a]	67	141	302
Westfal C	11	x_{max}	59	511	3528	4094
		$\bar{x}$	41	228	1545	1811
		s	16	133	1211	1304

n: number of drillings; x_{min}: minimum value; x_{max}: maximum value; x: average values; s: standard deviation; BM: sum of base metals

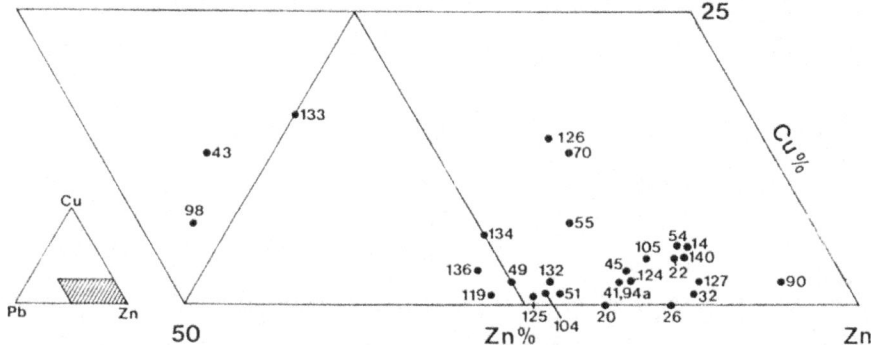

Fig. 6. Detail of the Cu-Pb-Zn triangle diagram showing the average base metal ratios in the Kupferschiefer of the investigated drillings (after Diedel and Friedrich, 1986)

molecule, three different stereoisomeres (diastereomeres) can occur (RR, SS, RS = SR).

The application of sophisticated capillary gas chromatography to organic sediment extracts allows the separation of the diastereomers of pristane into two peaks. One represents the RR and SS isomers, and the other the RS isomer, which is equivalent to the SR isomer. Only the RS isomer, which is present in the second eluting peak, is found in recent sediments as a stereospecific product of phythol degradation during early diagenetic processes. During maturation of sediments the first eluting peak (RR and SS isomers) is formed by the temperature-induced transformation of the recent isomer (Gassmann 1981). The value of "fossilized" to "recent" pristane is expressed in the ratio of diastereomers. Increasing temperature in the sediment causes an increase of this ratio until in the final stage an equilibrium value of 1 is reached. In the Kupferschiefer horizon of the Lower Rhine Basin the ratio of 1 is reached at a depth of burial of about 1000 m. Consequently, temperature effects in this sediment could be investigated by the diastereomeric ratio of pristane only up to 1000 m (e.g., drilling 124).

The maturation of sediments is primarily dependent on depth of burial. In this case the isograms of maturation should run parallel to isobaths. Isobaths of Kupferschiefer in the Lower Rhine Basin generally run SW-NE in the western and eastern part of the investigation area, and SSW-NNE in the middle part. The mentioned parallelism is given in the middle and eastern part. In the Krefelder Gewölbe, anomalous maturation values dominate (Fig. 7). In this area the isograms of maturation cut the isobaths perpendicularly. For this reason, an exclusive dependence of maturation on depth cannot be stated. Additionally, the maturation decreases going northeast from drilling 67. The latter intersects a fault in the Kupferschiefer.

For quantifying the maturation anomalies, a factor F has been developed, basing on the fact of an increasing ratio of diastereomeres of 0.1 for each 100 m of depth and using the parameters P (ratio of diastereomeres of pristane) and D (depth). The equation $D/P = F$ leads to a value of $F = 1000$ for the Kupferschiefer in the Lower Rhine Basin, if maturation is only influenced by depth (e.g., drilling 41:

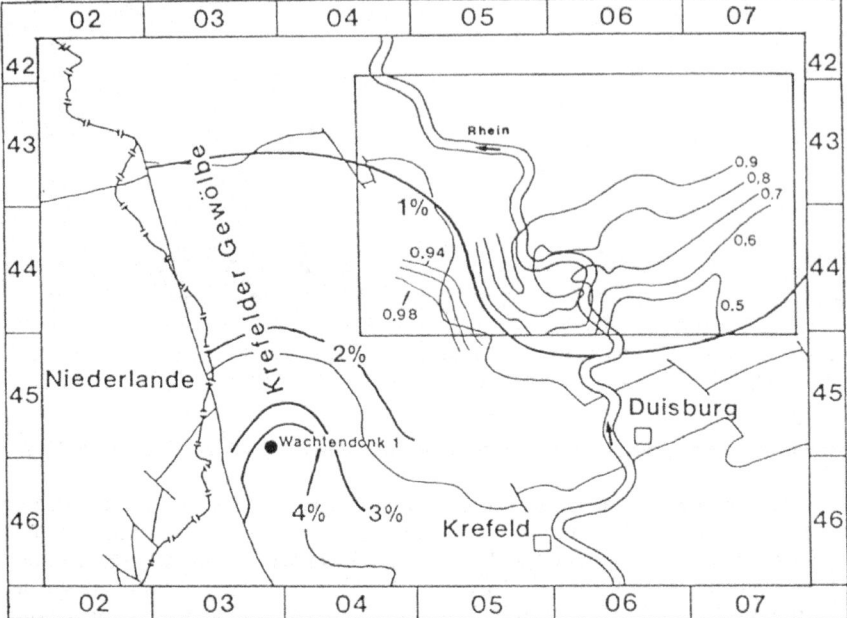

Fig. 7. Isograms of maturation, represented by the ratio of diastereomeres of pristane for the Kupferschiefer (*thin lines* values from 0.5 to 0.98) and for the Carboniferous (*thick lines* vitrinite reflectance Rm, values from 1 to 4%) (after Teichmüller et al., 1979); Wachtendonk 1: location of borehole in the center of the Krefelder Gewölbe

D = 752 m, P = 0.75, F = D/P = 1003; Table 3). Deviations of F ± 200 are caused by vertical tectonic movements, so all values < 800 can be defined as anomalous.

In the southeastern part of the investigated area a basic pluton beneath the Krefelder Gewölbe caused a coalification anomaly in the Upper Carboniferous formations (Westfal A). Here, the isograms of maturation of Kupferschiefer are oriented strictly parallel to those of the rocks of the Westfal (Fig. 7). Together with the fact that the isograms of maturation perpendicularly cut the isobaths of Kupferschiefer, a correlation between the maturation of Upper Carboniferous and Kupferschiefer, caused by an anomalous heat flow, is assumed.

Organic Carbon Content

The content of organic carbon has been determined in 22 drillings at the base of the Kupferschiefer (lowermost 10 cm). The values of organic carbon vary from 2.10 to 5.90% (Table 4). The low value of 0.51% in drilling 133 is caused by a local basement ridge, which led to an increased carbonate sedimentation under oxidizing conditions.

Table 3. Ratio F of diastereomers of pristane P and depth D (F = D/P)

Drilling	P	D	F	BM
14	0.98	685	699	1658
19	0.53	339	640	n.a.
20	0.67	538	795	1162
26	0.50	562	1124	1386
32	0.58	527	907	4094
41	0.75	752	1003	2492
45	0.67	583	870	2435
49	0.80	741	926	3331
51	0.37	446	1205	1725
54	0.72	524	727	1382
55	0.77	642	834	1191
64	0.50	683	1366	n.a.
67	1.00	671	671	n.a.
70	0.92	632	687	791
72	0.83	623	751	191
76	0.53	486	972	n.a.
94a	0.68	727	1069	3683
97	0.93	585	629	n.a.
116	0.78	801	1027	n.a.
119	0.81	847	1046	2157
123	0.90	615	683	n.a.
124	0.98	1012	1033	3539
125	0.72	472	656	680
126	0.99	562	568	1381
127	0.88	750	852	3353
132	0.95	384	404	4922
136	0.88	803	913	1807

Bm: sum of base metals; n.a.: not analyzed

Table 4. Content of organic carbon (C_{org}) in the Kupferschiefer of selected drillings and statistical parameters

Drilling/ specimen no.	C_{org} (%)	Drilling/ specimen no.	C_{org} (%)
19/11	3.40	94a/ 8	2.50
20/11	5.05	97/13	3.40
26/14	2.15	116/ 8	5.70
32/ 5	4.16	119/16	5.68
41/ 6	5.00	123/ 6	5.90
49/14	2.10	124/12	2.80
51/12	5.35	125/17	3.00
64/ 8	3.54	127/ 7	3.00
70/12	3.65	132/ 7	3.74
72/11	5.35	133/20	0.51
76/10	3.25	136/20	3.60
$x_{min} = 0.51$	$x_{max} = 5.90$	$\bar{x} = 3.77$	$s = 1.36$

x_{min}: minimum value
x_{max}: maximum value; x: average value; s: standard deviation

Additionally, the distribution pattern of organic carbon in a vertical profile of drilling 26 has been investigated within the lithostratigraphic units of Upper Carboniferous (Westfal C), Zechstein-Konglomerat, Kupferschiefer, and Zechsteinkalk (from bottom to top). In the Zechsteinkalk, which was deposited in oxidized facies, the average content of organic carbon is low (0.47%). The content of organic carbon in the Kupferschiefer increases from top (0.70%) to bottom (4.00%), with simultaneously decreasing carbonate content (Table 5). Significantly, the highest values of organic carbon occur 20 cm above the transition to the Zechstein-Konglomerat and decrease again in the direction of this transition. This phenomena may be explained by groundwaters circulating in the transition zone of Kupferschiefer and Zechstein-Konglomerat, depleting organic matter in the very bottom part of the Kupferschiefer. In the Zechstein-Konglomerat and in the Westfal C only low contents (< 0.1%) of organic carbon have been observed.

Table 5. Distribution of organic carbon (C_{org}) in the Kupferschiefer, in the Zechsteinkalk (hanging wall), in the Zechstein-Konglomerat and the Upper Carboniferous (footwall) of drilling 26

Specimen no.	$C_{org}(\%)$	Stratigraphic units	Specimen no.	$C_{org}(\%)$	Stratigraphic units
1	0.52		11	2.70	
2	0.41		12	3.40	Kupfer-
3	0.50	Zechstein-	13	4.00	schiefer
4	0.38	kalk	14	2.15	
5	0.54		15	0.09	Zechst.-Konglomer.
6	0.43		16	0.05	Upper
7	0.53		17	0.05	Carboniferous
8	0.70	Kupfer-	18	0.05	(Westfal C)
9	0.90	schiefer	19	0.08	
10	1.70				

Discussion

Base Metal Mineralization and Regional Tectonics

Block faulting is a characteristic feature of the Lower Rhine Basin and responsible for a system of NW-SE striking faults. These faults can be divided into two major types: type 1 represents the fault planes of the large horsts and grabens, the younger type 2 is situated within the horsts and grabens, causing step faults as in the southwestern part of the Lower Rhine Basin. High base metal contents in the Kupferschiefer are related to faults of type 1 (e.g., drillings 32, 51, 90). All drillings with low base metal contents are situated at a larger distance to type 1-faults. In contrast, the type-2 faults control only traces of the mineralization.

These results allow the conclusion that deep-seated faults of type 1 could have been the channels for ascending, metal-bearing solutions. A further indication for this hypothesis are the hydrothermal vein-type lead and zinc deposits in the Upper Carboniferous of the so-called Erzprovinz Ruhrgebiet, which occur in the same type of faults, only some kilometers away in the east and southeast.

Base Metal Mineralization and Maturation of Organic Material

Positive correlation between Kupferschiefer base metal occurrences and maturation is well known in Lower Silesia and southwest Poland (Speczik 1985b). In the Lower Rhine Basin two districts of anomalous base metal contents in the Kupferschiefer were recognized, one in the southwestern and another in the northern part of the investigated area. The Kupferschiefer of both districts is characterized by nearly the same stage of increased maturation and base metal contents. Different processes have been responsible for the high maturation. In the southwestern area (e.g., drilling 132, Table 3) the maturation is influenced by the anomalous heat flow in the Krefelder Gewölbe, in the northern part the same stage of maturation is related to "normal" depth of burial (about 1000 m). The low metal contents of the Kupferschiefer in the area of the Krefelder Gewölbe are due to the different facial development of the Kupferschiefer. Clayey-bituminous facies favors metal accumulation, whereas sandy-argillaceous and carbonate-rich facies causes metal dispersion. The latter is supported by drilling 126, which is situated nearly in the center of the highest maturation anomaly. Because of the high content of carbonates in the Kupferschiefer ($CaCo_3 + MgCO_3 = 55\%$), the base metal content is very low.

Base Metal Mineralization and Organic Carbon

No correlation has been found between base metal content and organic carbon content. This coincides with data given by Wedepohl (1964) for the Kupferschiefer of northwestern Germany. In contrast, Haranczyk (1972) found a positive correlation in the basal Zechstein of southwest Poland. On the other hand, the Kupferschiefer bottom seems to be leached by formation- or groundwaters circulating in the permeable transition zone between Kupferschiefer and underlying Zechstein-Konglomerat. Organic carbon and metals were mobilized, the latter of which are again redeposited in the Zechstein-Konglomerat. These processes would explain the high zinc content (0.5%) in the Zechstein-Konglomerat of drilling 26 (Diedel 1986). Similar descendent processes led to the formation of ore-bearing sandstones of the Lubin and Polkowice mines, southwest Poland (Banas et al. 1982).

Conclusions

It could be shown that high base metal contents in the Kupferschiefer of the Lower Rhine Basin occur in only two districts where increased maturation of the organic material was observed. Both districts show nearly the same stage of maturation. In the northern district the maturation depends only on depth of burial (about 1000 m). In contrast, the southwestern district is characterized by a high stage of maturation and by a depth of burial of only 400 m. Here, the maturation is caused by an anomalous heat flow in the roof of the pluton beneath the Krefelder Gewölbe.

Numerous factors are responsible for base metal enrichment in the Kupferschiefer horizon of the Lower Rhine Basin: (1) availability of thermal energy; (2) composition of basement rocks; (3) regional tectonic movements, and (4) facial development of the Kupferschiefer itself. It is suggested that anomalous high heat flow, resulting from the final stage of the geodynamical development of the Mid-European Variscides regionally led to a high geothermal gradient between 35° and 100°C (Buntebarth et al. 1982; Oncken 1984; Teichmüller et al. 1979; Weber and Behr 1983). This process enabled leaching of metals from the basement rocks and upward transporting in oxidized chlorine brines. Channels for the metal-bearing solutions have been both deep-seated faults and porous and permeable sandstones of the Upper Carboniferous. The clayey-bituminous Kupferschiefer, discordantly overlying the folded Carboniferous rocks, acted as a "geochemical trap", accumulating metals from the metalliferous brines. The metal accumulation represents the final stage of intraformational processes. In the Lower Rhine Basin the distribution of metals in the Kupferschiefer reflects the metal distribution (Zn, Pb, Ba) of the basement. The absence of anomalous copper values in the Kupferschiefer of the Niederrheinische Bucht is due to the missing of Rotliegendes sediments.

Acknowledgments. The authors are indebted to Prof. Dr. G. Friedrich, director of the Institute for Mineralogy and Economic Geology, for his support and the permission to publish data of his research project IV B 4 – FA 9828 "Sulfidmineralisation vom Typ Kupferschiefer in der Niederrheinischen Bucht", that had been financed by the Ministry of Science and Research of Nordrhein-Westfalen. Support of this study by Prof. M. Wolf is gratefully acknowledged. The thoughtful comments and suggestions of Dr. P.M. Herzig and the unknown reviewers were helpful in the clarification of certain points, and are most appreciated. The authors like to thank Mr. E. Barth for final drawing of figures.

References

Banas M, Salamon W, Piestrzynski A, Mayer W (1982) Replacement phenomena of terrigenous minerals by sulphides in copper-bearing Permian sandstones in Poland. In: Amstutz GC (ed) SGA Special Publ Vol 2. Springer Berlin Heidelberg New York
Buntebarth G, Michel W, Teichmüller R (1982) Das permokarbonische Intrusiv von Krefeld und seine Einwirkung auf Karbon-Kohlen am linken Niederrhein. Fortschr Geol Rheinl Westfalen 30:31-45
Buschendorf F, Richter M, Walther HW (1957) Die Blei-Zink-Erzvorkommen des Ruhrgebietes und seiner Umrandung. Beih Geol Jahrb 28:163
Diedel R (1985) Mineralogisch-geochemische Untersuchungen am Kupferschiefer der mittleren Niederrheinischen Bucht. Fortschr Miner 63:1, 47
Diedel R (1986) Die Metallogenese des Kupferschiefers in der Niederrheinischen Bucht. PhD Thesis Rheinisch-Westfälische Technische Hochschule, Aachen
Diedel R, Friedrich G (1986) Buntmetall- und Schwerspatmineralisation im basalen Zechstein (Kupferschiefer) der Niederrheinischen Bucht. Fortschr Geol Rheinl Westfalen 34:221-241
Gassmann G (1981) Chromatographic separation of diastereomeric isoprenoids for the identification of fossil fuel contaminants. Mar Pollu Bull 17:78-84
Geologische Karte des Ruhrkarbons 1:100 000, Krefeld (1982). Geologisches Landesamt Nordrhein-Westfalen
Haranczyk C (1972) Ore mineralization of the Lower Zechstein euxinic sediments in the Fore-Sudetic Monocline. Arch Miner 30:13-173 (in Polich)
Helgeson HC (1964) Complexing and hydrothermal ore deposition. Pergamon, New York
Knitzschke G (1966) Zur Erzmineralisation, Petrographie, Hauptmetallund Spurenelementführung des Kupferschiefers im SE-Harzvorland. Freib Forsch HC 207
Niemöller B, Stadler G, Teichmüller R (1973) Die Eruptivgänge und Naturkokse im Karbon des Steinkohlenbergwerks Friedrich Heinrich in Kamp-Lintfort (Linker Niederrhein) aus geologischer Sicht. Geol Mitt 12:197-218
Oncken O (1984) Zusammenhänge in der Strukturgenese des Rheinischen Schiefergebirges. Geol Rundsch 73:2, 619-649

Patience RL, Rowland SJ, Maxwell JR (1978) The effect of maturation on the configuration of pristane in sediments and petroleum. Geochim Cosmochim Acta 42:1871-1875
Pilger A (1961) Die Blei-Zink-Erzvorkommen des Ruhrgebietes und seiner Umrandung. Beih Geol Jahrb 40:385
Speczik S (1985a) Metallogeny of pre Zechstein basement of the Fore-Sudetic Monocline SW Poland. Geol Sudetica (Warsaw) 20:1 (in Polish)
Speczik S (1985b) Relation of Permian base metal occurrences to Variscan paleogeothermal field of the Fore-Sudetic Monocline. Results of fluid inclusion studies and vitrinite rank determination. Fortschr Miner 63:1, 222
Speczik S, Skowronek C, Friedrich G, Diedel R, Schumacher C, Schmidt FP (1986) The environment of generation of some Zechstein base metal occurrences of Central Europe. Acta Geol Pol 35:1-35
Teichmüller M, Teichmüller R, Weber K (1979) Inkohlung und Illit-Kristallinität. Fortschr Geol Rheinl Westfalen 27:201-276
Weber K, Behr HJ (1983) Geodynamic interpretation of the Mid-European Variscides. In: Martin H, Eder FW (eds) Intra continental fold-belts. Springer, Berlin Heidelberg New York, pp 427-469
Wedepohl KH (1964) Untersuchungen am Kupferschiefer in Nordwestdeutschland; Ein Beitrag zur Deutung bituminöser Sedimente. Geochim Cosmochim Acta 28:305-364

The Origin of the Kipushi (Cu, Zn, Pb) Deposit in Direct Relation with a Proterozoic Salt Diapir. Copperbelt of Central Africa, Shaba, Republic of Zaire

I. DE MAGNÉE[1] and A. FRANCOIS[2]

Abstract

The Kipushi deposit (Zn, Cu, Pb, Cd, Ag, Ge, Ga) is located in the southeastern part of the copper belt of Shaba. It is remarkable for its mineralogical wealth and its economic importance. This is also the case of the uranium deposit of Shinkolobwe (U, Ni, Co, Au, R.E.E., Pd, Pt), which is situated in the central part of the copper belt.

Both are clearly epigenetic and are located in the sedimentary rocks of the Katanga System (Upper Proterozoic). Its lower part, the Roan Supergroup, is mainly formed by alternating dolostones, siltstones, black shales, and chlorito-dolomitic siltstones. Its upper part, the Lower and the Upper Kundelungu Supergroups, is mainly detrital (tillitic conglomerates, sandstones, shales), but contains also limestones.

The Roan sediments are often of the sabkha or lagoonal type and at several levels of the stratigraphic column show "discontinuities" underlined by interstratified collapse breccias, formed probably by the dissolution of evaporites.

In the copper belt, the rocks of the Katanga System are intensely folded into an arcuate structure (Lufilian Arc), of dysharmonic Jurassic style. The cores of many anticlines are formed by megabreccias containing huge blocks of lower Roan dolomitic rocks of every size up to several kilometers. Many large clasts contain strata-bound Cu-Co sulfides and are actively mined. The matrix is formed by dolomitic siltstones, which may be residual sediments. They contain gypsum and anhydrite pseudomorphs.

The megabreccias often cut abruptly through the flank of an anticline and reach the Upper Kundelungu, with the result that blocks of this formation are mingled with those of the Roan. This is the case of Kipushi.

Anomalous cross-cutting megabreccias also exist along important strike-slip faults which cut obliquely across the folds (extrusion faults). The Shinkolobwe ore bodies are hosted inside such a structure.

Diapiric ascent of evaporites, followed by a long period of selective dissolution and the resulting collapse brecciation seem to be the only possible explanation of such generalized dislocations.

In our opinion, the saturated brines resulting from this dissolution played an essential role in the base metal mineralization, in particular for the formation of post-tectonic epigenetic deposits such as those considered in this paper. They are

[1]Free University of Brussels, 50, Av. F. Roosevelt, 1050 Bruxelles, Belgium
[2]21, Av. des Petits Champs, 1410 Waterloo, Belgium

Base Metal Sulfide Deposits
G.H. Friedrich, P.M. Herzig (Eds.)
© Springer-Verlag Berlin Heidelberg 1988

able to leach, transport, and deposit heavy chalcophile metals, as is well documented by present-day geothermal fields (Salton Sea, Phlegrean Fields).

To form economic ores, still other conditions are necessary: presence of sulfur and organic matter, high thermal gradient, micro- or macro- permeability, favorable host rock, and structures tending to channel the ascending hydrothermal solutions into a rather narrow space over a long period.

At Kipushi, this condition is created by a huge slab of shaly Kundelungu rocks sunken alongside the wall of the diapir and forming the impermeable roof of the ore bodies. Their footwall is formed by the dolostones of the Lower Kundelungu.

At Shinkolobwe, the rich uranium lodes and stockwerks are hosted by a heavily tectonized block of the Mine Series (lower Roan), which is in abnormal contact with Kundelungu rocks.

Introduction

Base metal deposits mined in the copper district of Shaba are all located in the Katanga System of Upper Proterozoic age (Fig. 1).

In the lower part of this system, one observes peculiar stratigraphic disconti-nuities and tectonic structures which have not yet been explained in a satisfactory and coherent manner.

The current general consensus on the genesis of the numerous stratiform and pre-tectonic copper-cobalt deposits of Shaba and Zambia is that they belong to the syndiagenetic Kupferschiefer type, but there are differences of opinion as to the genetic processes involved.

The purpose of this article is to explain the stratigraphic and tectonic peculiar-ities of the Shaba copper district by the presence of evaporites. It is also to discuss the origin of the post-tectonic and clearly epigenetic deposits, particularly the Kipushi orebody, which is one of the largest and richest deposits of its type in the world. The direct association of this ore deposit with a breccia-filled diapir suggests that the highly concentrated brines resulting from the solution of evaporites were also the mineralizing hydrothermal fluids.

Stratigraphy of the Katanga System

Stratigraphic Succession

In the Shaban copper district, the Katanga System constitutes a probably concord-ant sedimentary succession. Its rests on a Middle-Proterozoic or Archean basement (François 1974).

We divide the Katanga System into three supergroups, from top to bottom (Fig. 2):

The Upper Kundelungu Supergroup (Ks). A basal thin tilloid (mixtite) called Petit Conglomérat is overlain by pelitic-arenaceous formations with a dolomitic matrix. Total thickness: 2000 to 3000 m.

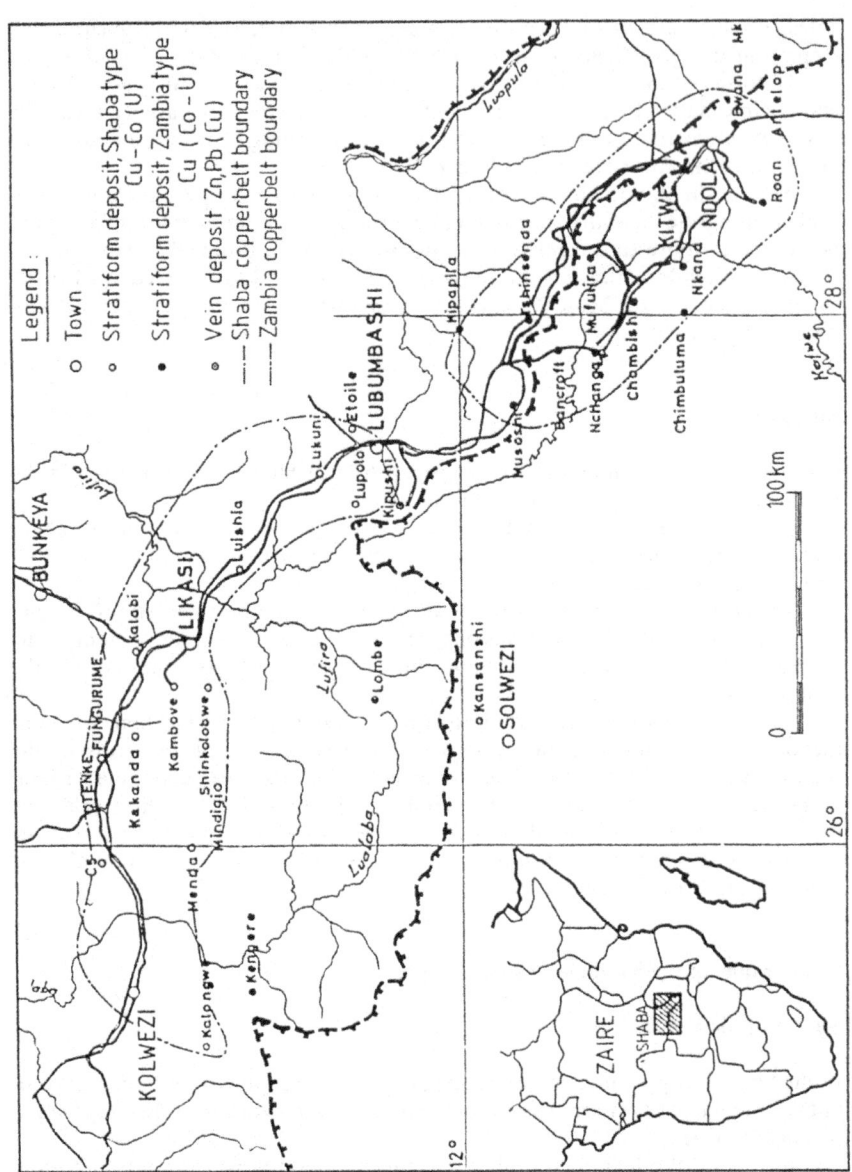

Fig. 1. Geographical map. (J. Cailteux)

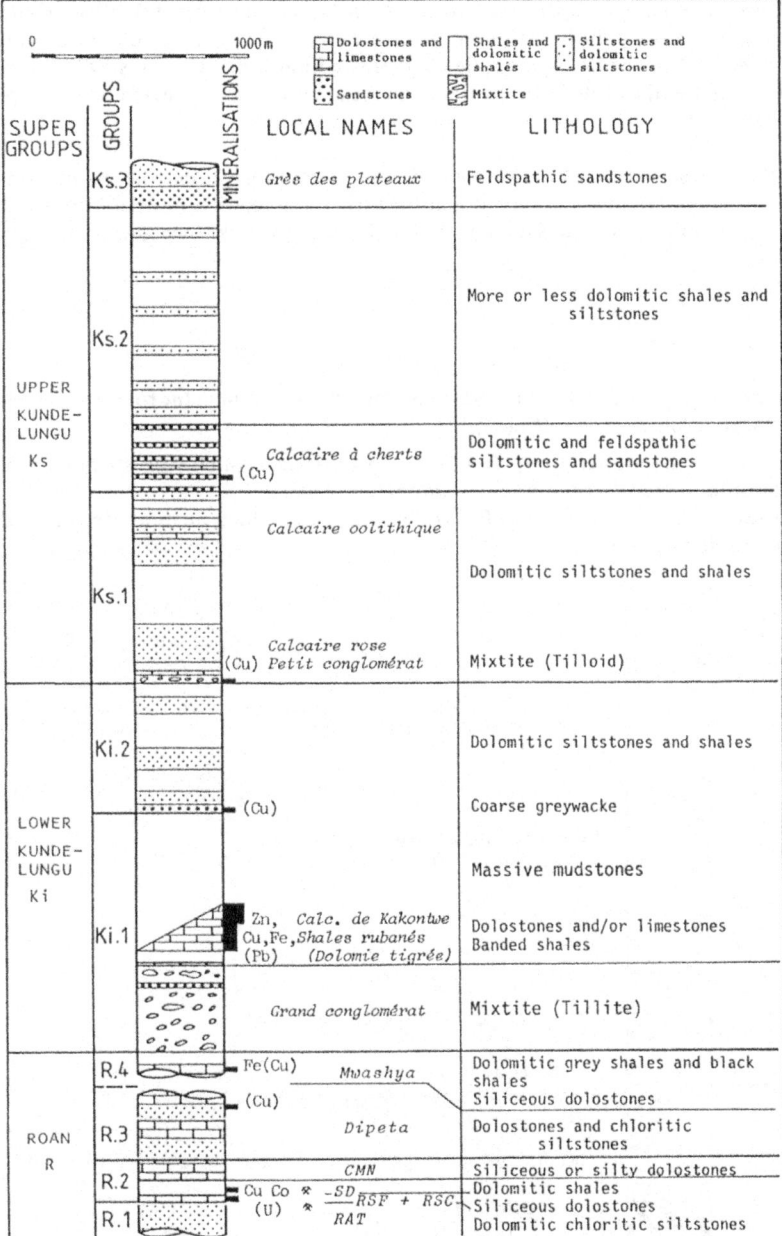

Fig. 2. Stratigraphy of the Katanga System (southern facies). (A. Francois)

The Lower Kundelungu Supergroup (Ki). A thick basal tillite (mixtite) called Grand Conglomérat is overlain by pelitic-arenaceous formations with a dolomitic component. A dolomitized limestone called Kakontwe Limestone (François 1973–1974) occurs just above the tillite on the southern fringe of the district. Total thickness: 1400 to 3500 m.

The Roan Supergroup (R). Consists of an alternation of detritic formations (chlorito-dolomitic siltstones) and dolomitic formations, consisting of dolostones as well as dolomitic shales and sandstones. Total thickness: unknown, approximately 1200 m.

Stratigraphic Anomalies

Stratigraphic gaps or discontinuities marked by breccias occur at the following levels of the stratigraphic column (Fig. 3):

a) Just below the upper group R.4 (Mwashya Group), which consists mainly of dolostones, grey dolomitic shales and black shales.
b) In several levels of the group R.2 (Dipeta Group), which includes detritic and dolomitic formations.

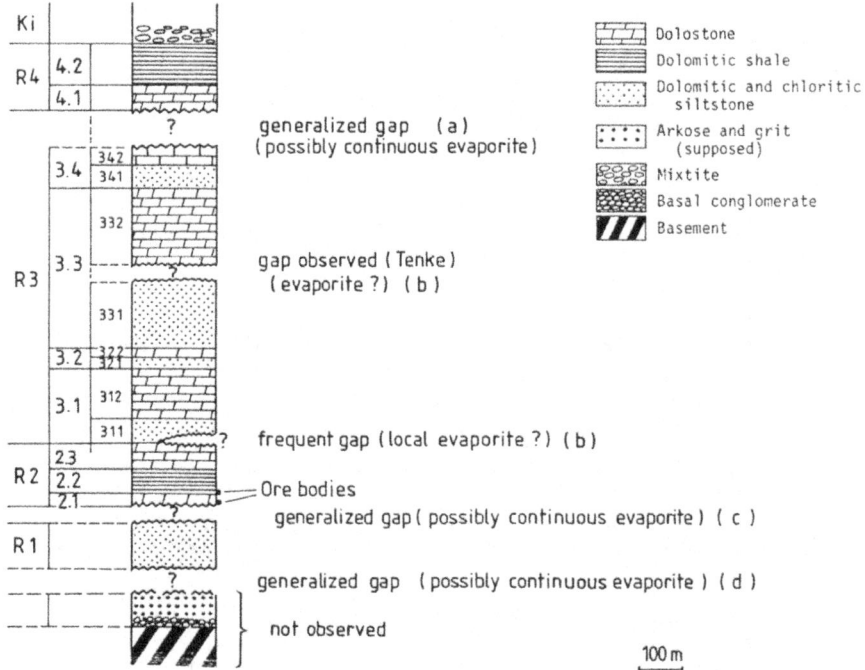

Fig. 3. Stratigraphy of the Shaban Roan Supergroup. (A. Francois)

c) Just below the mainly dolomitic group R.2 (Mines Group), which contains the stratiform Cu-Co ore bodies.
d) Below the lower group R.1, a detritic one, called RAT or Roches argilotalqueuses.

The a, b, and d discontinuities are regional. The discontinuities located in the R.3 group are poorly known, and could be either local or regional.

The breccias consist of subangular to rounded small fragments (0.2 to 5 cm in diameter) of chlorito-dolomitic siltstones, belonging to R.1 or R.3 Groups, cemented by the same rock finely crushed and by recrystallized dolomite.

Despite hundreds of drill holes, the pre-Katanga basement has never been reached (below the Katanga System) in the Shaban copper district. For this reason, the lower part of R.1 is unknown. It could consist of sandstones, grit, and conglomerate, as it does in the copper district of Zambia.

Tectonic Evolution

Tectonic Structures and Phases

A first tectonic phase produced folds overturned toward the north. The southern flanks of some anticlines are thrusted over the northern flanks and overlap them. The horizontal component of the displacement can reach 10 km. It reaches about 70 km for an anticline located in the western part of the copper district (Kolwezi nappe, François 1981).

A second tectonic phase produced folds overturned toward the south. The northern flanks of some anticlines are in this case thrusted over the southern flanks. The horizontal component of the slip can also reach up to 10 km.

A third tectonic phase produced important "oblique" faults which cut across folds at a low angle. The longest one follows a sinuous course over a distance of 170 km from Kalongwe in the extreme west to the Shinkolobwe "extrusion" in the center of the copper belt (Fig. 4). It is a dextral strike-slip fault (Demesmaeker et al. 1962; François 1973, 1974).

Tectonic Anomalies

First Kind of Anomaly.   The upper dolomitic R.4 Group remains everywhere linked to the younger Kundelungu formations. In contrast, the other dolomitic formations of the Roan, belonging in particular to the R.2 (Mines Group), occur everywhere as blocks of every size, up to several km., surrounded by the brecciated detritic formations described above. Whatever the structure in which these rocks outcrop (axial zones of regular anticlines, lower part of overthrusted anticlinal flanks, Kolwezi nappe), they always form what can be named Roan megabreccia (Fig. 5).

Second Kind of Anomaly.   Locally, one of the flanks of an anticline is breached along steep abrasion surfaces and the resultant void is filled with Roan megabreccia. This

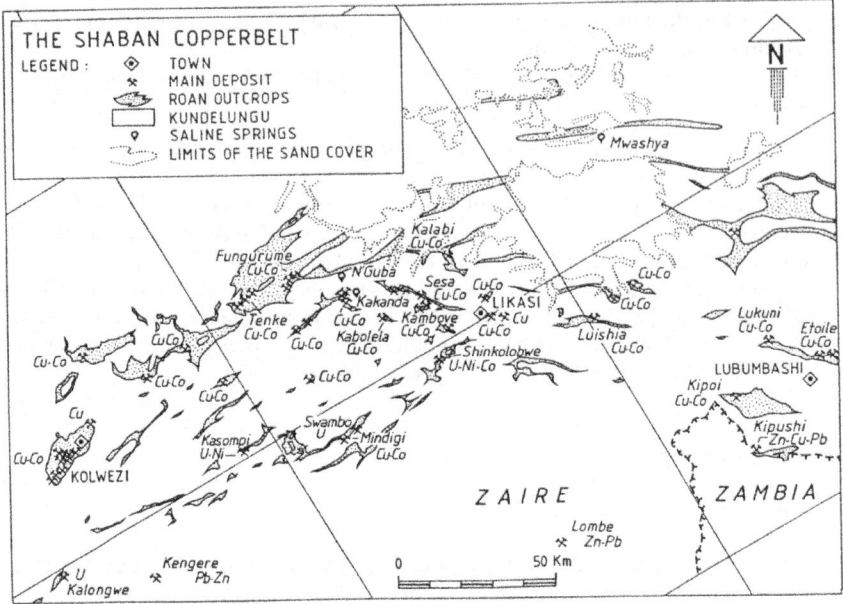

Fig. 4. Outcrops of the Roan in the Shaban Copperbelt. (A. Francois)

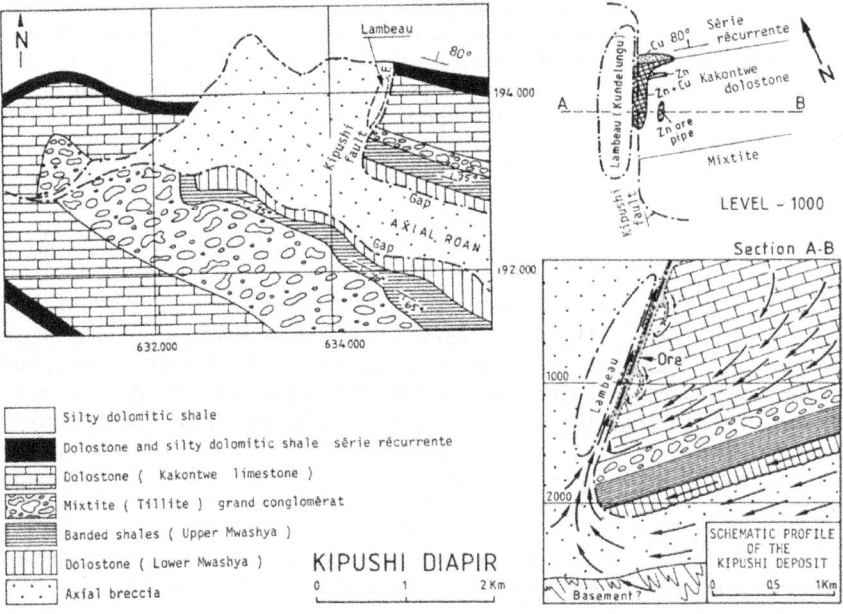

Fig. 5. Western part of the Kipushi diapiric anticline, schematic plan and profiles. (A. Francois)

is the case in the Kipushi structure (Fig. 6), where the breccia fill almost reaches the upper Kundelungu.

Third Kind of Anomaly. The late "oblique faults" are locally filled by Roan megabreccia, which has clearly risen from depth between two steep walls of nondisturbed Kundelungu. The ascent of Roan rock can reach up to 5 km, hence the local name extrusion faults is common. Large slabs of collapsed Kundelungu rock are "floating" in the same fill.

The Shinkolobwe uranium ore bodies (Fig. 6) are located inside a large extrusion which shows, at the same level, blocks and slabs of both the Mine Group (R.2) and the Kundelungu (Ki and Ks) (Demesmaeker et al. 1962).

Interpretation of the Stratigraphic and Tectonic Anomalies. The Evaporitic Nature of Some Roan Formations

To interpret the stratigraphic and tectonic anomalies, the common explanation was that the RAT and similar formations of the Roan Supergroup are supposed to be very plastic and incompetent (Schuiling 1947). Their resinous luster in breccias or along slickensided surfaces may have led to this convenient notion, as well as the occasional presence of talc and sepiolite. However, in unweathered form, it is a compact rock of normal density ($\pm$ 2.64), not significantly less coherent and dense than dolostones and shales.

In consequence, the accumulations of Roan debris squeezed upward and piercing Kundelungu formations, along the oblique faults (extrusions), cannot easily be explained without the isostatic uplift of low density halite, dragging the competent rocks of the Roan upward. This diapiric halokinesis explains also the mechanical breaching of some anticlinal flanks and in general the disharmonic nature of the tectonic structure.

At first sight, there is up to now no direct proof: no halite has been found, even in the deepest boreholes, but anhydrite occurs at several stratigraphic levels. Primary gypsum crystals and anhydrite nodules are pseudomorphosed by dolomite and quartz (Cailteux 1983).

In the copper belt of Zambia, in detritic formations which are correlated with the R.2 and R.3 groups of Shaba, several hundreds of meters of anhydrite and anhydritic dolostones are known (Garlick et al. 1972).

For additional evidence we refer to the report of Guilloux (1982), which gives a complete geochemical and mineralogical description of the proterozoic RAT facies, not only in Central Africa, but also in the other parts of the dislocated Gondwanaland.

The creep of the evaporites under gravitational and tectonic stress, and the later dissolution of thick salt formations, can easily explain the "gaps" observed every-where in well-defined horizons of the stratigraphic column, as well as the generalized dislocation and brecciation of the R.1, R.2, and R.3 Groups. The massive nonstratified parts of the R.1 and R.3 Groups are probably residual rocks, the result of solution collapse. The stratified parts (laminites, algal mats, dolostones containing anhydrite nodules) strongly suggest a coastal sebkha facies, as advocated by Cailteux (1983).

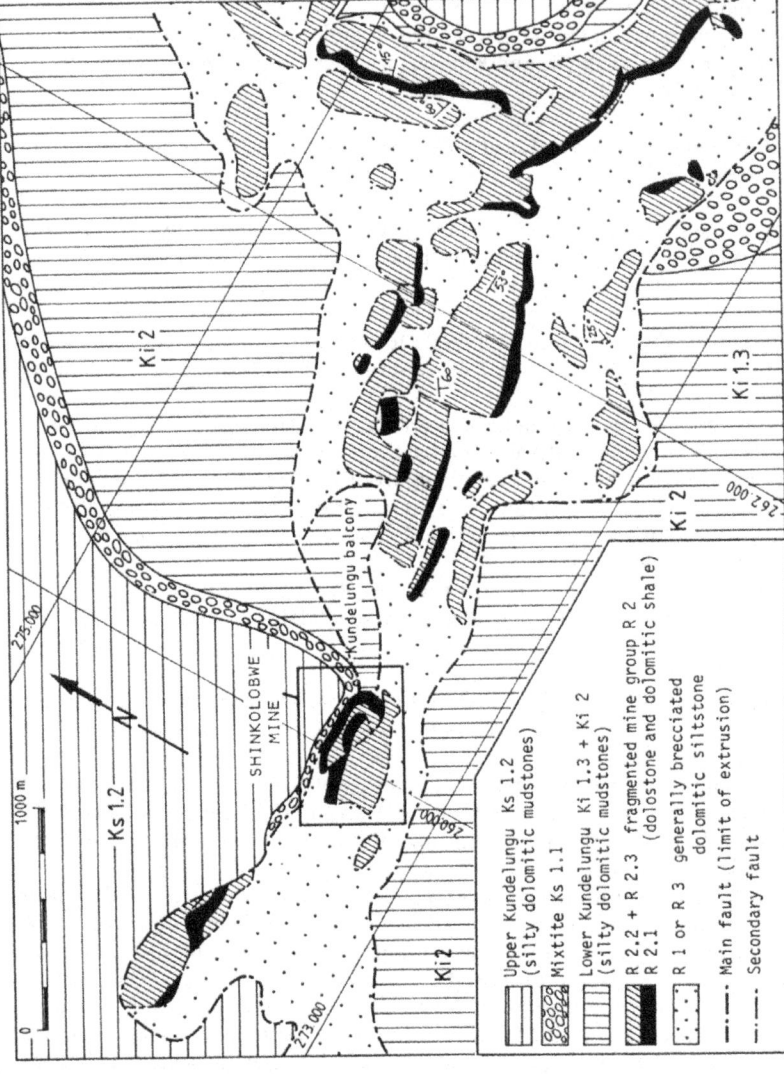

Fig. 6. Western part of the Shinkolobwe diapir anticline. (A. Francois)

The extrusions are diapiric structures modified by the vanishing of the evaporitic component. The residual collapse breccia contains locally large clastic blocks of Kundelungu rocks (Ki and K2) torn from the walls or the roof of the diapir (Fig. 5).

The dissolution process is still going on today on a modest scale: saline springs are worked for NaCl production near the rivers Kafila and Kiandamu, as well as at Nguba and Mwashya, always near R.4 outcrops (Fig. 4). Here we agree with the hypothesis of Buffard and Grujenshi (1979) concerning the relation of these springs with a saline horizon which existed just below the R.4 Group (gap a, Fig. 3).

Still other evidence is given by the liquid inclusions of the stratiform and the younger epigenetic ores. In the important strata-bound ore deposits of Kamoto (Fig. 4), the authigenous dolomite crystals contain microcrystals (Pirmolin 1970). Recently, at Shinkolobwe, Raman microsonde analysis has proven the presence of $MgCl_2$, $CaCl_2$, NaCl and KCl (Audéoud et al. 1984). In the epigenetic deposit of Kipushi, Intiomale (1982) describes inclusions containing a single cubic microcrystal, presumably halite. However, this evidence is still scanty and does not exclude connate brines.

Admittedly, largely evaporitic basins are scarce in the Proterozoic and unknown in metamorphic rocks, except as remnants such as anhydrite and scapolite (Moine et al. 1981). However, in the center of Australia remains the Amadeus evaporite basin, with its assorted diapirs (Mc Naughton et al. 1968). The diapir of Goyder Pass has the same size and elongated but irregular form as some extrusions of the Shaba district. Rock salt has been cored in a few of the deepest holes drilled for oil and gas exploration.

Base Metal Deposits in the Shaba Copper District and Their Connection with Evaporitic Formations

The Katanga System in the Shaba copper district, contains a large number of base metal deposits.

Stratiform Deposits of Copper and Cobalt with Subsidiary Uranium

Most of the ore bodies found in Shaba are stratiform or at least strata-bound. They are located in the lower part of the R.2 Group (Mines Group), a few meters above an extensive discontinuity which may result from the leaching of a widespread evaporitic horizon, as mentioned before.

It is generally admitted that these stratabound Cu-Co-(U) ores belong to the syndiagenetic Kupferschiefer type. Compaction, dehydratation of gypsum and conversion of the clay minerals into chlorite created a slow migration of hypersaline brines under lithostatic pressure. They selectively leached the R.1 (red RAT) and the basement which lost part of their original Cu and Co content. They reached, at least locally, the overlying dolomitic, carbonaceous, and pyritic R.2 rocks.

It is not our purpose to substantiate this concept, this paper being limited to the post-tectonic epigenetic deposits.

The Epigenetic Zinc-Copper-Lead-Cadmium-Germanium-Silver Deposit of Kipushi

Geological Structure. The diapiric anticline of Kipushi is an elongated dome about 25 km long, whose axial plane is slightly inclined to the south. Its axial core is filled by a megabreccia of Roan rocks: massive or fractured chloritic silstones, enclosing clasts of silicified or talcose dolostones. They probably belong to the R.3 Group (Intiomale and Oosterbosch 1974).

In the western part of the dome, a huge block of the northern flank, some 1800 m long and more than 2000 m deep, has been torn away and replaced by a megabreccia, probably similar to the axial breccia (Fig. 5).

This breccia is limited toward the east by a discontinuity surface, dipping 70°C west, called Kipushi Fault. It cuts orthogonally the steep northern flank of the anticline. From south to north it exposes the normal stratigraphic succession, from the upper Roan (Mwashia Group R.4) to the lower Kundelungu Group (Ki). Figure 5 represents this series in a simplified form.

From a metallogenetic point of view, the important formations are the dolomitized Kakontwe Limestone (Ki.2) and the overlying "Série Récurrente" (local name for a banded alternation of dolostone and silty dolomitic shale). Both are invaded by the mineralization and form the footwall of the ore bodies.

Another important feature is their roof, which is formed by a gigantic slice of stratified shales, silty shales, and sandstones with occasionally irregular bedding. It is called Grand Lambeau and considered by Intiomale (1982) to belong to the Upper Kundelungu.

This Lambeau, up to 250 m thick, forms the hanging wall of the Kipushi Fault on a distance of about 500 m. It is continuous between levels −160 and −1800 m. Its stratification is nearly parallel to the bedding of the adjoining Kakontwe dolomites, which dip 70° to 80° NE.

Mineralization. The Kipushi ore body is formed by three types of ore:

The main ore, rich in both Cu and Zn (about 20%), located along and on both sides of the Kipushi Fault, with a horizontal length varying from 200 to 500 m.

Relatively poor Cu ore (about 2% Cu) which penetrates inside the Série Récurrente, forming offshoots which attain a horizontal length of 180 m, starting from the mineralized Kipushi Fault (Figs. 5 and 9).

Very rich Zn ore (about 40% Zn) forming elliptic pipes (diameters 5 to 20 m). They penetrate deeply into the Kakontwe dolostone, and are connected with the main ore, either laterally or upwards (Fig. 5). The rich sphalerite core of the pipes is often surrounded by a pyritic sheath.

The shape of the ore lenses is illustrated by a vertical profile (Fig. 7) and three horizontal sections corresponding to different levels of the mine (Figs. 8, 9, 10).

Two long horizontal drillholes, at the levels −240 and −1150 m, penetrated into the axial breccia and crossed slabs of saussuritized "gabbro" (dolerite), dolomitic clasts and also important empty cavities (Fig. 7). However, this breccia remains poorly known.

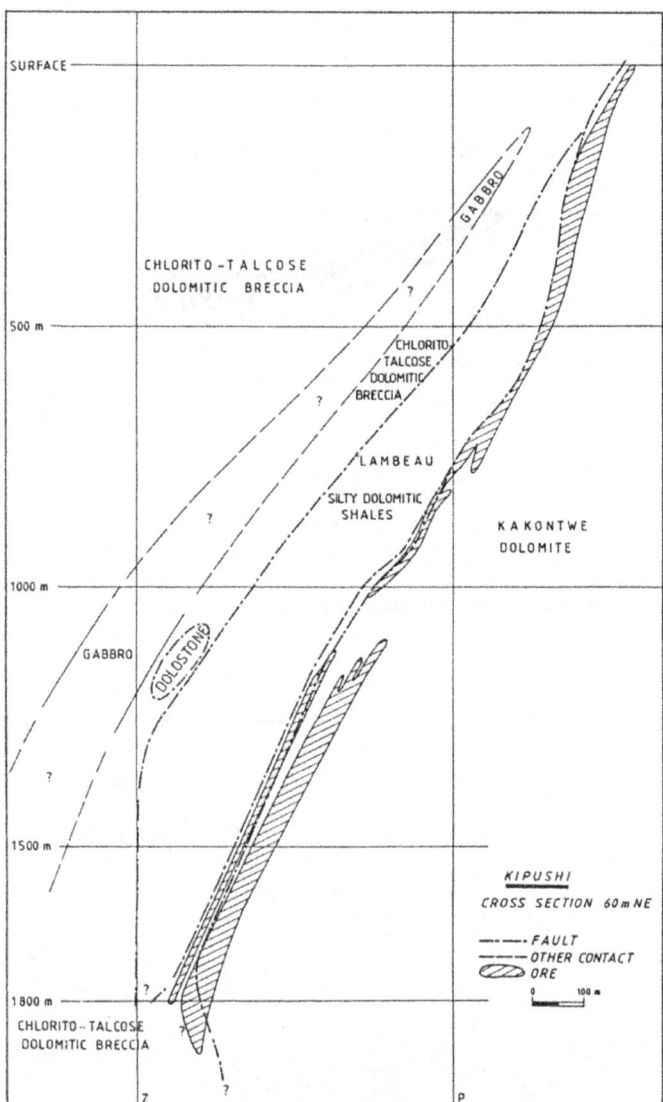

Fig. 7. Vertical profile perpendicular to the Kipushi ore bodies. (M. Intiomale)

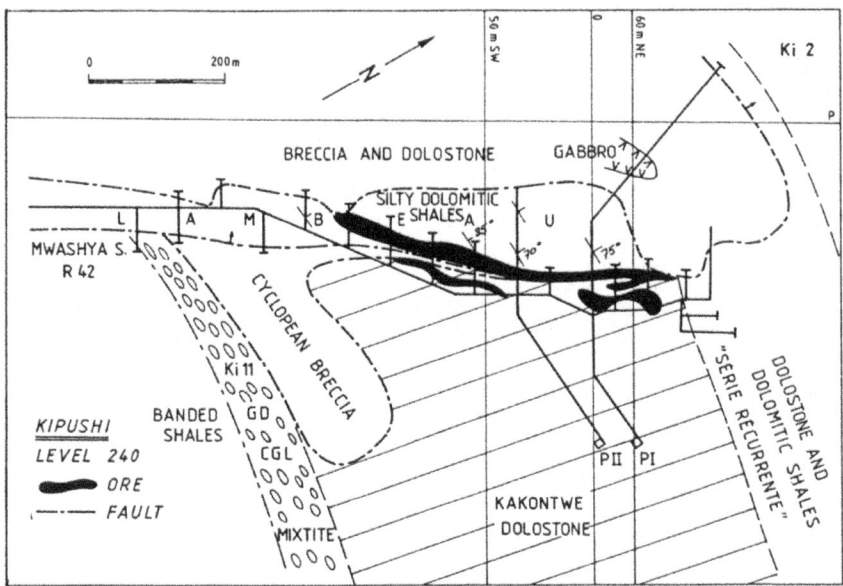

Fig. 8. Horizontal section, Mine level — 240 m. (M. Intiomale)

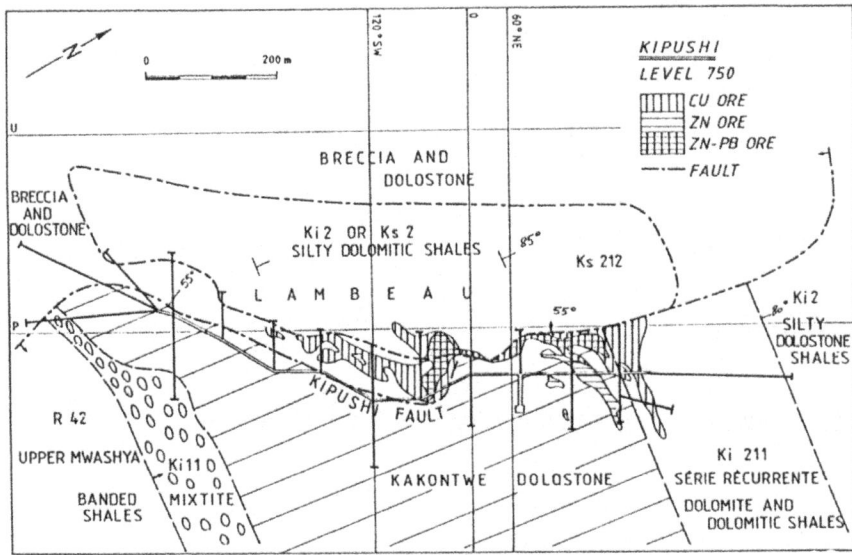

Fig. 9. Horizontal section, Mine level — 750 m. (M. Intiomale)

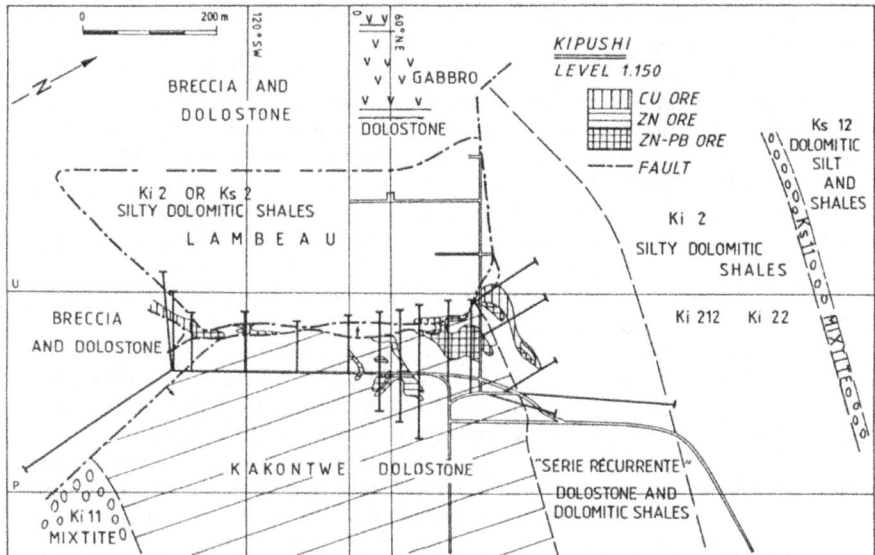

Fig. 10. Horizontal section, Mine level — 1150 m. (M. Intiomale)

Numerous publications describe the mineral wealth of the Kipushi ores (Dimanche 1974; De Vos et al. 1974).

Metallogenetic Model. The Kipushi ore body coincides, at every level of the mine below −160 m, with the extension of the Lambeau forming its hanging wall. Both seem to pinch out at the depth of about 1800 m (Fig. 7).

The mineralization, being in part metasomatic, invades not only the breccia and the footwall dolostones but also locally the shaly hanging wall. Reopenings were frequent during the mineralization period, as shown by multiple (polyphased) ore breccias.

The direct association with the Kipushi Fault is obvious. Geologists who have studied Kipushi unanimously consider this feature as the channel along which rose the mineralizing fluids.

It remains difficult to imagine how more than 10×10^6 tons of heavy metals could accumulate in such a limited space during a post-tectonic period, when the country rocks had already lost most of their connate water.

Solution of the diapiric evaporites requires an immense volume of water, that only infiltration of ground water during a long period could provide.

One may object that thermal convection cells, known in every geothermal system, transport the dissolved elements upward without need of much additional water (Marinelli 1976), but this cannot explain the vanishing of enormous quantities of soluble chlorides and sulfates. Outlets at the surface of the lithosphere are indispensable. Natural saline springs are known in most geothermal fields and are still active in the Shaban Copperbelt, albeit on a modest scale.

On the vertical profile Fig. 5 the arrows indicate schematically the supposed circulation, convection cells included.

The pipe-like offshoots of the main ore body mentioned before are considered by the local geologists (Intiomale and Oosterbosch 1974) to be karstic features mineralized per descensum, in contrast with the main ore. This hydrothermal circulation pattern suggests a convective cell whose descending cooler branch completed the mineralization.

Another requisite is that this hydrothermal system should drain and leach a vast area of deep-seated rocks, including the lower Roan and the weathered part of the old basement. This basement, in addition to granite and gneiss, contains also gabbros. In Zambia, it shows copper anomalies and even minable concentrations in many places.

Magmatic additions cannot be ruled out, but no clear indications are observed.

One should consider that the Kakontwe dolomitized limestone is a permeable rock and could easily provide the needed meteoric water. It does it at present: pumping at the -1150 m level of the mine delivers about 2000 m^3/h.

When the thermal gradient returned to normal, the water circulation changed: cold oxygenated phreatic water reached the deepest levels of the mine, as proven by the presence in the fissures of secondary minerals, such as limonite, copper and zinc carbonates.

Limonite underlines "collapse faults" cutting the Kipushi breccia. They tend to be funnel-shaped. In the horizontal plane, they are more or less elliptical (Intiomale 1982).

The northern part of the Kipushi Fault itself turns sharply from N 20°E to E-W (Fig. 6), suggesting also a collapse fracture.

Comparable Mineralizations and Geothermal Systems. In their important paper of 1985 about the Pb-Zn mineralization of the Eastern Maghreb, Rouvier et al. (1985) propose a model for "Peridiapiric metal concentration". It invokes a similar circulation of the fluids, with ascending "channelization" along the contacts of a piercing diapir.

In Tunisia the salt is Triassic, only superficially dissolved, and the diapirs pierce the upper Tertiary.

We reproduce the general map (Fig. 11) for the sake of comparison with the map of the Copper Belt of Shaba (Fig. 4). Their analogy is striking: both display a typical example of halotectonic disharmonic folding.

In some salt diapirs of the Gulf Coast (Texas), the cap rock is mineralized and contains uneconomic concentrations of sphalerite and galena (Price et al. 1983). The same dispersed mineralization occurs at the Reitbrook salt dome, near Hamburg (Lietz 1951).

Carpenter et al. (1974) described the Pb- and Zn-rich oil field brines of Central Mississipi.

To understand the hydrodynamic and geochemical aspects of the process involved, it seems appropriate to compare this peridiapiric model with some presently active mineralized geothermal systems.

Salton Sea, in the Imperial Valley of California, is well documented. Its deep brines contain up to 970 mg/l Zn, 104 mg/l Pb, 10 mg/l Cu and 185 g/l Cl (Table 1).

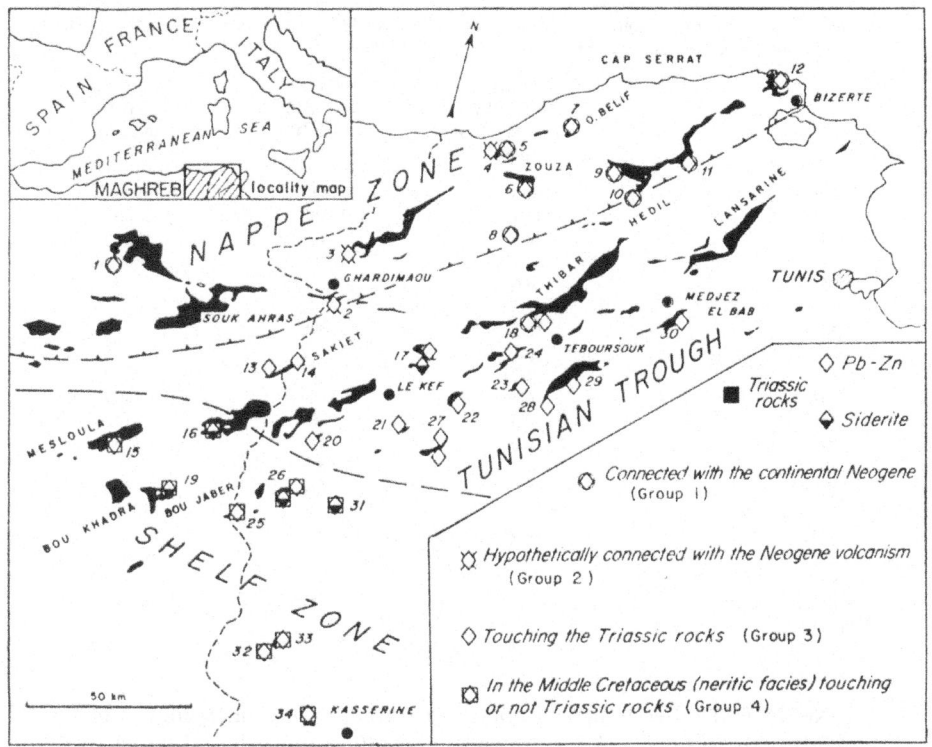

Fig. 11. The triassic diapirs and associated Pb-Zn deposits of the Eastern Maghreb. (H. Rouvier, V. Perthuisot, A. Mansouri).

Nevertheless it is proven that the fluid is simply water of the Colorado River which infiltrates the deltaic host rocks (White 1981).

Recent drillholes in the Phlegrean Fields, north of Napoli, revealed a comparable nydrothermal system, established in a thick, mainly volcanic pile resting upon a deep basement of Triassic limestones, dolostones and evaporites. The lavas and tuffs are mainly latites.

In the region of Mofete, three permeable geothermal reservoirs are superposed. The fluid of the deepest one is heavily chlorinated and mineralized, as shown in Table 1. Its down-hole temperature is 350°C, at a depth of ±2500 m. The second reservoir (depth 1250–1600 m) is much less mineralized, although its temperature is about 300°C. Its rocks contain disseminated sphalerite, pyrite, and pyrrhotite.

The authigenic metamorphic minerals form successive zones, well defined by the isotherms. At ±150°C, the montmorillonite-rich upper zone changes into the illite-chlorite zone, with mixed-layer clay minerals, K-feldspar and albite.

The 250°C isotherm defines the lower boundary and the top of the Ca-Al silicate zone, characterized by epidote and a sharp increase of feldspar.

Table 1. Chemical composition of the fluids of two wells of the Phlegrean geothermal fields compared with fluids of Salton Sea. (White 1981)

	Mofete well MF1 (550–896 m) (mg/l)	Mofete Well MF5 (2450–2650 m)	Salton Sea Well II D1 (< 1000 m)
Na	9316.5	71439	51000
K	1143.7	36398	25000
Li	23.5	401	215
Ca	523.5	45268	28000
Mg	6.0	25.9	
Ba	3.3	249	
Sr	32.5	1098	
Cu	0.13	2.7	8
Fe	1.7	4287	2290
Mn	6.6	2500	1400
Pb	0.5	368	102
Zn	0.09	436	540
B	111.3	1107	
SiO_2	384.0	176	400
Cl	16591	254359	155000
SO_4	44.1	Trace	5
HCO_3	74.2	Trace	690
CO_2	11250	n.d.	n.d.
Temperature in the reservoir	240°C	350°C	340°C

At 325°C the thermometamorphic zone starts with the emergence of biotite, amphibole, and scapolite. At a still higher temperature, diopside and garnet are the index minerals.

All these authigenic minerals are known at Kipushi, except garnet and diopside. However, they are rather sporadic, except the ubiquitous magnesian chlorite. In addition, one finds muscovite (phengite), tourmaline, and sepiolite-palygorskite.

This comparison confirms earlier estimates of the formation temperature of the Kipushi ores: 200° to 350°C.

Many high enthalpy geothermal fields are vapor-dominated. Abundant bubbling initiates two-phase convection which is much more effective than one-phase thermal convection circulation. Flashing in such a system occurs when a leak opens suddenly, causing a pressure drop in the reservoir.

The highly mineralized hot brines are of course self-sealing, and narrow fissures are rapidly filled.

If similar hot brines were the mineralizing fluids at Kipushi, boiling could also play an important role, explaining the abnormally high metal concentration in a limited space.

After all, solution collapse is not a continuous process, but proceeds by stages and intervals, and opens new fissures crossing older ones already sealed (eventually by sulfides). We cite Elders and McKibben (1985): "A history of multiple fracturing and sealing is common in geothermal systems, as well as frequent earthquake swarms". A pulsating regime of the hydrothermal brine circulation could explain the zonation of the sphalerite of Kipushi (Katanga 1982).

Organic Carbon Association with the Kipushi Ore Deposit. Both walls of the ore body are blackened by carbon, as observed by Katanga (1982). The copper sulfides themselves contain residual inclusions of anisotropic carbon (shungite) displaying odd forms and even structured algae (Francotte and Jedwab 1963). This high carbon anthraxolite is also present in the dolomitic clasts of the breccia and infills fissures and pores.

This fact strongly suggests that part of the carbonaceous matter, during maturation, was in liquid form before or during the mineralization.

The same relation between mineralization and migration of hydrocarbons seems to exist at the famous Tsumeb Mine (Namibia), where it has been described as graphitization of the country rocks (Hughes 1979).

At Tsumeb the mineralization is peripheral to a cylindrical pipe of collapse breccias that transect the dolostone of the Otavi Group, which is part of the Damara Supergroup, correlated with the Katanga System.

Tsumeb has been called the sister deposit of Kipushi, the main reason being their identical complex mineralogy, inclusive rare minerals such as germanium and gallium sulfides.

Uraninite Vein Deposits in the Mines Group R.2. Shinkolobwe Type

Scattered black uranium oxides are located in the gray RAT's, at the base of the lower stratiform ore horizon. This kind of ore, which seems to be strata-bound, has not been dated. The epigenetic uraninite, 520 to 620 m.y. old, is located in narrow cracks and brecciated zones which cross the Cu-Co ore bodies. Both types of uranium ore are present in slabs of the Mines Group belonging to overthrusted masses of Roan (Kamoto Principal, Kambove Ouest, Luishia, etc.) (Fig. 4).

More concentrated masses of uraninite, 620 to 670 m.y. old, form veins in isolated slabs of R.2. floating inside diapiric breccias ("extrusions" for Kalongwe, Swambo and Shinkolobwe, anticline with an extrusive core for Kasompi) (Fig. 4).

Uranium is accompanied by precious metals (Au, Pt, Pd) and also by non-radioactive monazite and by sulfides and seleno-sulfides of Ni, Co, Cu, Fe, and Mo.

Except for the scattered black oxides, the ores are vein shaped, and clearly epigenetic. They probably were formed by remobilization of a syn-diagenetic protore. It has been suggested that hot brines, formed by leaching of halite beds, were the mineralizing fluids (Audéoud et al. 1984).

The remarkable association of precious metals with uraninite or pechblende is also known in the Zechstein Kupferschiefer of Germany and Poland (Kucha 1982).

In the copper deposits scattered along the borderline between Shaba and Zambia, the ages of the uranium occurrences are much younger and variable (Meneghel 1981). Many are dated at 520 to 536 m.y. which is the age ascribed to the thermal event advocated by Cahen et al. (1984).

This may be explained by the very long period necessary to dissolve the evaporites down to depths of more than 2000 m.

In this context we mention that the isotopic model age of the lead of Kipushi, repeatedly measured, is ± 425 m.y. This is also true for Tsumeb and Kombat (Namibia). It is possible that this model age represents the absolute age of these ore bodies.

Conclusions

The former presence of halite-rich evaporites in the Upper Proterozoic Katanga
System of the Shaban copper district explains many stratigraphic and tectonic
peculiarities. Their selective solution could have generated the metal-rich brines
which formed the hydrothermal and post-tectonic Cu-Zn-Pb and U deposits. They
are geographically and genetically linked to halocinetic structures.

A thermal event dated isotopically to 500-525 m.y. explains the post-tectonic
thermal and hydrothermal evolution and the active circulation of mineralized hot
brines.

Leaching, transport, and precipitation of the heavy metals are mechanisms
which are known in several presently active geothermal fields, also linked with old
evaporites.

Kipushi is far from being the only ore body directly located along the flank of
a salt-diapir or dome, but it is certainly the largest and the richest (total production:
about 10 million tons of metals).

One would be tempted to consider salt diapirs and domes as being good targets
for exploration, with the hope of finding some other Kipushi; but we should
remember that the exceptional size and concentration of this deposit seems to be the
result of a quite exceptional circumstance: the presence of a nearly impervious slab
of shaly rocks alongside the flank of a diapir, which guided and localized the
ascending mineralizing fluids.

We realize that the proposed genetic model is far from well established. It needs
criticism and additional observations, such as isotopic studies, Raman spectroscopy
of the liquid inclusions, etc.

Acknowledgments. We thank Dr. J. Cailteux and Dr. M. Intiomale for permission to reproduce several
illustrations of their memoirs which are unfortunately unpublished. We acknowledge the fact that Dr.
Intiomale suggested that the Lambeau forming the hanging wall of the Kipushi ore bodies is a solution
collapse feature and emphasized its role in the formation of a very rich hydrothermal deposit. We also
thank the AGIP Company and its geologists, MM. W. Chelini, M. Rossi and A. Sbiana, for permission
to publish some date given in their private report: *Geology of the Phlegrean Fields* (1985).

References

Audéoud D, Moine B, Poty B (1984) Minéralisations uranifères et milieux confinés du Shaba. Greco 52
    SGA, Paris
Buffard R, Grujenshi C (1979) Les sources salines de l'Arc du Shaba méridional. Leur relation probable
    avec l'existence d'une assise salifère de la couverture katanguienne. Ann Soc Géol Belg T
    102:285-294 .
Cahen L, Snelling NJ, Delhal J, Vail Jr (1984) The geochronology, and evolution of Africa. Clarendon,
    Oxford, p 512
Cailteux JLH (1983) Le "Roan" shabien dans la région de Kambove (Shaba, Zaïre). Thèse Univ Liège
Carpenter AB, Trout ML, Pickett EE (1974) Preliminary report on the origin and chemical evolution of
    Pb- and Zn-rich oil-field brines in Central Mississippi. Econ Geol 69:1131-1206
Demesmaeker G, Francois A, Oosterbosch R (1962) Gisements stratiformes de cuivre en Afrique.
    Symposium coordonné par Lombard J et Nicolini P, 2ème partie Tectonique Lusaka ASGA, pp
    47-115

De Vos W, Viaene W, Moreau J (1974) Minéralogie du gisement de Kipushi Shaba Zaïre. Centenaire de la Soc Géol de Belgique, pp 165–183

Dimanche F (1974) Paragenése des sulfures de cuivre dans les gisements du Shaba Zaïre, pp 186 201

Elders WA, McKibben MA (1985) Fe-Zn-Cu-Pb mineralizations in the Salton Sea geothermal system. Imperial Valley California. Econ Geol 80:539–559

Francois A (1973) L'extrémité occidentale de l'arc cuprifère shabien. Etude géologique. Edité par le Département Géologique de la Gécamines Likasi République du Zaïre, pp 1–65

Francois A (1973-1974) Le niveau du Calcaire de Kakontwe et ses facies au Shaba. Académie Royale des Sciences d'Outre-Mer, Bulletin des Séances, pp 845–867

Francois A (1974) Stratigraphie, tectonique et minéralisations dans l'arc cuprifère du Shaba (République du Zaïre). Centenaire de la Société Géologique de Belgique. Gisements stratiformes et provinces cuprifères, pp 79–101

Francois A (1981) La couverture katanguienne entre les socles de Nzilo et de la Kabompo, Zaïre. Musée Roy de l'Afrique Centrale. Annales 87:1–50

Francotte J, Jedwab J (1963) Traces d'organites (?) dans la Shungite de Kipushi. Bull Soc belge Géol T LXXII fasc 3:393–398

Garlick WG, Fleischer VD (1976) Sedimentary environment of Zambian copper depositions. In: Outline of the geology of Zambia. Geol en Mijnbouw 51:277–298

Guilloux L (1982) Etude chimique des séries porteuses de quelques grands gisements du type Kupferschiefer. Mèm no 43 Sciences de la Terre, Nancy, p 659

Hughes MJ (1979) Some aspects of the genesis of the Tsumeb ore-body, SW Africa, and of its subsequent deformation, part 1. 18th Congress of the Geological Society of SA, pp 200–206

Intiomale M (1982) Le gisement Zn-Pb-Cu de Kipushi (Shaba Zaïre). Etude géologique et métallogénique. Thèse de Doctorat. Univ de Louvain-la-Neuve

Intiomale M, Oosterbosch R (1974) Géologie et Géochimie du gisement de Kipushi Zaïre. Centenaire de la Soc Géol de Belgique, pp 123–164

Katanga WK (1982) Contribution á la connaissance de la structure, de la lithologie et de quelques types particuliers de minéralisation du gisement de Kipushi (Shaba Zaïre). Univ de Liège, Fac des Sci Appl, Institut de Géologie

Kucha H (1982) Platinum group metals in the Zechstein copper deposits of Poland. Econ Geol 77:1578–1592

Lietz J (1951) Sulfidische Klufterze im Deckgebirge des Salzstockes Reitbrook. Mitt Geol Staatsinst Hamb H 20:110–119

Marinelli G (1976) Géothermie et théories métallogénétiques. Deuxièmes Journées de l'Industrie Minérale. Univ libre de Bruxelles, pp 1066–1074

Mc Naughton, Quinlan et al. (1968) The evolution of salt anticlines and salt domes in the Amadeus Basin. Central Australia. Am Assoc Pet Geol, Spec Pap 88:229–247

Meneghel L (1981) The occurrence of uranium in the Katanga System of Northwestern Zambia. Econ Geol 76:56–68

Moine B, Sauvan P, Jarousse J (1981) Geochemistry of evaporite-bearing series: a tentative guide for the identification of meta evaporites. Contrib Miner Pet 76:401–402

Pirmolin J (1970) Inclusions fluides dans la dolomite du gisement stratiforme de Kamoto. Ann Soc Géol Belg T 93:193–202

Price PE, Kyle JR, Wessel GR (1983) Salt dome-related Zn Pb deposits. Dept of Geol Sci. Univ of Texas Austin

Rouvier H, Perthuisot V, Mansouri A (1985) Pb-Zn Deposits and Saltbearing Diapirs in Southern Europe and North Africa. Econ Geol 80(3):666–687

Schuiling H (1947) La tectonique des gîtes de cuivre du Katanga. Centenaire de l'AILg Liège, pp 309–315

White D (1981) Active geothermal systems and hydrothermal ore fluids. Econ Geol 75th Anniv Vol:392–423

Sulfide Mineralization of Paleozoic Rocks in the Northern Eifel, F.R.G.

V. Scheps[1] and S. Keyssner[1]

Abstract

In the northern Eifel, Paleozoic rocks were investigated for base metal sulfides in a drilling program.

An increase of base metals was noted in Lower Paleozoic black slates at the Cambrian/Ordovician (Revin/Salm) transition zone. Near Lammersdorf/Konzen the magnetic anomaly at the southern flank of the Venn anticline is due to a pyrrhotite-bearing sequence of black slates and siltstones of this transition zone. Minor amounts of chalcopyrite and sphalerite and traces of galena are associated with the pyrrhotite enrichment.

An increase of base metals, predominantly Zn, was also noted in Upper Devonian (Frasnian) black shales and nodular limestones of the Inde syncline. These sediments were deposited in a restricted environment under euxinic and partly evaporitic conditions in a swell and basin facies of the shelf platform.

The acid intrusives of the Venn anticline, the "Lammersdorf tonalite" and several "tonalite porphyrites", as opposed to the "Hill tonalite" in adjacent Belgium, do not contain any sulfide mineralization of interest.

Numerous vein-type Pb-Zn deposits, mined near Aachen and Stolberg until the beginning of this century, are related to NW and N striking faults cutting Middle to Upper Devonian and Lower Carboniferous carbonate sequences.

A model for metallogenesis of the epigenetic ore deposits in the northern Eifel is introduced including the Triassic sandstone hosted Pb deposits of Maubach-Mechernich. The metal content is suggested to be derived from Paleozoic sedimentary rocks by leaching and lateral secretion. The potential for metals would be increased if a model of overthrusting is adopted to repeat the stratigraphic sequence in the northern Eifel.

Introduction

Two areas of Paleozoic rocks are regarded as having potential for syngenetic base metal sulfides on the basis of metal anomalies in rocks and stream sediments (Scheps 1982; Scheps and Friedrich 1983):

---

[1]Institut für Mineralogie und Lagerstättenlehre, RWTH Aachen, Wüllnerstr. 2, 5100 Aachen, FRG

Base Metal Sulfide Deposits
G.H. Friedrich, P.M. Herzig (Eds.)
© Springer-Verlag Berlin Heidelberg 1988

- Lower Paleozoic black slates of the Revin and Salm (Upper Cambrian/Lower Ordovician) of the Venn massif and associated tonalite and tonalite porphyrite intrusives;
- Frasnian (Upper Devonian) nodular limestones, marly shales, and black shales of the Inde syncline.

Both rock groups are poorly exposed, so that a drilling program was required to collect samples for detailed and systematic geological and mineralogical investigations.

Epigenetic vein-type Pb-Zn deposits in Devonian and Lower Carboniferous carbonates of the Inde syncline have been mined to a depth of about 130 m until the beginning of the century. As the mines are no longer accessible, studies of the orebodies were possible by exploration drilling only.

Diamand core drilling was carried out in different phases from 1982 to 1985.

Methods

Along the southeastern limb of the Venn massif, north of Lammersdorf, 19 shallow holes of 30 m each were drilled in black slates, sandstones, and quartzites of Revin and Salm formations and intercalated tonalite porphyrites. Two deeper boreholes were sunk to the Revin/Salm transition zone V-21 (Venn-21, 300 m) and V-22 (470 m). Three boreholes intersected the "Lammersdorf tonalite": V-4 (60 m), V-4a (100 m) and V-23 (146 m) (Figs. 1, 2, 3).

In the area of the Inde syncline, cores of four holes drilled into Upper Devonian sedimentary rocks were investigated: I-1 (Inde-Mulde 1, 342 m) and I-3 (210 m) at the southeastern flank of this structure, I-2 (125 m) and I-4 (204 m) at its northwestern flank (the core of I-2 was kindly placed to our disposal by Kaiserbrunnen AG, Aachen).

At two ancient mining sites of fault controlled vein-type Pb-Zn deposits (Albertsgrube and Roemerfeld) near Hastenrath/Gressenich (Stolberg) two drillings each (AG-1 220 m, AG-2 210 m; RF-1 112 m, RF-2 127 m) were sunk below the deepest former mining levels. Further shallow drillings were carried out on vein-type barite occurrences along the same tectonic structures (BA-1 to BA-13).

The drill cores were sampled equidistantly at intervals of 1 m. The attempt was made to collect samples of the black shales without any epigenetic fissure filling mineralization to gain characteristic data of the intersected rocks.

Chemical analyses were carried out by X-ray fluorescence. Selected samples were investigated by X-ray diffraction, microscopy of thin and polished sections and electron microprobe. Data processing was performed using the MAX software package (Kottrup and Rehder 1983).

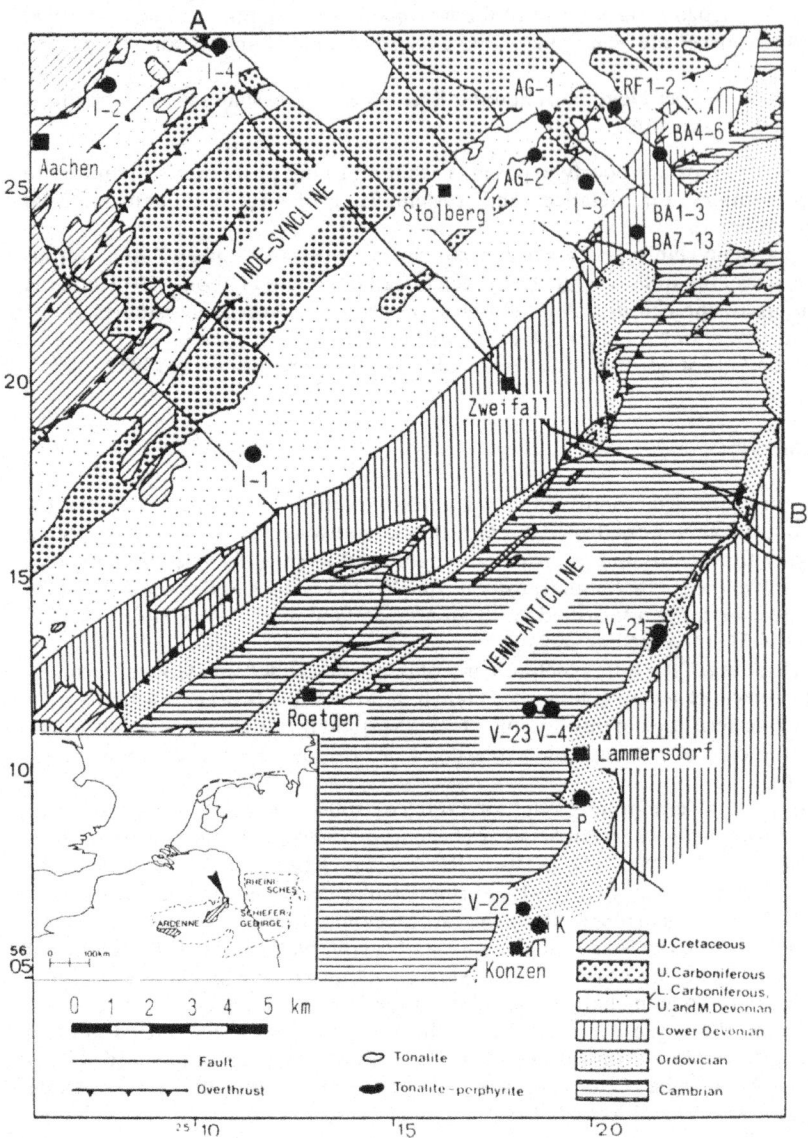

Fig. 1. Geological map of the northern Eifel (after Knapp 1980) with project drill sites I-1 to I-4, AG-1, AG-2, RF-1, RF-2, BA-1 to BA-13 and V-4, V-21 to V-23. Further drill sites: *K* Konzen (Geol. Inst. RWTH Aachen); *P* Paustenbach (Mannesmann AG). For section along A-B see Fig. 2

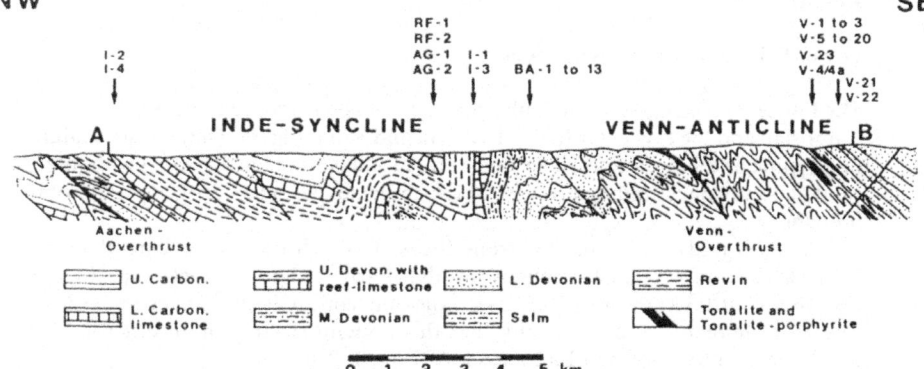

Fig. 2. Geological section crosscutting the northern Eifel (cf. Fig 1 – after Knapp 1980) and relative positions of drill sites

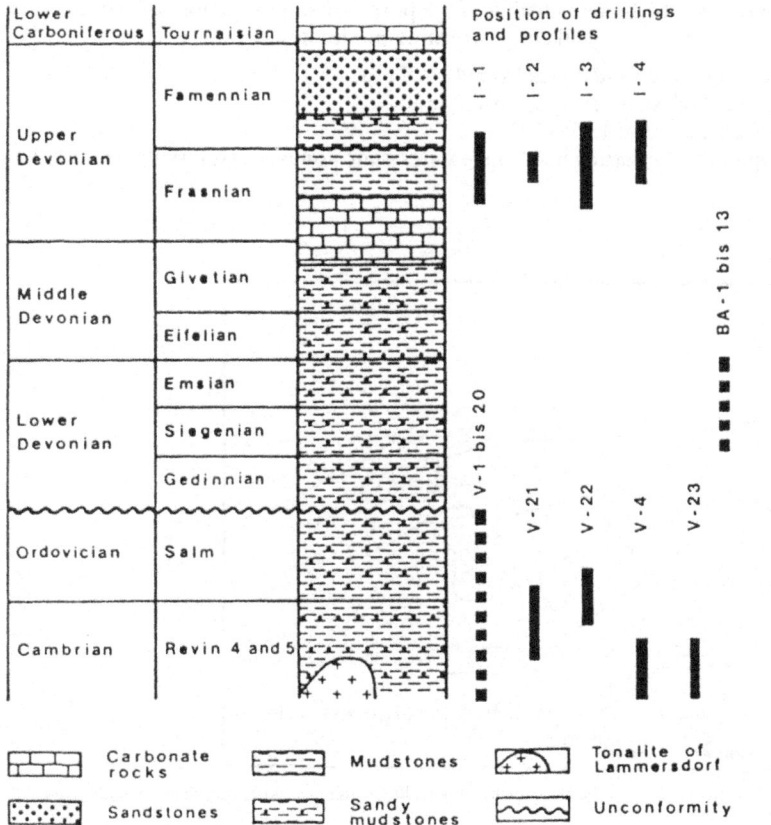

Fig. 3. Stratigraphic positions of drillings

Results

Lower Paleozoic Sedimentary Rocks

The Lower Paleozoic sediments with a thickness of about 1500 m comprise the Upper Cambrian and the Lower Ordovician (Revin and Salm, Fig. 3). They consist mainly of phyllitic and bituminous slates and banded slates (Baenderschiefer) with intercalated sandstones and quartzites which indicate periodic terrigenous detrital influence.

Analytical data of samples from the shallow drillholes indicate that the highest base metal concentrations in the Early Paleozoic rocks are in the Cambrian/Ordovician (Revin/Salm) transition zone (Vogtmann et al. 1986a). Therefore borehole V-21 was sunk into this domain starting in an outcrop of a tonalite porphyrite north of Lammersdorf (Figs. 1, 2, 4).

The transition zone in that area corresponds to an SP-anomaly (H. Jödicke, pers. commun.). According to drill core studies, this anomaly is caused by metaanthracite-bearing black slates (Jödicke 1985).

Apart from the hanging wall tonalite porphyrite, four sections of sedimentary rocks occur in the V-21 drill core, consisting predominantly of:

1. sandy slates and sandstones from 47 to 71 m;
2. phyllitic black slates from 71 to 130 m;
3. banded slates from 130 to 233 m;
4. a frequently alternating bedding of slates, sandstones, and quartzites from 233 to 300 m.

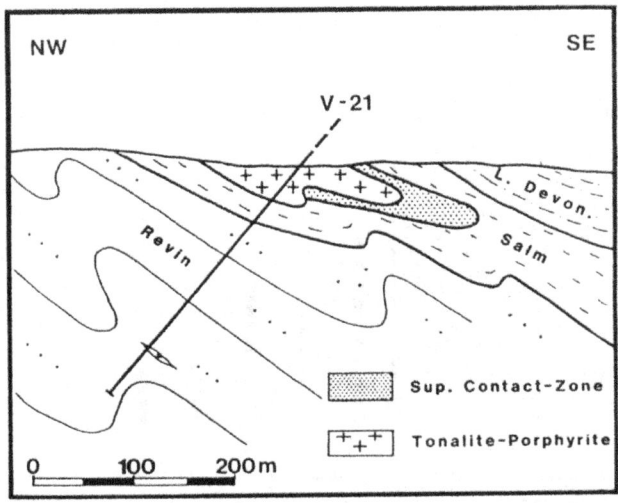

Fig. 4. Geological section of the area north of the Kalltal water reservoir (after Schmidt 1956) with drill site V-21

Several intrusions of tonalite porphyrite with a maximum thickness of 3 m occur in the latter. According to lithology the uppermost section (1) of the V-21 core is part of the lowermost Salmian, unit (2) consists of Revinian 5 slates.

The most important result of V-21 is the increase of Zn and Pb from 70 to 140 m depth (section 2) with maximum values of 2750 ppm Zn and 500 ppm Pb. In this part of the core numerous sphalerite grains were found ranging within a few millimeters in diameter. Their occurrence as well in sedimentary rocks as in surrounding fissures and veins points to sulfide mobilization by lateral secretion.

Similar results were obtained by Hack et al. (1985) who investigated the distribution of ore minerals in the Konzen borehole, a 400 m pre-investigation drilling of the German Continental Drilling Program (KTB) (Walter and Wohlenberg 1985a).

Echle et al. (1985) found additional evidence of lateral secretion by comparing the mineralogical and chemical composition of quartz-carbonate veins and surrounding sedimentary rocks of the Konzen drill core.

At the southern limb of the Venn anticline near Lammersdorf/Konzen a strong magnetic anomaly corresponds with the Salm/Revin transition zone (Franken et al. 1985). A deeper drill (Fig. 5, V-22, 470 m) was carried out to study the coincidence of geochemical and geophysical anomalies on the same stratigraphic level. Drilling has shown that the geophysical anomaly is caused by a more than 250-m-thick pyrrhotite-bearing sequence of black slates and siltstones (Friedrich et al. 1985). Pyrrhotite occurs lens-shaped up to some centimeter in length, in layers a few millimeters thick, intimately disseminated and on small fissures. Minor amounts of chalcopyrite, sphalerite, and traces of galena are associated with the pyrrhotite enrichment.

In hanging wall rocks up to the surface, pyrite and marcasite occur instead of pyrrhotite, so that the source of the magnetic anomaly was detectable only by drilling.

In all drill cores studied, the Cambro-Ordovician strata generally comprise large amounts of pyrite which often contains small inclusions of sphalerite, galena, and chalcopyrite. High C_{org} (up to 4.6 wt.%) and V concentrations and extraordinary low CaO and MnO concentrations (Table 1) indicate a sedimentary or early diagenetic

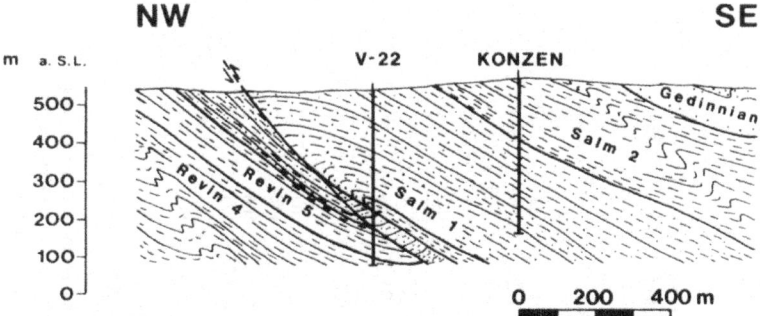

Fig. 5. Geological section across the Revin/Salm transition zone at the southeastern rim of the Venn-Anticline near drill sites V-22 and Konzen

Table 1. Main and trace element content of representative lithological units of Paleozoic sedimentary rocks and magmatic rocks intersected by drillings (main elements in weight-%, trace elements in ppm)

Lithol. unit	1	2	3	4	5	6	7	8	9	10	11	12	13	14	15
n	27	37	174	46	13	22	70	22	229	18	18	37	19	45	14
SiO_2	16.96	47.73	31.30	47.09	2.84	57.81	58.81	61.30	62.92	55.54	69.38	68.03	68.11	70.30	70.54
Al_2O_3	5.85	20.85	12.46	18.99	<1.00	21.36	20.32	20.28	19.60	21.82	18.37	16.53	16.84	17.79	15.64
CaO	37.56	6.08	23.31	7.19	52.22	<0.50	<0.52	<0.50	<0.69	<0.52	<0.50	2.37	1.92	<0.50	2.00
MgO	1.70	2.26	2.18	2.27	0.90	2.27	2.10	1.66	1.70	2.24	1.52	1.37	1.61	0.56	1.50
Na_2O	<0.10	<0.10	<0.10	<0.10	<0.10	0.69	0.58	0.66	0.92	0.51	0.29	2.83	3.21	1.50	2.72
K_2O	1.30	4.51	2.66	4.15	0.23	3.77	3.75	3.68	4.32	4.12	1.83	1.57	2.10	2.31	2.11
TiO_2	0.25	0.85	0.49	0.77	0.06	0.94	0.95	0.94	0.92	0.96	0.56	0.51	0.51	0.64	0.45
Fe_2O_3	2.22	7.89	4.49	7.03	0.52	8.40	8.44	7.16	6.66	8.41	3.83	2.74	3.34	3.46	3.40
MnO	0.08	0.08	0.10	0.09	<0.04	<0.04	<0.04	<0.04	0.13	0.17	0.05	0.08	0.07	0.06	0.09
P_2O_5	0.08	0.12	0.08	0.08	0.02	0.10	0.11	0.11	0.10	0.12	0.11	0.11	0.11	0.12	0.12
SO_3	0.68	0.55	0.25	0.75	0.31	n.b.	<0.20	<0.20	0.20	<0.20	<0.20	<0.20	0.43	<0.20	<0.20
Pb	<25	<21	<22	35	<20	39	35	<29	43	36	<26	<21	<20	<20	<20
Zn	<23	64	40	178	<29	143	143	108	162	137	101	−23	58	32	50
Cu	<20	<26	<25	<20	<20	33	37	<25	76	46	237	157	66	61	59
Ni	<20	50	30	39	20	53	44	41	50	51	<20	<20	<20	<20	<20
Co	<20	<23	<20	<20	<20	<25	<24	<22	<24	<25	<20	<20	<20	<20	<20
Cr	39	104	66	95	21	111	101	98	103	105	26	23	20	22	25
V	56	139	90	118	35	185	159	152	124	181	85	80	71	90	62
Ba	122	506	303	526	44	600	572	579	1243	942	462	500	438	447	296
Sr	458	173	260	156	323	114	112	117	137	113	69	480	577	241	175
Rb	50	202	109	178	30	182	179	173	201	175	48	35	50	74	100
Pb+Zn+Cu	68	111	87	233	69	215	215	162	281	219	364	201	144	113	129

1 = I-1 Frasnian nod. limestones
2 = I-1 Low. Frasn. shales
3 = I-1 Frasnian nod. limestones and shales
4 = I-1 Upp. Frasn. shales (Matagne-Schiefer)
5 = I-1 Frasn. reef carbonates

6 = V-21 Revinian slates
7 = V-21 Revin. banded slates
8 = V-21 Revin. slates. banded slates, quartzites
9 = V-22 Salm I banded slates
10 = V-21 Salm I banded slates

11 = V-4 altered Lammersdorf tonalite
12 = V-4A "fresh" tonalite
13 = V-23 "fresh" tonalite
14 = V-21 tonalite porphyrite
15 = V-21 apophyses of tonalite porphyrite

n = number of analyses

environment favorable for sulfide formation. Compared with Cambrian and Ordovician sedimentary rocks and in respect of a later mobilization and leaching of these strata (see metallogenetic model) the Revin/Salm transition zone and the Lower Salm are significantly enriched in Pb, Zn, Cu, Ni, and Co. These sections, in particular, are characterized as having potential for syngenetic sulfides.

Additionally, Lower Salmian strata are significantly enriched in Ba bound to silicates, thus representing a potential source for Ba in younger sedimentary rocks.

Acid Intrusives

The Tonalite of Lammersdorf (see also Scherp 1959; Van Wambeke 1955a,b) is characterized by intensive weathering causing feldspar decomposition at its top zone (Fig. 6). The geochemistry of unweathered rocks from deeper levels of the Lammersdorf intrusion corresponds to granodioritic composition (see Table 1).

The following ore minerals, in order of decreasing frequency, were determined: pyrite, pyrrhotite, rutile/anatas, chalcopyrite, marcasite, galena, sphalerite. Smaller accumulations of chalcopyrite were found in marginal zones of the intrusive. The content of Mo, Sn, and W in rocks was lower than 10 ppm. Generally the Lammersdorf tonalite does not show such a complex ore mineral association as described by Weis et al. (1980) for the analogous Hill tonalite in Belgium.

The increasing pyrrhotite content toward depth and related high susceptibility values (Franken 1984) may point to an eventual relationship between magmatism and the magnetic anomaly at the southeastern rim of the Venn anticline.

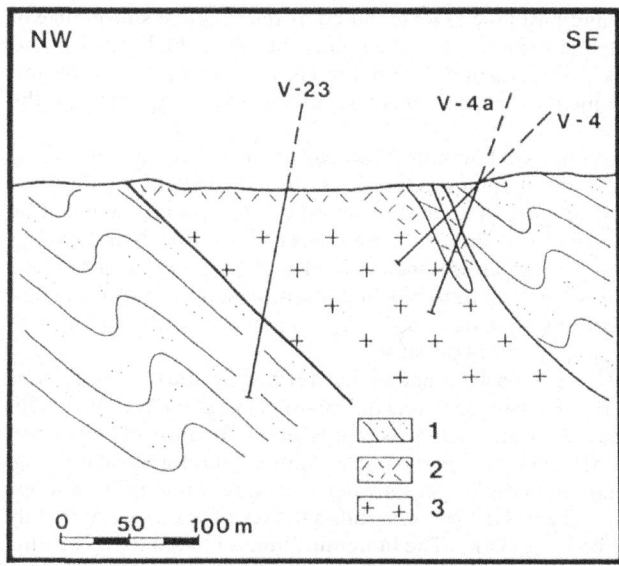

Fig. 6. Geological section of the Lammersdorf tonalite, drill sites V-4, V-4a and V-23; *1* Revinnian slates and quartzites; *2* tonalite (weathered); *3* tonalite (fresh)

Different tonalite porphyrite occurrences were intersected by several shallow drillings and borehole V-21 (Fig. 4). Phenocrysts of plagioclase are embedded within a fine-grained sericitized matrix of quartz, albite, and chlorite. Accessory minerals are apatite, zircon, sphene, epidote, and carbonate (Scherp 1959; Van Wambeke 1955c). There is only a small amount of opaque minerals (pyrite, pyrrhotite, galena, chalcopyrite). Up to now the porphyrites are interpreted as "hydrothermal branches" of the tonalite (Knapp 1980). However, intercalations of thin beds of pelites, host rock xenolithes with diameters up to 3 cm, sharp contacts with foot wall and hanging wall sedimentary rocks, and the lack of contact metamorphic patterns in surrounding rocks may indicate a pyroclastic origin (tuffs).

Main and trace element distribution reveal a relatively similar geochemical composition of the tonalite and tonalite porphyrite. Influences of supergene alteration become evident by a relative decrease in mobile (Na, Ca, Sr) and increase in immobile or less mobile elements (Ti, Zr, Cr). Hypogene metasomatic processes could be distinguished neither laterally nor vertically due to the restricted extension of the magmatic bodies.

Upper Devonian Sedimentary Rocks

At the end of the Middle to Upper Devonian reef growth, the Hercynian basinal facies transgressed onto the shelf platform of the Old Red Continent. Bituminous shales, marls and nodular limestones were deposited in a restricted environment under euxinic and partly evaporitic conditions. The carbonates were derived by erosion from older reefs or from reefs persisting on submarine elevations. The development of a swell and basin facies was favored by the irregular subsidence of the Devonian carbonate platform. A regressive phase began in the Early Famennian: the Frasnian black shales and nodular limestones were overlain by sandy and mica rich sediments which form the transition to the thick sequence of the Condroz-Sandstones.

The stratigraphic sequences of borehole I-1 and I-3 are shown in Fig. 7, revealing a good correlation with other profiles along the southeastern flank of the Inde syncline. Lithogeochemical correlation and stratigraphic indications are given by Co, SO_3, and certain base metal distribution patterns (Table 1, Fig. 7) and also by C_{org} values. These results suggest the presence of swells and basins within a distance of few kilometers, possibly forming traps in which stratiform sulfides accumulated (third-order basins according to Large 1983). A correlation between the northwestern and southeastern flank is not possible.

In addition to local metal enrichment found during previous investigations (Scheps and Friedrich 1983), recent borehole information (Vogtmann et al. 1986b) from the Frasnian reveals increased sulfide contents, predominantly of sphalerite, concentrated in Frasnian hanging wall parts which comprise bituminous shales (Fig. 7). Sphalerite is of a light, anhedral type, showing aggregates ranging from a few millimeter up to 1 cm in diameter. The longitudinal axes of crystals are mainly oriented parallel to the bedding plane. The mineralization is related to the euxinic black shale environment. Recrystallization and redeposition of low temperature sphalerite (low Fe content) may occur as a result of a late diagenetic dissolution of evaporites.

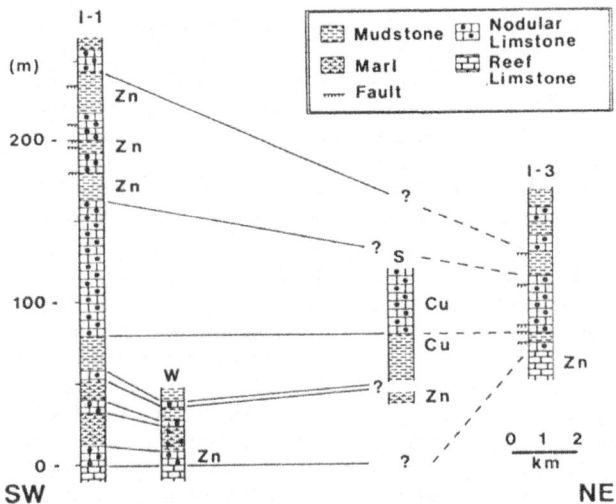

Fig. 7. Borehole sections showing Frasnian sedimentary rocks and indications of increased base metal content; correlation of the profile I-1 with two profiles (*W* Walheim; *S* Schleicher-Stollen) along the southeastern flank of the Inde syncline and relative position of core drilling I-3

Evidence for an evaporitic environment during the formation of the Frasnian black shale facies includes the positive correlation between Sr and SO_4^{2-} in marly black shales and nodular limestones. The elevated contents of metals in the Matagne-Schiefer (unit 4 in Table 1) indicates a close relation between saline facies and metal enrichment (av. 0.75% SO_4^{2-} up to 0.2% Zn, av. 178 ppm). In marly shales of the Schleicher Stollen profile (Fig. 7) pseudomorphs after halite occur at a stratigraphic level corresponding to foot wall black shales of I-1 (Scheps 1982).

The highest content of Zn (0.5%) is observed in Frasnian dolomites and limestones from borehole I-4. This section with generally high Zn values also contains elevated Mn contents averaging 0.19% MnO, which may point to a proximal manganese halo.

The metal accumulation occurs on the same stratigraphic level as the galena-sphalerite-barite-pyrite mineralization near Chaudfontaine in Belgium (Dejonghe 1979).

According to our results, further exploration targets for a syngenetic base metal accumulation are in the central part of the Inde syncline. However, in this area the black shales are buried under a sediment cover of more than 500 m.

Epigenetic Mineralization

Pb-Zn Deposits of Aachen/Stolberg Type. The epigenetic vein-type Pb-Zn deposits of Aachen/Stolberg show many similarities with low-temperature carbonate hosted Pb-Zn deposits corresponding to the Mississippi Valley type after Sangster (1976). They are structurally and lithologically controlled; the ore with collomorph textures fills cavities within Devonian and Lower Carboniferous carbonate sequences.

Former mining activity generally reached levels of only about 130 m because of engineering problems caused by high ground water currents in the carbonate rocks.

Boreholes AG-1 and AG-2 drilled in the vicinity of the former Albertsgrube (Fig. 2) were expected to intersect two of the mineralized fault systems beneath the deepest mining level (Fig. 8). While AG-2 met a blind fault with only calcite filling, AG-1 intersected a 7-m-thick fault system mineralized with calcite, dolomite, sphalerite, and galena.

Boreholes RF-1 and RF-2 at the Roemerfeld location were planned to meet vein-type mineralization underneath the former mining level. Both drillings were stopped in tectonically completely disturbed and karstified carbonate rocks. Fissure fillings of galena and sphalerite occur, as well as traces of disseminated galena with inclusions of pyrite, bravoite (vaesite), and chalcopyrite (Krahn 1985).

Knowledge of the Aachen/Stolberg type deposits is based on limited information derived from former mining activities and studies of material collected from dumps (Gussone 1964). The drilling program has provided more information on host rock and lode filling for detailed geological and mineralogical studies. Microscopic and geochemical investigations of the cores reveal a close genetic association of dolomitization and Pb-Zn-mineralization. Sole relicts in form of halite occur within the fault breccia of the AG-1 ore body. They are bound to gangue

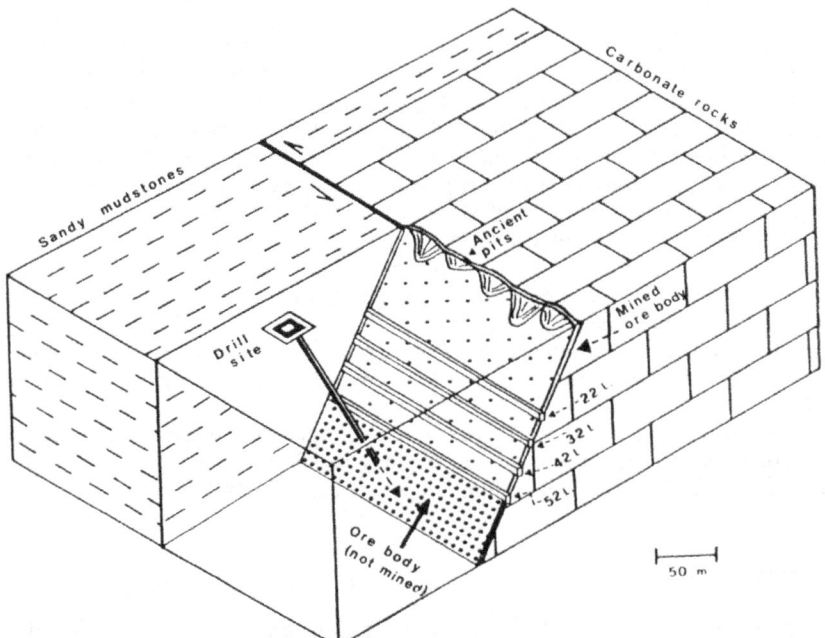

Fig. 8. Block diagram showing geological setting of the vein-type deposit Albertsgrube, ancient pits, not-mined ore body at depth and drill site; mining levels in Lachter (1 L. = 2.05 m)

carbonates encrusted by younger ore minerals and indicate that highly saline solutions were active during ore formation (Krahn 1985; Krahn et al. 1986).

In spite of these results, considerable ore reserves may still be found in relative shallow depth within other former mines, e.g., the Breiniger Berg, and under an about 100-m-thick cover of Tertiary sediments in the northeastern extension of the Inde syncline.

Barite Occurrences. East of Stolberg (Fig. 1), two buried barite occurrences near Krewinkel and Gressenich were delineated by surface geochemistry (Ba and Hg) and subsequently studied by shallow core drillings (Scheps et al. 1986).

The mineralization follows the long-lived northwest-southeast striking major tectonic structures Sandgewand and Zittergewand (Holzapfel 1910). Along the same fault systems carbonate complexes of the Inde syncline represent the host rocks of the Aachen/Stolberg Pb-Zn deposits.

The vein-type mineralizations have a horizontal extension of about 150 to 200 m. Vein filling varies in thickness from a few centimeters to several meters and is accompanied by traces of chalcopyrite and sphalerite. Host rocks and barite are strongly brecciated and mylonitized.

The structural control and a distinct positive Cu-Ba correlation point to a relation to those Cu-bearing barite veins occurring in the southeastern extension of the studied fault zones in the Mechernich Trias anticline (Schröder 1938/1979). A relationship to barite veins of the southeastern Eifel near Ürsfeld (Weisser 1963) is not evident.

Conclusions – Metallogenetic Model

The genetic interpretation of the vein-type Pb-Zn deposits of Aachen/Stolberg type is still being discussed.

Present knowledge indicates that the metal content of epigenetic deposits in the northern Eifel is derived from sedimentary rocks of the Paleozoic basement. Models of metal mobilization from older sedimentary rocks by hydrothermal activities, especially during tectogenesis, have been developed, e.g., by Dejonghe (1979). Additional indications (e.g., sphalerite in sedimentary rocks and surrounding fissures) have been found in this study.

Concerning metallogenetic aspects the following model for intraformational ore deposition is proposed (Fig. 9; for details see Krahn et al. 1986; Vogtmann et al. 1986a; Vogtmann et al. 1986b):

— During Cambro-Ordovician time (A) base metals are syngenetically accumulated in a black shale environment suitable for sulfide deposition.
— During Paleozoic time the release of metal-bearing interstitial waters is due to tectonic stress and progressive compaction. Syn- and epigenetic enrichment occurs as well in Upper Devonian black shales (B) as in joints and remaining pore volume of the Paleozoic sedimentary rocks (phases Ia and Ib).
— During late Paleozoic and Mesozoic age (Upper Carboniferous to Upper Cretaceous) epigenetic metal enrichment takes place along post-Variscan fractures (C, carbonate complexes of Aachen/Stolberg) and in sedimentary rocks south-

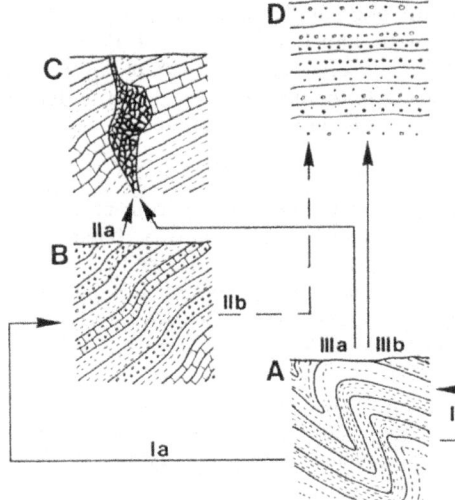

Fig. 9A-D. Metallogenetic model concerning the northern Eifel. **A** Cambro-Ordovician black slates, siltstones, sandstones, and quartzites. **B** Devono-Carboniferous carbonate rocks, nodular limestones, black shales, and sandstones. **C** Vein-type Pb-Zn deposits in Devonian and Lower Carboniferous carbonate complexes (Aachen/Stolberg type). **D** Pb-Zn impregnations in the Buntsandstein Lower Triassic; type Maubach/Mechernich; for explanations of phases I, II, and III see text

east of the Venn anticline containing a high pore volume (D, Triassic sandstones of Maubach/Mechernich). The most important accompanying mechanisms are:
1. mobilization from late Paleozoic sediments (phase II) by highly saliniferous solutions probably supported by decomposition of evaporitic rock complexes;
2. mobilization by processes of lateral secretion from the early Paleozoic basement.

The early Paleozoic formations with their high metal content have a total thickness of several hundred meters. In contrast Upper Devonian black shales, also enriched in base metals, are comparatively thin with their thickness of only a few tens of meters.

The Pb-isotope correspondence of galena from vein-type deposits of Aachen/Stolberg and the mineralization at Chaudfontaine/Belgium (Dejonghe 1979; Dejonghe et al 1982) should be the result of a single source (Krahn et al. 1986). More investigations of samples from Maubach/Mechernich and Lower Paleozoic series are planned.

Younger Devonian and Carboniferous sedimentary rocks most likely were not effective as a metal source for Maubach/Mechernich (dotted line in Fig. 9) because the ore-hosting clastic rocks of Triassic age are underlain by Lower Devonian and older beds.

The acid igneous rocks of the northern Eifel (Keyssner et al. 1986) with an age of 385 m.y. (Kramm and Buhl 1985) did not play any role concerning the formation of deposits known so far. They may have influenced and accelerated processes in phases Ia and Ib by heat supply.

A wide-range overthrust in the northern Eifel (Meissner et al. 1981; Walter and Wohlenberg 1985b) would lead to duplication of beds and would thus represent an important increase in the potential of metal supplying rocks (Friedrich et al. 1981).

Different forms of ore development and appearance in different localities are controlled by the carbonate content and cavity volume of the host rocks:
— Along fault zones cutting relatively carbonate-poor clastic rocks (e.g., clastic Upper Carboniferous rocks of Aachen), a comparatively slow increase in pH values takes place causing precipitation and crystallization of larger sulfide aggregates.
— Within coarse clastic sediments (Buntsandstein of the Mechernich Triassic triangle) ore deposition is orientated primarily by cavity volume.
— Coming across carbonate complexes along faults metal bearing hydrothermal solutions suffer a sudden increase in pH value. This leads to a quick precipitation of sulfides showing collomorph textures.

The minor pore volume of these carbonates complexes (Schneider 1977) is responsible for solutions migrating predominantly along faults and for metal enrichment taking place in fault zones only. Nevertheless the Aachen/Stolberg Pb-Zn-deposits are to be classified as of Mississippi Valley type.

Acknowledgments. The authors are indebted to Prof. Dr. G.H. Friedrich, RWTH Aachen, for supporting this paper and for the permission to publish these data. The project was funded by the Federal Minister for Research and Technology (BMFT-Project No. 03 R 242). We thank two anonymous reviewers for reading the manuscript and for helpful suggestions.

References

Dejonghe L (1979) Discovery of a sedimentary Ba (Fe, Zn, Pb) ore body of Frasnian age at Chaud-fontaine, Province of Liege, Belgium. Min Dep 14:15–20

Dejonghe L, Rye RO, Cauet S (1982) Sulfur isotopes of barite and lead isotopes of galena from the stratiform deposit in Frasnian carbonate and shale host-rocks of Chaudfontaine (Province of Liege, Belgium). Ann Soc Geol Belg T 105:97–103

Echle W, Plüger WL, Zielinski J, Frank B, Scheps V (1985) Petrography, mineralogy and geochemistry of the Salmian rocks from research borehole Konzen, Hohes Venn (West Germany). Neues Jahrb Geol Paläontol Abh 171:31–50

Franken D (1984) Die Magnetisierung des Kambriums und Ordoviziums im nordöstlichen Stavelot-Venn Massiv. Dipl Thesis, Rheinisch-Westfälische Technische Hochschule, Aachen

Franken D, Bosum W, Wohlenberg J (1985) A geological and geophysical interpretation of the magnetic anomaly of Lammersdorf, Hohes Venn (West Germany). Neues Jahrb Geol Paläontol Abh 171:363–375

Friedrich G, Scheps V, Gussone R (1981) Die nördliche Eifel als höffiger Bereich für tiefer liegende Blei-Zink-Erzlagerstätten. In: Walter R (ed) Vorstudie für eine übertiefe Forschungsbohrung im Massiv von Stavelot-Venn, Nordeifel. Report Geol Inst, Aachen (unpubl)

Friedrich G, Scheps V, Keyssner S, Vogtmann J, Bosum W (1985) Pyrrhotinführende Sedimentgesteine als Ursache der magnetischen Anomalie am Südrand des Hohen Venns. Fortschr Miner 63(1):66

Gussone R (1964) Untersuchungen und Betrachtungen zur Paragenesis und Genesis der Blei-Zink-Erzlagerstätten im Raume Aachen-Stolberg. — PhD Thesis, Rheinisch-Westfälische Technische Hochschule, Aachen

Hack U, Friedrich G, Martin J (1985) Distribution of accessory ore minerals in the Lower Ordovician sediments of Salm 1 and Salm 2, research borehole Konzen, Hohes Venn (West Germany). Neues Jahrb Geol Paläontol Abh 171:51–62

Holzapfel E (1910) Die Geologie des Nordabfalls der Eifel mit besonderer Berücksichtigung der Gegend von Aachen. Abh Kgl Preuß Geol Landes Anst N F 66:218

Jödicke H (1985) A large selfpotential anomaly at the SE flank of the Venn-Stavelot anticline originating from metaanthracite-bearing black shales at the Salm/Revin boundary. Neues Jahrb Geol Paläontol Abh 171:387–402

Keyssner S, Scheps V, Friedrich G (1986) Zur Geochemie und Petrologie saurer Magmatite im Altpaläozoikum der nördlichen Eifel. Neues Jahrb Geol Rheinl Westfalen 171:187–206

Knapp G (1980) Erläuterungen zur geologischen Karte der nördlichen Eifel 1:100000. Geol Landesamt NRW, Krefeld

Kottrup G, Rehder S (1983) MAX Program Description. Fed Inst Geosci Nat Res (BGR), Hannover

Krahn L (1985) Geologisch-lagerstättenkundliche Untersuchungen der Bohrungen "Albertsgrube 1 und 2", Blei-Zink-Distrikt Aachen-Stolberg. Dipl Thesis, Rheinisch-Westfälische Technische Hochschule, Aachen

Krahn L, Friedrich G, Gussone R, Scheps V (1986) Zur Blei-Zinkvererzung in Karbonatgesteinen des Aachen-Stolberger Raumes. Fortschr Geol Rheinl Westfalen 34:133–157

Kramm U, Buhl D (1985) U-Pb zircon dating of the Hill Tonalite, Venn-Stavelot Massif, Ardennes. Neues Jahrb Geol Paläontol Abh 171:311–327

Large D (1983) Sediment-hosted stratiform lead-zinc-deposits: An empirical model. In: Sangster DF (ed) Short Course in Sediment-hosted stratiform lead-zinc deposits. Min Assoc Can (MAC) Victoria/BC:1–29

Meissner R, Bartelsen H, Murawski H (1981) Thin-skinned tectonics in the northern Rhenish Massif, Germany. Nature 290:399–401

Sangster DF (1976) Sulphur and lead isotopes in strata-bound deposits. In: Wolf KH (ed) Handbook of strata-bound and stratiform ore deposits, Vol 2. Elsevier/Amsterdam pp 219–266

Scheps V (1982) Geochemische Untersuchungen in der Eifel – Ein Beitrag zur Geochemie der Flußsedimente und der paläozoischen Schwarzschiefer und Karbonatgesteine des Venn-Massivs und der Inde-Mulde. PhD Thesis, Rheinisch-Westfälische Technische Hochschule, Aachen

Scheps V, Friedrich G (1983) Geochemical investigations of Paleozoic shales and carbonates in the Aachen region. Min Dep 18:411–421

Scheps V, Keyssner S, Sänger-von Oepen P, Friedrich G (1986) Gangförmige Barytvorkommen östlich von Stolberg, Nordeifel. Fortschr Geol Rheinl Westfalen 34:207–219

Scherp A (1959) Die Petrographie der Eruptivgesteine im Kambro-Ordovizium des Hohen Venns. Geol Jahrb 77:95–120

Schmidt Wo (1956) Neue Ergebnisse der Revisionskartierung des Hohen Venns. Geol Jahrb Beih 21:146

Schneider W (1977) Diagenese devonischer Karbonatkomplexe Mitteleuropas. Geol Jahrb D21:3–107

Schröder E (1938) with a contr. of Pfeffer P. Erläuterungen zu Blatt Zülpich – Geol. Karte Preussen und benachbarte deutsche Länder 1:25.000, Berlin, 2nd edn (1979). Geol Kt Nordrh Westf 1:25.000, Erl Bl 5305 Zülpich, Krefeld

Vogtmann J, Scheps V, Friedrich G (1986a) Zur Geochemie und Mineralogie der kambro-ordovizischen Sedimentgesteine an der Südostflanke des Stavelot-Venn-Massivs, Nordeifel. Fortschr Geol Rheinl Westfalen 34:159–185

Vogtmann R, Scheps V, Friedrich G (1986b) Geochemie und Sulfidführung oberdevonischer Schwarzschiefer und Knollenkalke der Inde-Mulde. Fortschr Geol Rheinl Westfalen 34:103–131

Walter R, Wohlenberg J (eds) (1985a) Geology and geophysics of the northeastern Hohes Venn area – Report on a joint scientific venture. Neues Jahrb Geol Paläontol Abh 171:467

Walter R, Wohlenberg J (1985b) Proposal for an ultra-deep research borehole in the Hohes Venn area (West Germany). Neues Jahrb Geol Paläontol Abh 171:1–16

Wambeke L van (1955a) Compositions minéralogiques et chimiques des tonalites de la Helle et de Lammersdorf (Hautes-Fagnes). Bull Soc Belge Géol Paléontol Hydrol 64:477–509

Wambeke L van (1955b) La minéralisation des tonalites de la Helle et de Lammersdorf et leurs relations avec les autres minéralisations. Bull Soc Belge Géol Paléontol Hydrol 64:534–581

Wambeke L van (1955c) Note sur les Venn Porphyres des Hautes Fagnes allemandes. Bull Soc Belge Géol Paléontol Hydrol 64:510–521

Weis D, Dejonghe L, Herbosch A (1980) Les associations des minéraux opaques et semi-opaques de la roche ignée de La Helle. Ann Soc Géol Belg 103:15–23

Weisser D (1963) Tektonik und Barytgänge in der SE-Eifel. Z Dtsch Geol Ges 115:33–68

The Significance of Diagenesis in Emplacement of Strata-Bound Zn-Pb Mineralization in Carbonate Sediments

H. KUCHA[1]

Abstract

An emplacement of base metal sulfides in subsurface sedimentary environment is controlled by:

   1. Sediment inherited porosity-permeability. Carbonates with high initial porosities such as intertidal sediments (Navan, Moate), Waulsortian limestones with stromatactics (Ballinalack, partly Tynagh), and gravity-induced deposits on Waulsortian (Silvermines, Tynagh) and submarine slopes (Navan in places) act as favorable host to strata-bound mineralization to be emplaced in the subsurface environment.

   2. Retention and recreation of porosity-permeabilty. Controlling factors are polymorphic transformations of carbonates, recrystallization of carbonates, introduction of base metals in the structure of carbonates, stylolitization, tectonic fracturing, and brecciation and dissolution of carbonate host. Replacement of base metal carbonates reduces volume and releases large amounts of CO_3^{2-}. As a result, host carbonates are leached and collapse breccias are produced (Upper Silesia, Navan). A retention of porosity-permeability is favored by the existence of two separate solutions, one with metals and another with sulfur.

Introduction

In carbonate-hosted strata-bound Zn-Pb deposits there are several stages of mineralization distinctly separated in time yet occurring within the same limited space of the ore zone. Massive ore bodies are formed by the repetition of mineralization within the same space. It appears that the creation and retention of porosity and permeability of the ore zone during emplacement of an epigenetic mineralization may be as important as is the metal source. The expected metal concentration in the mineralizing solutions is below 100 ppm (Bischoff et al. 1981) and therefore an enormous volume of fluids has to pass through the ore zone to produce a high-grade ore. The aim of this paper is to discuss how a high permeability-porosity can be retained and recreated during emplacement of the ore in intertidal carbonates (Navan, Moate), Waulsortian reefs (Ballinalack, Tynagh), gravity flow deposits generated on reef slopes (Silvermines) and strata-bound dolomites (Upper Silesia).

[1]H. Kucha, Afdeling fysico-chemische geologie, Celestijnenlaan 200 C, B-3030 Heverlee, Belgium

Base Metal Sulfide Deposits
G.H. Friedrich, P.M. Herzig (Eds.)
© Springer-Verlag Berlin Heidelberg 1988

Mineralization Hosted in Intertidal Carbonates

The Navan ore body, Ireland, consists of five stacked tabular lenses hosted in shallow water carbonates composed of bioclastic and oolitic calcarenites and pelsparites of the Lower Carboniferous (Andrew and Ashton 1982; Kucha and Wieczorek 1984). The ore lenses are conformable to the bedding and are surrounded by a halo composed of dispersed Fe-dolomite, Zn-dolomite, ankerite, pyrite, and silicification (Kucha 1987). The main ore minerals are sphalerite and galena. The Zn:Pb ratio is about 5:1. Associated minerals are pyrite, marcasite, melnikovite and barite. Quartz overgrowths, Fe-dolomite, and Zn-dolomite may be abundant in places.

The Navan ore exhibits five macrotextures:

1. Massive and finely banded, paralleling the bedding. Microtextures are represented by sulfide rims on bioclasts, ooids, pellets (Fig. 1), clasts of barite, sulfides infilling interparticle porosity and sulfides replacing bioclasts, ooids, and carbonate microclasts (Kucha 1987). These microtextures ensure that macrotexture "1" follows a sedimentary fabric of the host.

2. Sulfide breccia and microbreccia recemented by later sulfides, metal-bearing carbonates and barite. Microtextures are banded sulfide and carbonate cements, fillings, and replacements.

3. Host rock breccia cemented with sulfides. Microtextures consist of banded sulfides with preserved relics of banded Zn-calcite (Fig. 2).

4. Cavity infills. Microtextures consist of banded sulfides and carbonates. Sulfides replace carbonates.

5. Veins and veinlets including horizontal interlayer veins with sulfide stalactites. Microtexture is represented by sulfide bands. The earliest sphalerite at Navan forms rim cement on bioclasts and ooids.

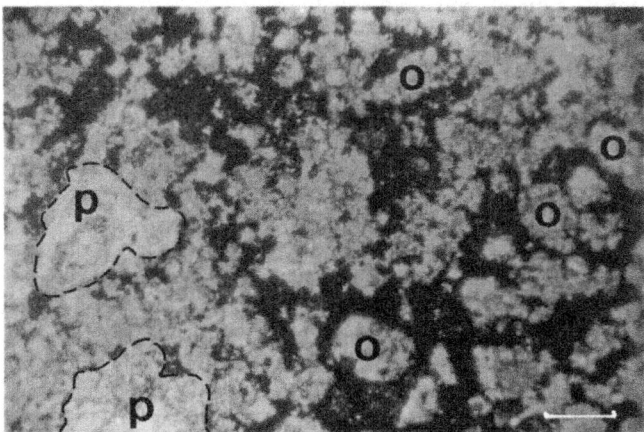

Fig. 1. Sphalerite (*black*) cementing oolitic calcarenite. Sphalerite is present around ooids (*O*), pisoids (*P*, encircled by *dashed line*) and as infill of interparticle porosity. Navan, Ireland. Stained thin section, transmitted light, bar = 150 μm

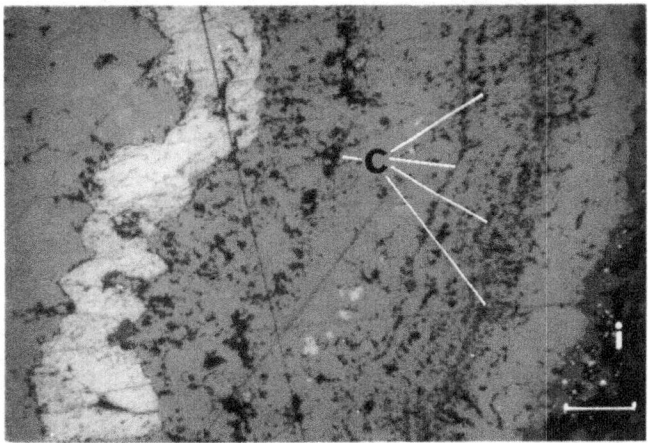

Fig. 2. Banded sphalerite interfingered with relic bands of Zn-calcite (*C*). Galena (*white*) and internal sediment (*i*) are also present. Navan, Ireland, carbonate breccia in the wall rock. Reflected light, bar = 150 μm

Zn-Pb mineralization in the Moate (Moyvoughly) prospect, Ireland, is hosted on oolitic calcarenites of the Moyvoughly Beds, Lower Carboniferous. Mineralized calcarenites are truncated by supratidal micrites preferentially replaced by post-ore Fe-dolomite. Fe-dolomite and silicification form a dispersed halo surrounding sulfide mineralization. Main ore minerals are sphalerite and pyrite-melnikovite. Associated minerals are barite, galena, and celsian constituting in places 10 vol.% of the rock. Zn:Pb ratio is about 7:1. Sulfides typically replace carbonate allochems, cements, and barite.

Macrotextures of the Moate ore are as follows:

1. Massive and finely banded, paralleling the bedding. Microtextures are represented by sulfide rims on carbonate and silicate allochems, sulfides as infill of interparticle porosity, sulfides replacing carbonate allochems (Fig. 3), silica overgrowing detrital quartz and celsian overgrowing detrital feldspars. Macrotexture "1" controls about 60% of the mineralization. The original fabric of the host carbonate is typically preserved.

2. Fracture infill. Microtexture is formed by sulfide bands. Macrotexture "2" controls about 40% of the mineralization.

3. Stylolite guided sulfide infills and replacements.

4. Finely disseminated sulfides in carbonate matrix.

5. Nodular. Nodules are several cm in diameter with external parts cemented with sphalerite and interior parts cemented with barite and Fe-calcite.

A typical ore macrotexture at Moyvoughly is a combination of "1" and "2", i.e., fracture infill with massive mineralization spreading out laterally into porous and permeable oolitic calcarenites. Single mineralized zones extend vertically from 3 to 30 m and laterally up to 200 m.

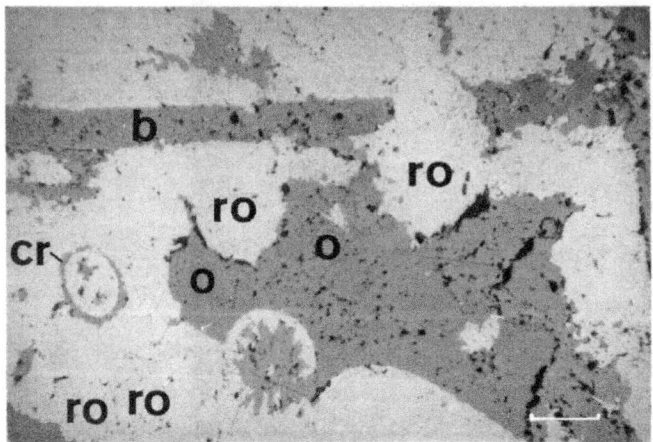

Fig. 3. Replacement of ooids by sphalerite. A radial texture of sphalerite suggests that replacement took place during transformation of the original aragonitic ooid into low Mg-calcite polymorph. *cr* rim cement composed of low Mg-calcite; *b* barite; *ro* replaced ooid; *o* unreplaced ooid. Moate, Moyvoughly Beds, Ireland. Reflected light, bar = 400 μm

Mineralization Hosted in Waulsortian Reefs

At Ballinalack, Ireland, sphalerite, pyrite, melnikovite, and barite form infills, linings of stromatactis cavities (Fig. 4), and replacements of the host Waulsortian limestones. Associated minerals are: galena, Zn-dolomite, and Fe-dolomite, which in places may constitute as much as 20 vol.% of the host rock.

Three major ore macrotextures can be distinguished at Ballinalack:

1. Stromatactis infillings subdivided into:
   a) stromatactis linings composed of sphalerite, pyrite, melnikovite and minor barite. Microtextures are finely banded linings and replacements of banded metalliferous carbonates;
   b) internal sediments in stromatactis composed mainly of zoned rhombs of Zn-Fe-dolomite replaced by sphalerite. Microtextures are represented by banding related to the diffusion of H_2S through the internal sediments in the stromatactis cavity system.
2. Stylolite guided infills and replacements.
3. Fracture infills.

Ore mineralization at Tynagh, Ireland, is hosted mainly in Waulsortian limestones (85–90%) and less common in off-reef gravity-induced deposits (10–15%). Waulsortian hosted mineralization at Tynagh (Fig. 5) has been ac-cumulated in three stages:

1. Synsedimentary-early diagenetic. Sulfides, mainly pyrite-melnikovite and minor sphalerite form linings and infills of stromatactis cavities. They constitute 1–4

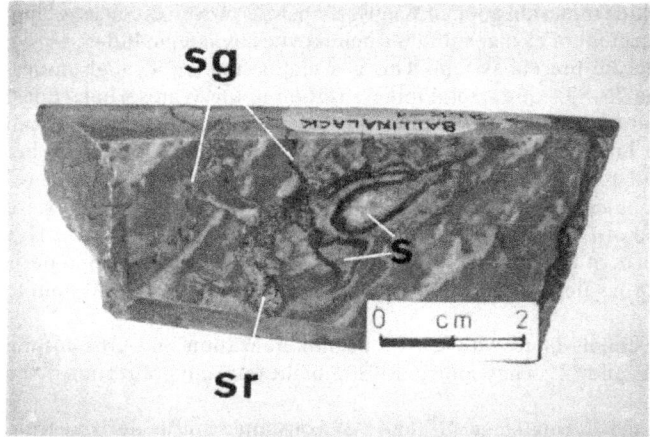

Fig. 4. Mineralization in Waulsortian at Ballinalack, Ireland, as stromatactis infill (*s*) by radiaxial fibrous mosaic calcite, sphalerite, pyrite, barite and galena linings, as stylolite guided infills (*sg*), and as stylolite guided replacements (*sr*)

Fig. 5. Mineralization in Waulsortian at Tynagh, Ireland. *s* stromatactis infilled by radiaxial fibrous mosaic calcite and Fe-calcite and sphalerite linings; *f* dilatant fracture with symmetrical infill composed of radiaxial fibrous mosaic calcite, sulfides, barite and Zn-bearing carbonates

vol.% of the total sulfide mineralization in Waulsortian. Microtextures consist of fine banding and replacement of earlier sulfides (melnikovite) by later sulfides.

2. Dilatant fracture-breccia system. This is a major stage of Tynagh mineralization controlling 75–80% of the total mineralization in the Waulsortian (Boast et al. 1981). Dilatant fracture-breccia is infilled by banded linings and by internal sediments. Banded linings are composed of radiaxial fibrous mosaic calcite, Fe-calcite, sulfides, and barite. Microtextures are composed of fine bands and replacements of carbonates and early melnikovite by sulfides. Internal sediments were originally composed of Fe-, Zn- and Ba-dolomites, converted by diffusion of H_2S into intimate mixtures of Fe-dolomite + sphalerite or nodules with external parts built up of sphalerite or galena and interior parts built up of a mixture of Fe-dolomite + barite.

3. Veins and veinlets filled with Cu-Pb-Ba mineralization and crosscutting deposits of stages "1" and "2". They control 15–20% of the total mineralization in the Waulsortian.

Sulfides deposited during stage "1" and "2" consumed sulfur derived from marine sulfates (Boast et al. 1981), i.e., metals and sulfur came from separate sources. Mineralization of stage "3" indicates an igneous source of sulfur (Boast et al. 1981), i.e., metals and sulfur were transported in the same solution.

At Tynagh 10–15% of the total mineralization is hosted in gravity-induced deposits in off-reef facies. Macrotextures follow the sedimentary fabric of debris-flows; microtextures are sulfide rims on bioclasts, infills and replacements of allochems and carbonate spar.

Mineralization Hosted in Off-Reef Gravity-Flow Deposits

Zn-Pb-Ba mineralization at Silvermines, Ireland, is hosted in gravity deposits induced on Waulsortian slopes. Two types of mineralization may be distinguished: synsedimentary barite and sulfides in back-reef facies (Tylor 1984). The back-reef facies consist of large clasts of dolomitized Waulsortian and clasts of barite. Small clasts are formed by bioclasts, ooids, rhombs of Fe- and Zn-dolomite, siderite, and clasts of sulfides. Gravity flows were activated three or four times as suggested by cements rimming, filling and recementing intraclastic porosity (Fig. 6).

There are four major macrotextures of mineralization:

1. Massive, parallel to the bedding. Bedding is underlain by irregular patches of galena enclosed in "massive" sphalerite (Tylor 1984). This massive sphalerite appears as an intimate mixture of sphalerite, Fe-, Zn-dolomite and siderite. Patches of galena and sphalerite were probably formed by diffusion of H_2S through a metalliferous carbonate sediment.

2. Cements and fillings in debris-flow breccia (Fig. 6). Clasts are cemented (rimmed) and replaced by sphalerite and intraclastic porosity is filled with a mixture of sphalerite, galena, barite, and metalliferous carbonates. Dominant microtextures are sphalerite rims and replacements together with fillings of metalliferous carbonates.

3. Three generations of veinlets crosscutting debris-flow breccias and containing banded sulfides.

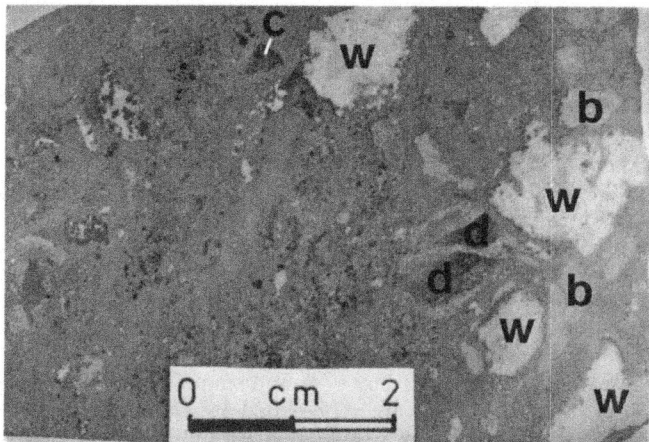

Fig. 6. Mineralized debris-flow breccias in Silvermines, Ireland. *w* clasts of dolomitized Waulsortian surrounded by sphalerite rims; *b* barite clasts surrounded by rims as above; *c* Waulsortian clasts with dispersed nodular galena; *d* clasts composed of Fe-dolomite rhombs cemented by galena. Matrix is composed of fine mixture of sulfides, barite, Fe-dolomite, and siderite

4. Finely dispersed sulfides.

The main minerals at Silvermines are pyrite, melnikovite, marcasite, sphalerite, barite, galena (Tylor 1984) and Fe-, Zn-dolomite, and siderite.

Mineralization Hosted in Strata-Bound Dolomites

The MVT-type mineralization in Upper Silesia, Poland, is hosted in dolomites, close to limestone transition. The ore-bearing dolomites are a neosome developed by dolomitization of Triassic carbonates deposited at the basin margin separated from an open sea by a barrier of Devonian islands (Sass-Gustkiewicz et al. 1982). Originally, these sediments probably consisted of shallow water carbonates truncated by supratidal dolomitized micrites. The main minerals are sphalerite, galena, pyrite, marcasite, Zn-dolomite, smithsonite, and melnikovite. Earlier nonstoichiometric sulfides are typically replaced by younger stoichiometric varieties. Furthermore, metalliferous carbonates are replaced by sulfides (Kucha and Czajka 1984). The Zn:Pb ratio varies from 2:1 to 5:1 within the ore district. Sulfide mineralization is enveloped by a thin halo composed of metalliferous carbonates. The metal concentration in the halo decreases rapidly with increasing distance from the sulfide orebody.

Four major macrotextures can be distinguished:

1. Finely dispersed texture in the ore-bearing dolomite outside of the sulfide orebodies. Small sulfide grains are dispersed in a recrystallized carbonte matrix. Typical microtextures are framboids of pyrite, sphalerite and galena, and re-

placement of carbonates. These sulfides may control up to 2.5 wt.% of Zn + Pb in the carbonate matrix.

2. Massive texture, parallel to the bedding consisting of singular or multiple sphalerite layers with numerous vugs and empty spaces separating individual layers. The overall thickness of the sphalerite layers is from a few centimeters to a few meters; they may extend laterally up to 200 m. They are assumed to have been formed by a metasomatic replacement (Sass-Gustkiewicz et al. 1982). Sphalerite rims around carbonate grains, sphalerite filling intergranular porosity and sphalerite replacing carbonate grains are typical observed microtextures (Fig. 7). The distribution of this ore type probably follows an original distribution of carbonate grainstones characterized by high initial porosities and permeabilities. This ore type was probably introduced early below the redox interface before an essential period of recrystallization of the carbonate host.

3. Breccia macrotexture, consisting of host-rock and sulfide clasts cemented with colloform sulfides. Microstructures are represented by colloform fillings, cements, and replacements of metalliferous carbonates by sulfides (Kucha and Czajka 1984).

4. Fracture fillings surrounded by a metasomatic halo (Kucha and Czajka 1984).

5. Sulfides in disaggregated (delithified) dolomite. This ore type was probably introduced after an essential period of recrystallization of the carbonate host. The distribution of ore probably follows an original distribution of Zn-dolomite which can be considered as a precursor of this ore macrotexture. Replacement (sulfidization) of Zn-dolomite reduces the volume remarkably (Kucha and Czajka 1984) and may be responsible for disaggregation (delithification) of the dolomite host.

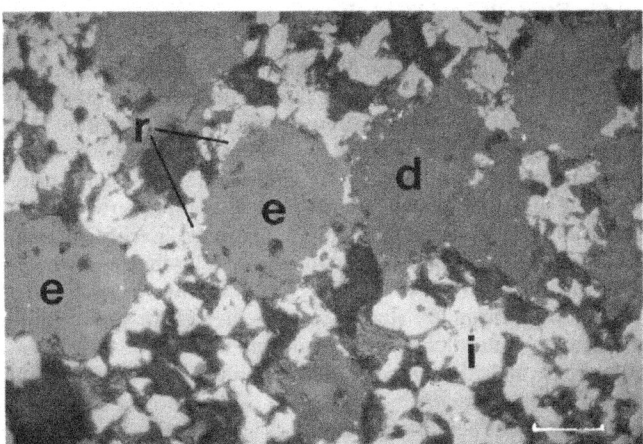

Fig. 7. Sphalerite (*white*) as rim (*r*) and filling cement (*i*) in dolomite (*grey*) with preserved fabric of the original carbonate grainstone. *e* echinoderm composed of single dolomite crystal; *d* overgrown dolomite rhomb. Pomorzany mine, Upper Silesia, Poland. Reflected light, bar = 150 μm

Porosity-Permeability of the Carbonate Host as a Factor Controlling Emplacement of Strata-Bound Sulfides

Intertidal Sediments

Intertidal sediments have a high porosity and permeability (Bathurst 1971). The original sedimentary porosity is reduced by consolidation defined as dewatering in response to burial stress. Consolidation reduces the initial porosity by 30–40% of its original value (Jones et al. 1984). When cementation starts before consolidation, as seems to be the case in Navan, Moate, and Upper Silesia, the rock may preserve a greater porosity.

Another process reducing the primary porosity is cementation. It can take place early at or near to the sediment-water interface, where the source of cement is marine water and therefore these cements are originally Mn- and Fe-free. A major cementation is usually generated by pressure solution. These cements are generated in sufficient amounts at the depth exceeding 1000 m (Neugebauer 1973). Unbroken rim cements on carbonate allochems at Navan, Moate, and Upper Silesia suggest that intertidal sediments at these three locations underwent a first cementation under light overburden, close to the sediment-water interface. If shallow water carbonates are preserved under shallow burial they may remain unlithified even for 30–40 millions of years (Schlager and James 1978). Therefore such sediments constitute an ideal potential host for the Zn-Pb mineralization to be introduced as cements in a subsurface environment.

At Moate, Navan and Upper Silesia the first rim cements on carbonate allochems are sphalerite rims (Figs. 1, 3, 7) without preceding calcite rims. This may suggest that the first mineralization at these three locations was introduced early and close to the sediment-water interface. A lack of earlier calcite rim cement may suggest rapid burial below the redox interface due to storm deposits. Further sulfides at these three locations are infill cements and replacements of carbonate allochems.

Waulsortian Reefs

Waulsortian limestones form another type of carbonate host. Recently, accumulation of calcite muds and growth of mud mounds is ascribed entirely to the algal activity (Pratt 1982). Recent muds have a porosity of 40–80% (Bathurst 1971; Enos and Sawatsky 1981). However, the permeability of mud is very low. During early stages of mud diagenesis due to dewatering and recrystallization, a high initial porosity is transformed into a highly permeable system of stylolites and stromatactis cavities. Such a hydraulic system represents a favorable place for the emplacement of sulfides, when protected against oxidation. The best example of such a mineralization is Ballinalack (Fig. 4), where stromatactis cavities are filled by sulfide linings and laminated internal sediments composed of sulfides and metalliferous carbonates. The original fabric of the stromatactis filling is transformed by a diffusion of H_2S through the cavity system. Also at Tynagh the first stage of mineralization is formed as stromatactis fillings; only 1–4% of the total Waulsortian mineralization appear in this form.

Gravity Flow Deposits Generated on Reef Slopes

Gravity-induced deposits on reef and submarine slopes are an efficient way of recreating the porosity and permeability of partly lithified sediments (Enos and Moore 1983). Such sediments are usually graded and bedded. At a distance they transgress into turbidites, which may be the most remote indicators of gravity flows. At Silvermines, most sulfide mineralization appears as rim cement on clasts of dolomitized Waulsortian, barite (Fig. 6), nodules of Fe-dolomite cemented with galena and bioclasts. Sulfides, together with siderite and metalliferous carbonates, form a breccia matrix and they also form infill cement and replacements of gravity flows and therefore the ore has a sedimentary appearance in hand specimens.

At Navan, small-scale gravity flows probably produced clasts of synsedimentary barite (Kucha 1987) and generated carbonate clasts following in their shape and size geometrical figures produced by uneven cementation rates.

Retention and Recreation of Porosity and Permeability

The retention and recreation of porosity and permeability in carbonate rocks are controlled by:

1. Polymorphic transformations of carbonates. The most suitable rock types are rocks with an original aragonite or high Mg-calcite mineralogy. They are metastable during diagenesis and convert to calcite or dolomite or are leached out leaving moldic porosity filled by later sulfides (Figs. 1, 3, 7). The appearance of ooids at Moate (Fig. 3) may suggest that only aragonitic varieties are replaced by sphalerite contemporaneously with their transformation to calcite. The original low Mg-calcite rim cement (CR, Fig. 3), not subjected to polymorphic transformation, is preserved in the unchanged form. Ooids without radial texture (O, Fig. 3) are not replaced by sphalerite. As replacement of ooids progresses, the Fe content in the carbonate structure decreases from 0.35 to 0.12 wt.% and the Zn content increases from 0.37 to 0.68 wt.%.

Dolomitization reduces the volume of limestone by 10.5% and may promote circulation of mineralizing fluids (Kucha and Czajka 1984).

2. Introduction of metals in the structure of carbonates and recrystallization of carbonates. During recrystallization carbonates may incorporate Zn, Pb, and Fe into their lattices (Kucha and Czajka 1984; Kucha 1987). Such carbonates can be easily and efficiently replaced by sulfides (Kullerud 1967) and their relics can be found even in massively banded sphalerites (Figs. 2, 8).

3. Stylolitization. Stylolites develop during dewatering of carbonate sediments (Jones et al. 1984). They increase the permeability and guide sulfide mineralization at Navan, Moate, and Ballinalack (Fig. 4).

4. Tectonic fracturing and brecciation. Tectonic fractures are observed at all mineralization sites discussed but are most clearly developed at Tynagh as a dilatant fracture-breccia system (Boast et al. 1981). The Waulsortian mound exposed at least several tens of meters over the sea floor was probably isostatically unstable and, when cut by the north Tynagh fault, split into a dilatant fracture-breccia system filled by cement linings (Fig. 5) and internal sediments (Boast et al. 1981). Internal sediments consist of a very fine-grained mixture of Zn- Fe-dolomite and sulfides

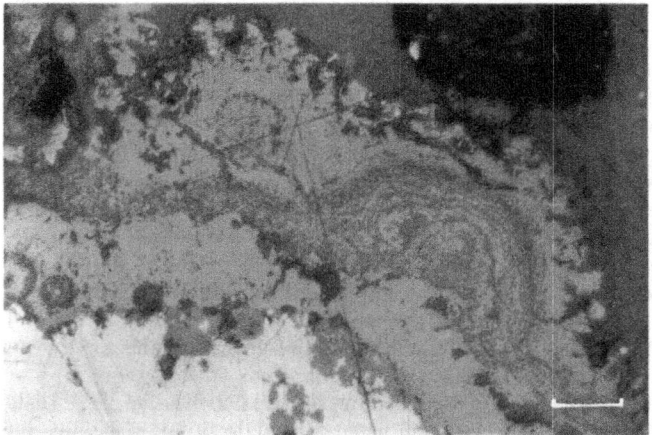

Fig. 8. Interfingerings of banded sphalerite (*grey*) with smithsonite (*dark grey*) overgrowing galena (*white*). Olkusz mine, Upper Silesia, Poland. Reflected light, bar = 150 μm

(mainly sphalerite). Metals and sulfur were derived from different sources and a diffusion of H_2S through the dilatant fracture-breccia system produced disturbances in the original lamination of internal sediments such as overgrowths of barite-albite nodules, overgrowths of barite-sulfide nodules, and overgrowths of celsians.

　5. Dissolution of the carbonate host. The dissolution of carbonates (karst, etc.) may produce vast empty spaces. Sulfur may be provided by groundwaters and metals may be provided by warm ascending solutions. Such mineralizing systems will repeatedly produce metalliferous carbonates replaced by sulfides by addition of sulfur from a separate source. In this way a high permeability and porosity of the ore zone can be retained by the reduction in volume and leaching of carbonates by released CO_3^{2-} (Kucha and Czajka 1984). The sulfidic sulfur was probably produced by bacterial reduction. Framboids of pyrite, sphalerite, and galena abundant in places provide some evidence of this process.

Discussion

An initial emplacement of sulfides in the deposits discussed uses porosity and permeability of the original carbonate sediment and may control 30–50% of the total mineralization. As a result, a typical sedimentary macrotexture is produced, but microtexture such as sulfide rims on allochems and sulfide cement provide evidence of epigenetic emplacement. Further ore generations are metasomatically emplaced in the host carbonates. The replacement process is so closely connected with collapse breccias that some authors suggest a "hydrothermal karst" as an explanation of these phenomena (Sass-Gustkiewicz et al. 1982). Relics of Zn-calcite (Fig. 2) containing 0.70–1.20 wt.% Zn and smithsonite (Fig. 8) even in colloform massive sulfides suggest that base metals may have been partly introduced as carbonates in environment free from sulfidic sulfur. Sulfidization of metal carbonates may reduce the volume as much as 58% (Kucha and Czajka 1984) and releases large volumes of CO_3^{2-}, causing extensive leaching which may be a reason of collapse breccia development. Sulfidization of metal carbonates releases such great amounts of CO_2 (Kullerud 1967) that leaching probably extends into the wall rock and nonmineralized collapse breccias may result.

Experimental carbonate-sulfide reactions provide a basis for the understanding of sulfidization process (Kullerud 1967). Reaction of metal carbonates with sulfur is rapid; at 200°C appreciable amounts of galena form in 1 h. In reaction of carbonates with sulfur, high gas pressures are produced, the main gas component being CO_2. The stability series of metal carbonates in the presence of sulfur determined experimentally is $MnCO_3 > FeCO_3 > ZnCO_3 > PbCO_3$, and is similar to that observed in nature (Kucha and Wieczorek 1984). Numerous relics, pseudomorphs, and preserved idiomorphic crystals of metal and metalliferous carbonates have been found at all the six deposits discussed.

The mechanisms and reactions discussed above suggest that massive, large-scale strata-bound base metal deposits can form when carbonate-sulfide reactions, retaining a high permeability of the ore zone, are involved. This process requires that the metal and sulfur sources be, at least partly, separate.

References

Andrew CJ, Ashton JH (1982) Mineral textures; metal zoning and ore environment of the Navan orebody, Co. Meath, Ireland. In: Brown AG (ed) Mineral exploration in Ireland. Irish Assoc Econ Geol: 35–45

Bathurst RGC (1971) Carbonate sediments and their diagenesis. Elsevier, Amsterdam, 620 pp

Bischoff JL, Radtke AS, Rosenbauer RJ (1981) Hydrothermal alteration of graywacke by brine and sea-water: roles of alteration and chloride complexing on metal solubilization at 200 and 359°C. Econ Geol 76:659–676

Boast AM, Coleman ML, Halls C (1981) Textural and stable isotopic evidence for the genesis of the Tynagh base metal deposit, Ireland. Econ Geol 76:27–55

Enos P, Moore CH (1983) Fore-reef slope environment. In: Scholle PA, Bebout DG, Moore CH (eds) Carbonate depositional environments. Am Assoc Pet Geol: 508–537

Enos P, Sawatsky LH (1981) Pore networks in carbonate sediments. J Sediment Pet 51:961–985

Jones ME, Bedford J, Clayton Ch (1984) On natural deformation in the chalk. J Geol Soc (Lond) 141:675–683

Kucha H (1987) Carbonate and silicate precursors of the sulfide mineralization in the Navan Zn-Pb deposit, Ireland. Miner Pet (in press)

Kucha H, Czajka K (1984) Sulphide-carbonate relationships in Upper Silesian Zn-Pb deposits (Mississippi Valley type), Poland, and their genesis. Trans Inst Miner Metall Sect B 93:12–22

Kucha H, Wieczorek A (1984) Sulfide-carbonate relationships in the Navan (Tara) Zn-Pb deposit, Ireland. Miner Dep 19:208–216

Kullerud G (1967) Sulfide studies. In: Research in geochemistry, Vol 2. Wiley, New York, pp 286–327

Neugebauer J (1973) The diagenetic problem of chalk: the role of pressure solution and pore fluid. Jahrb Geol Paleontol Abh 143:223–245

Pratt BR (1982) Stromatolitic framework of carbonate mud mounds. J Sediment Pet 52:1203–1277

Sass-Gustkiewicz M, Dzulynski S, Ridge JD (1982) The emplacement of zinc-lead sulfide ores in the Upper Silesian district – a contribution to the understanding of Mississippi Valley-type deposits. Econ Geol 77:392–412

Schlager W, James NP (1978) Low-magnesian calcite limestone forming at the deep-sea floor, Tonge of the Ocean, Bahamas. Sedimentology 25:675–702

Tylor S (1984) Structural and paleogeographic controls of lead-zinc mineralization in the Silvermines orebodies, Republic of Ireland. Econ Geol 79:529–548

Genetical Significance of Saline Relics in Carbonate Host Rocks of Alpine Pb-Zn Deposits

R. WOLTER and H.-J. SCHNEIDER[1]

Abstract

Geochemical research on saline relics in Alpine Pb-Zn deposits established important indications as to the composition and concentration of ore-bearing solutions. The investigations proceed from the principle that saline solutions are commonly responsible for the formation of ore deposits. Therefore the mere existence of saline relics in Alpine Pb-Zn deposits had to be proved initially. For this purpose a leaching method was developed and applied to barren and mineralized rocks of the Bleiberg-Kreuth (Austria) and Raibl (Italy) Pb-Zn deposits. Predominantly Ca^{2+}, Mg^{2+}, SO_4^{2-}, accompanied by generally minor amounts of Na^+, K^+, and Cl^- were found in the leachates. Only Na^+, Cl^-, and Mg^{2+} (Ca^{2+}) can be assigned to fluid inclusions and salt crystals, here referred to as "saline relics".

The concentration of the entrapped solutions was estimated by the Na/Cl ratio of the leachates. For barren rocks, these ratios frequently indicate solutions saturated with respect to NaCl (halite). In strongly mineralized domains the salinity is significantly increased. Here the existence of $MgSO_4$ saturated brines seems to be probable.

Additionally Na/Cl ratios of fluid inclusions and oil field brines from Mississippi Valley type deposits are very much in line with our results and emphasize the reliability of our method of investigation. The high salinity of entrapped solutions (saline relics) indicates that formation water itself might be of major importance for transport, precipitation, and mobilization of ore matter.

Introduction

Saline solutions are of great importance for the formation of many types of ore deposits. Fluid inclusions, for instance, are considered as relics of ore-bearing solutions. Extensive research on fluid inclusions in Mississippi Valley-type deposits by Roedder (1976) and others revealed salinities that usually exceed 15 wt.% NaCl-equivalent. Additionally, the occurrence of salt crystals (daughter crystals), for example NaCl, indicates that the solutions were saturated with respect to these salts.

Physiochemical calculations of the solubility of metals in solutions (Anderson 1973; Barret and Anderson 1982) demonstrated an increasing ability of brines to carry metal ions with rising salinity.

---

[1]Institut für Angewandte Geologie, Freie Universität Berlin, Wichernstr. 16, D-1000 Berlin 33, FRG

Base Metal Sulfide Deposits
G.H. Friedrich, P.M. Herzig (Eds.)
© Springer-Verlag Berlin Heidelberg 1988

122   R. Wolter and H.-J. Schneider

Some authors (Newhouse 1932; Carpenter et al. 1974) suggest that brines expelled from sediments during compaction (e.g., oil field brines) are responsible for the formation of the Mississippi Valley-type Pb-Zn deposits. This is supported by analyses of many oil field brines which occasionally showed salinities of more than 35% of total dissolved solids and metal contents up to 100 ppm Pb and 500 ppm Zn (Collins 1975). The formation of metal enrichments by solutions was proved recently, when outflows of hot brines were discovered in the Red Sea and in the Pacific Ocean. Even if one takes into account that these mineralizations have little to do with the Mississippi Valley-type environment, they stress the importance of brines for the ore formation at the sea floor in general.

Apart from metal sulfides, $CaSO_4$ is also occasionally precipitated from some of these solutions (MacDonald and Luydendyk 1981; Haymon 1983), which are saturated with $CaSO_4$ at least after mixing with sea water.

The research on saline relics is not only of scientific interest, but contains essential economic aspects as well. The application of fluid inclusion research to the prospection on porphyry copper deposits is only one example of its economic importance.

Our investigations were carried out in the Pb-Zn deposits of Bleiberg-Kreuth (Eastern Alps/Austria) and Raibl (Southern Alps/Italy) (Fig. 1). They belong to the so-called Alpine Pb-Zn deposits which occur in the Eastern and Southern Alps. Other economically important deposits of this type are Mezica (Yugoslavia), Salafossa (Italy), and Gorno (Italy) (Fig. 1). The host rocks of Alpine Pb-Zn deposits are Triassic limestones and dolomites. The mineralization is strata-bound, some-

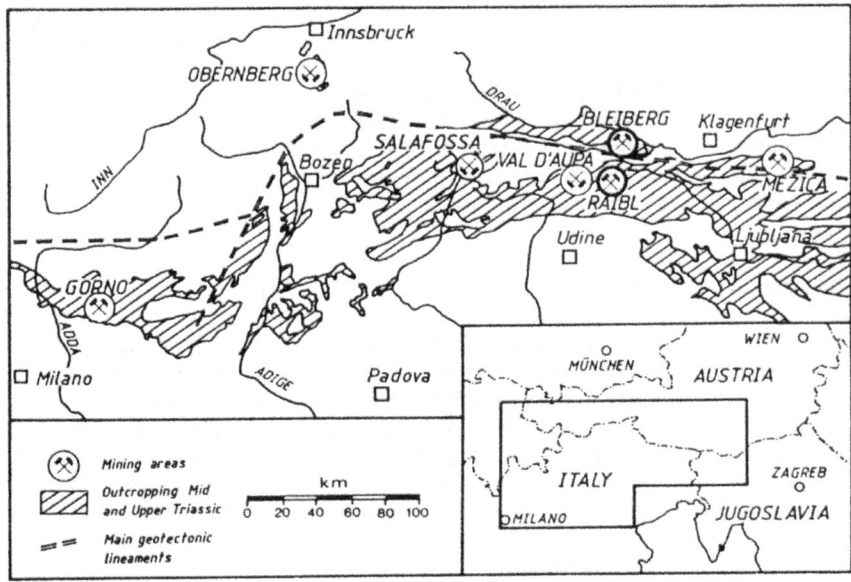

Fig. 1. Location map of important Alpine Pb-Zn deposits depicting the investigated deposits of Bleiberg-Kreuth and Raibl

times even stratiform, revealing sedimentary bedding and top-bottom textures. Mineralized veins and breccia bodies are nearly always present and occasionally linked with stratiform ores. Predominant ore minerals are sphalerite and galena accompanied by pyrite/marcasite (Brigo et al. 1977). Similarities between the Alpine and Mississippi Valley Pb-Zn deposits are obvious with respect to the ore minerals, ore textures, and the facies of the host rocks (Sangster 1983). These two types, scattered all over the world, are classed under the term carbonate hosted Pb-Zn deposits.

Analytical Method

For the analysis of saline relics we developed a leaching technique similar to the method described by Lamar and Shrode (1953). In Alpine Pb-Zn deposits fluid inclusions are generally very small and cannot be analyzed using the common methods. On the other hand, the bulk analysis of carbonate and ore minerals is not appropriate in order to recognize the small amounts of saline relics. Therefore a selective analytical method for brine relics, such as a leaching method, is the most appropriate technique.

Prior to leaching, the samples were cleaned and pulverized. Fifty grams of the pulverized samples were leached for 30 min in boiling tri-distilled water. After the separation of solid matter from leachate, the solution was analyzed for Ca^{2+}, Mg^{2+}, Na^+, K^+, Cl^-, and SO_4^{2-}. In addition, the HCl insoluble residue and the calcite/dolomite ratio of each sample were determined.

Results

Absolute Amounts of Leached Solids

The predominant ions in the leachates are Ca^{2+}, Mg^{2+}, and SO_4^{2-}, whereas in general Cl^-, Na^+, and K^+ are quantitatively subordinate (Table 1). Concerning the absolute amounts of leached solids our results correspond very well to those of Lamar and Shrode (1953) and Schneider (1969). This is an indication for the reliability of our leaching method.

Distribution of Leached Solids

As previously reported by Wolter and Schneider (1983; 1985) the amounts of Ca^{2+} (Mg^{2+}) and SO_4^{2-} increase significantly in mineralized sequences, particularly in the presence of sulfide minerals (e.g., pyrite/marcasite, sphalerite, and galena). The distribution pattern of Na^+ and Cl^- in the vicinity of orebodies is not homogeneous. In contrast to an almost unmineralized sequence with lower contents (Fig. 2 top), a significant enrichment of Na^+ and Cl^- is evident in the domain of a stratiform mineralization (Fig. 2 bottom). Furthermore, a correlation between mineralization and K^+ contents of the leachates is not observable.

Table 1. Average element contents (in ppm) of leachates from Bleiberg-Kreuth and Raibl. (St.dev. = standard deviation)

BLEIBERG-KREUTH (Austria)

	Limestones (n = 58)			Dolomite (n = 39)			Mineralization (n = 36)		
	Median	Mean	St.dev.	Median	Mean	St.dev.	Median	Mean	St.dev.
Na	10.1	11.7	5.8	17.8	21.2	12.6	12.3	24.1	21.3
K	4.75	6.7	6.0	5.4	15.9	32.8	6.8	29.4	49.9
Ca	44.0	71.6	108.0	27.0	82.5	226.2	62.4	101.4	174.3
Mg	5.6	6.6	4.6	48.0	77.0	103.0	42.4	53.7	68.5
Cl	26.9	29.4	16.7	43.5	51.4	30.3	41.6	62.1	54.8
SO$_4$	52.5	144.6	362.0	74.0	390.0	1280	164	403.2	960.8

RAIBL (Italy)

	Limestones (n = 19)			Dolomite (n = 42)			Mineralization (n = 31)		
	Median	Mean	St.dev.	Median	Mean	St.dev.	Median	Mean	St. dev.
Na	34.2	34.1	11.4	23.7	24.2	5.6	19.4	20.1	4.0
K	3.5	3.7	0.8	4.2	7.2	7.5	7.3	8.9	5.5
Ca	48	47.8	7.1	20	23.2	13.2	44	47.1	51.0
Mg	6.4	6.3	1.6	62	69.3	31.1	76.8	110	58
Cl	87.9	83	27.4	71.9	70.3	16.9	61.5	73.4	34.6
SO$_4$	39	43.6	19.6	78.5	113.4	137.2	198	287.7	205.7

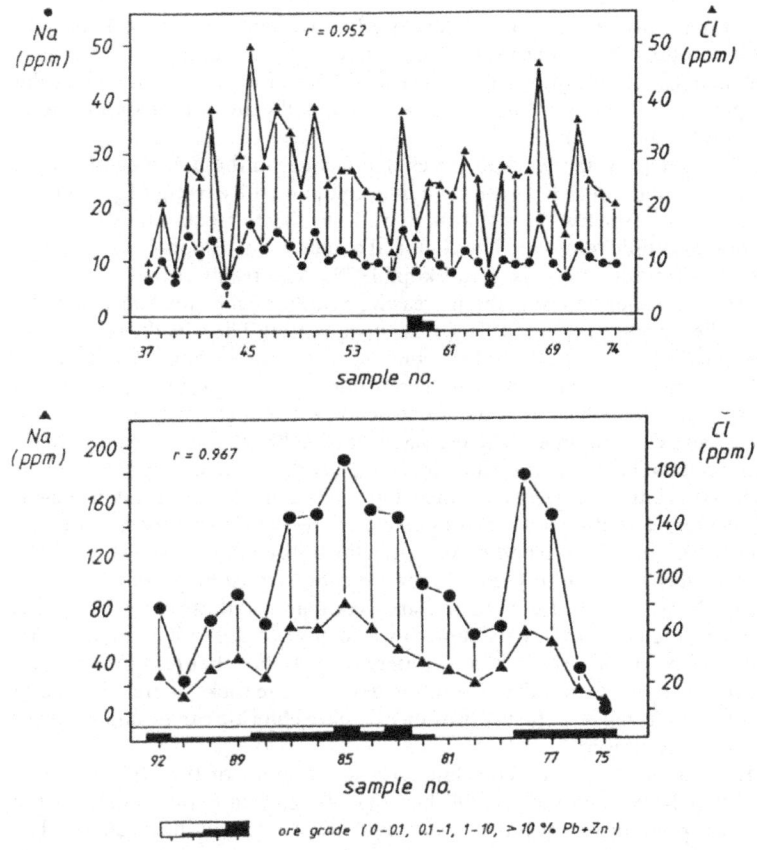

Fig. 2. Sodium and chlorine variations in a nearly barren (*top*) and in a mineralized (*bottom*) Wetterstein sequence of Bleiberg-Kreuth. Mineralizations are marked and quantitatively specified. Note different scale of y-axis. *r* correlation coefficient

Composition of Saline Relics

Generally four sources of leached solids can be distinguished (Wolter and Schneider 1985):

1. Partial dissolution of host rock minerals.
2. Fluid inclusions.
3. Individual salt crystals.
4. Ions adsorbed to clay minerals.

To determine the composition of saline relics, it is essential to determine the proportion of leached solids derived from the above sources. In this context, elements which are to be attributed to fluid inclusions or salt crystals are of major importance.

In general a positive correlation between Mg^{2+} contents of the leachates and the dolomite portion of the rocks is evident. Likewise the matching Ca^{2+} contents reflect the chemical composition of the host rock. Therefore, at least a part of the dominating ions Ca^{2+} and Mg^{2+} has to be attributed to the dissolution of limestone and/or dolomite respectively.

The SO_4^{2-} content of the leachates rises significantly in the presence of sulfide minerals. The increase of SO_4^{2-} is quite often accompanied by rising Mg^{2+} and/or Ca^{2+} contents. Experiments show that high SO_4^{2-} levels are frequently due to the disintegration of sulfide minerals. The chemical reactions correspond to a chemical "weathering" of sulfide minerals (Krauskopf 1979). As a result of these processes SO_4^{2-} and H^+ are generated and the increasing acidity of the solution raises the solubility of the carbonate minerals (e.g. calcite, dolomite). Thereby the concurrent increase of the SO_4^{2-}, Ca^{2+} and/or Mg^{2+} contents is easy to explain. In some cases abnormal Ca^{2+} and SO_4^{2-} values have to be attributed to the dissolution of gypsum or anhydrite which have been detected microscopically in a few samples.

The Na^+ and Cl^- amounts of the leachates have to be attributed to fluid inclusions or salt crystals. There are no indications for any other sources. Figure 3 depicts the excellent correlation between Na^+ and Cl^- ($r = 0.971$, $n = 58$) of limestone leachates from Bleiberg-Kreuth. However a significant excess of Cl^- is apparent after the conversion of Na^+ and Cl^- into halite. Assuming the bulk content of saline relics has been dissolved totally, some further chlorides or cations must be present.

The Ca/Mg ratios in the leachates are small compared to those of the host rocks (Wolter and Schneider 1983). Frequently in leachates of dolomites the absolute amount of Mg^{2+} even exceeds the Ca^{2+} content (Fig. 4). For that reason Mg^{2+} can be only partially derived from the dissolution of carbonate rocks, it must be bound in part to fluid or salt relics. In fact very often excessive chlorine can be balanced out with excessive magnesium.

The K^+ contents generally correlate with the amount of the HCl insoluble residue. Detailed investigations established that the clay mineral content of the samples — here predominantly illite — controls the K^+ content of the leachates. The amount of leached K^+ does not change significantly if a rising insol content is due to

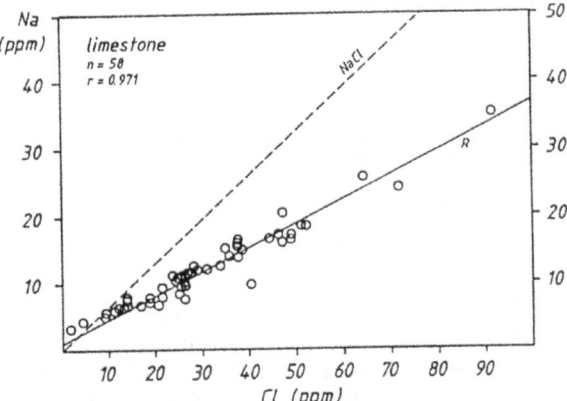

Fig. 3. Plot of sodium vs. chlorine in limestone leachates of Bleiberg-Kreuth. r correlation coefficient; R regression line

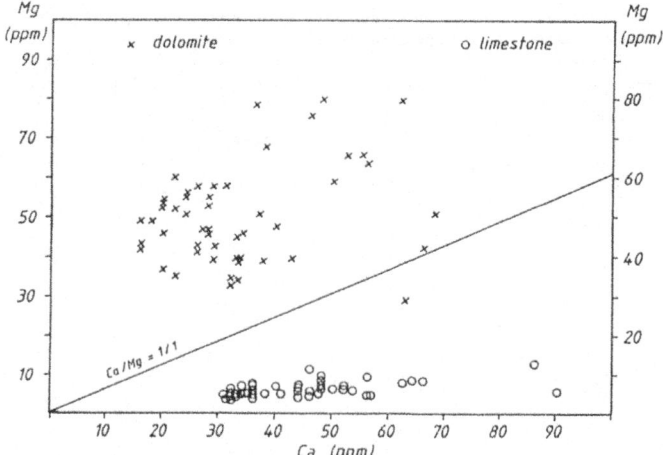

Fig. 4. Plot of calcium vs. magnesium contents of leachates from limestones and dolomites (Bleiberg-Kreuth). *Line* (Ca/Mg = 1 / 1) represents the composition of a stoichiometric dolomite

quartz or fluorite solely. Most probably, potassium was bound to the surface of clay minerals prior to leaching.

To sum up the major components of saline relics are Na^+, Cl^-, and Mg^{2+}. The Ca^{2+} and SO_4^{2-} concentrations of saline relics are not exactly determinable. Referring to the chemical composition of natural brines it is supposed that parts of Ca^{2+} and SO_4^{2-} are components of brine relics.

Concentration of Saline Relics

In order to estimate the (relative) concentration of saline relics the attempt was made to use the Na/Cl ratio of leachates. Most of the Na/Cl ratios are typical for natural brines saturated with halite.

During the carbonate sedimentation sea water was trapped as connate water. It changed its composition during the various stages of diagenesis due to mineral dissolution, recrystallization, and ion exchange reactions. In general an increasing salinity of brines is mainly caused by membrane filtration processes (Engelhardt 1973). Increasing salinities may result in the precipitation of salt minerals. Collins (1975) revealed the composition of solutions attaining saturation with $CaSO_4$, $NaCl$, $MgSO_4$, etc. (Table 2) which demonstrates Cl^-, Na^+, and Mg^{2+} being predominant. These ions are exactly the same as those we assign to saline relics. The absolute amounts of ions resulting from our leaching technique cannot be used to estimate the salinity. Therefore element ratios were calculated and related to the various saturation points (Fig. 5). Out of all the ratios plotted in the diagram, the Na/Cl ratio is the only suitable one because Na^+ and Cl^- derive from saline relics solely as shown above. These ratios can be used to determine the salinity of the brines analyzed.

Table 2. Concentration changes (mg/l) during evaporation of sea water and brine (according to Collins 1975). *Arrows* indicate the precipitation of the mineral concerned.

	Sea water	CaSO$_4$↓	NaCl↓	MgSO$_4$↓	KCl↓	MgCl$_2$↓
Sodium	11.000	98.000	140.000	70.000	13.000	12.000
Chlorine	19.000	178.000	275.000	277.000	360.000	425.000
Magnesium	1.300	13.000	74.000	80.000	130.000	153.000
Potassium	350	3.600	23.000	37.000	26.000	1.200

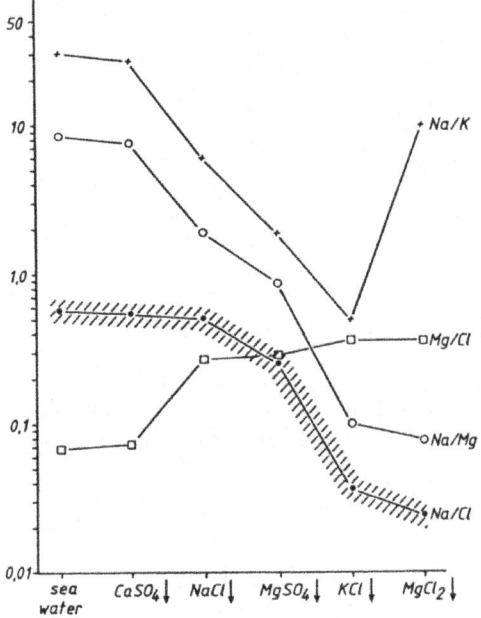

Fig. 5. Element ratios during the evaporation of sea water and brine (calculated according to Collins 1975) (see also Table 2). *Arrows* indicate the precipitation of the mineral concerned

Barren Host Rocks. The frequency distributions of Na/Cl ratios of barren lime-stones and dolomites from Bleiberg-Kreuth deposit (Fig. 6A, B) demonstrate a maximum Na/Cl ratio at about 0.4. Therefore, even in barren host rocks solutions saturated with halite seem to occur frequently. Additionally ratios below 0.25 and above 0.57 indicate MgSO$_4$ saturated brines respectively CaSO$_4$ saturated brines, sea water, and meteoric water with low salinities.

Mineralized Rocks. The frequency distribution of the Na/Cl ratios in mineralized samples is not significantly different compared to barren ones (Fig. 6), because most of the mineralizations are mixtures between various amounts of carbonate host rocks and ore minerals. Therefore the question arises which part of the saline relics has to be attributed to formation water and which part to ore-forming solutions. Generally the salinity of the leachates is higher in mineralized domains than in barren host rocks.

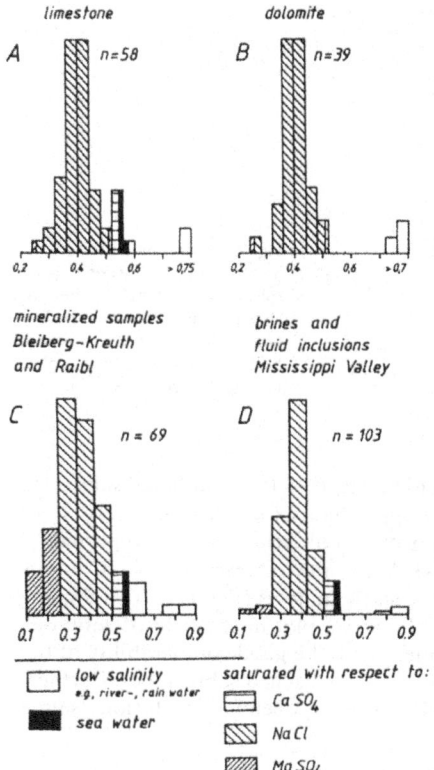

Fig. 6. Na/Cl frequency distribution of barren **A** limestone and **B** dolomite leachates from Bleiberg-Kreuth, **C** mineralizations from Bleiberg-Kreuth and Raibl, and **D** oil field brines and fluid inclusions from Mississippi Valley deposits. **D** calculated according to Roedder (1972) and Carpenter et al. (1974)

A definite correlation in any case exists between the ore grade (stated by the sum of galena and sphalerite) and the Na/Cl ratios in the leachates (Fig. 7). Obviously the Na/Cl ratio decreases with increasing metal content indicating an increase of excessive chlorine. The concentration of these brines seems to be significantly higher than that of formation water, which is underlined by a certain number of samples saturated with $MgSO_4$ (Figs. 5, 6C).

Probability of the Estimated Concentration Levels

Fluid inclusion research indirectly proved the occurrence of NaCl saturated brines in connection with mineral formation. Na/Cl ratios of oil field brines indicate the existence of solutions saturated with NaCl.

The salinity of fluid inclusions is generally determined by freezing temperature measurements. From these data the concentration of a pure NaCl solution is determined. Therefore salinities are generally expressed as NaCl-equivalent. Occasional freezing temperatures below –20.7°C cannot be explained by NaCl satu-

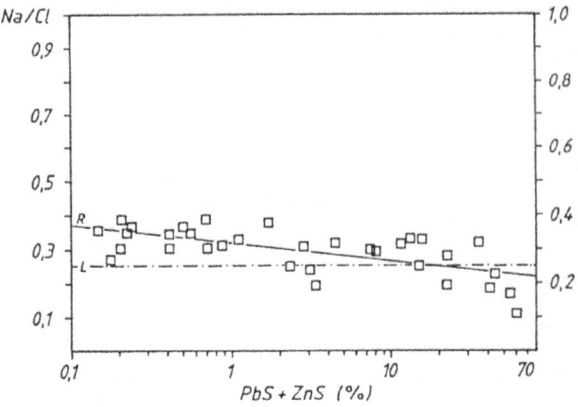

Fig. 7. Plot of the Na/Cl ratios in leachates vs. galena plus sphalerite contents in mineralized rocks of Raibl. Below line L (Na/Cl = 0.252) natural brines are saturated with respect to MgSO$_4$ R regression line

rated solutions solely. In these cases it has to be assumed that additional salts (e.g., CaCl$_2$) are important constituents of the entrapped fluids. Chemical analyses of fluid inclusions from the Mississippi Valley deposits and oil field brines are reported by Roedder (1972) and Carpenter et al. (1974). According to these data, Na/Cl ratios were calculated and compared with those of leach analyses (Fig. 6D). The similarities are striking because most of the Na/Cl ratios plot in the field of solutions saturated with NaCl. This finding underlines again the good comparability of the composition of fluid inclusions, oil field brines, and leached saline relics. Furthermore the chemical analogies indicate a relationship between fluid inclusions, saline relics, and ore-forming solutions.

Summary and Conclusions

Our investigations proved the existence of saline relics in barren as well as in mineralized sequences of the Alpine Pb-Zn deposits. Major components and genetically indicative constituents of saline relics are Na$^+$, Cl$^-$, and Mg^{2+} whereas K$^+$, Ca^{2+}, and SO$_4^{2-}$ are of minor importance. Regarding the Na/Cl ratio as a salinity indicator, most of the solutions are saturated with NaCl, while ore-forming solutions may occasionally have reached the level of MgSO$_4$ saturation. The high salinity of solutions entrapped in unmineralized sequences was amazing and indicates the presence of solutions capable of transporting metal ions everywhere in the host rocks. At least remobilization of ore minerals might be due to the migration of formation water since the earliest stages of diagenesis (Schneider 1969).

The genesis of the Alpine Pb-Zn deposits is indicated by some geological facts: The decline of the Mesozoic syncline established by high sedimentation rates caused subsequently a submersion of the carbonate host rocks in about 3000 m with a corresponding rise of pressure and temperature (Brigo et al. 1977). Furthermore, the tectonic stress and strain during the Alpine orogeny opened pathways and increased the mobility and chemical activity of the migrating brines. They might have been mixed locally with ascending geothermal solutions since the earliest stages of diagenesis. However, this contingency is not detectable by the reported analytical method.

At any rate, a genetic interpretation of Alpine Pb-Zn deposits has to take into account the composition and concentration of saline relics present everywhere during the geological history of the deposits and their host rocks.

Acknowledgments. The research program was financially supported by the Deutsche Forschungsgemeinschaft, which is gratefully acknowledged. We would also like to thank the Bleiberger Bergwerks Union especially Dr. I. Cerny (Bleiberg-Kreuth/Austria) and the staff of the Raibl Mine (Cave del Predil/Italy) for their kind assistance.

References

Anderson GM (1973) The hydrothermal transport and deposition of galena and sphalerite near 100°C. Econ Geol 68:480–492

Barret TJ, Anderson GM (1982) The solubility of sphalerite and galena in NaCl brines. Econ Geol 77:1923–1933

Brigo L, Kostelka L, Omenetto P, Schneider HJ, Schroll E, Schulz O, Strucl I (1977) Comparative reflections on four Alpine Pb-Zn deposits. In: Klemm DD, Schneider HJ (eds) Time- and strata-bound ore deposits. Springer, Berlin Heidelberg New York, pp 273–293

Carpenter AB, Trout ML, Pickett EE (1974) Preliminary report on the origin and chemical evolution of lead- and zinc-rich oil field brines in Central Mississippi. Econ Geol 69:1191–1206

Collins AG (1975) Geochemistry of oilfield waters. Elsevier, Amsterdam

Engelhardt WV (1973) Die Bildung von Sedimenten und Sedimentgesteinen. Schweizerbart, Stuttgart

Haymon RM (1983) Hydrothermal deposition on the East Pacific Rise at 21°N. J Geochem Explor 19:493–495

Krauskopf KB (1979) Introduction to geochemistry, 2nd edn. McGraw-Hill, New York

Lamar JE, Shrode RS (1953) Water-soluble salts in limestones and dolomites. Econ Geol 48:97–112

MacDonald KC, Luydendyk BP (1981) Tauchexpedition zur Ostpazifischen Schwelle. Spektrum Wiss 7:73–87

Newhouse WE (1932) The composition of vein solutions as shown by liquid inclusions in minerals. Econ Geol 27:419–436

Roedder E (1972) Composition of fluid inclusions. U S Geol Surv Prof Pap 440-JJ:JJ1-JJ164

Roedder E (1976) Fluid-inclusion evidence on the genesis of ores in sedimentary and volcanic rocks. In: Wolf KH (ed) Handbook of strata-bound and stratiform ore deposits, vol 2. Elsevier, Amsterdam, pp 67–110

Sangster DT (1983) Mississippi Valley-type deposits; a geological melange. In: Kisvarsanyi G, Grant SK, Pratt WP, Koenig JW (eds) International conference on Mississippi Valley type lead-zinc deposits. Proceedings Volume, University of Missouri-Rolla, Rolla, Missouri, pp 7–19

Schneider HJ (1969) The influence of connate water on ore mobilization of lead-zinc deposits in carbonate sediments. In: Zuffardi P (ed) Remobilization of ores and minerals. Mulas, Cagliari, pp 314–322

Wolter R, Schneider HJ (1983) Saline relics of formation water in the Wettersteinkalk and their genetical connection with the Pb-Zn mineralization. In: Schneider HJ (ed) Mineral deposits of the Alps and of the Alpine epoch in Europe. Springer, Berlin Heidelberg New York, pp 223–230

Wolter R, Schneider HJ (1985) Solerelikte in Erz und Nebengestein der Blei-Zink Lagerstätte Bleiberg-Kreuth. Arch Lagerst Forsch Geol B A Wien 6:201–208

Ore Mineralogy of the Tatestown Prospect, Ireland

E. TIJSKENS, W. VIAENE[1], P. VAN OYEN[1], J. CLIFFORD[2]

Abstract

The Tatestown prospect is a strata-bound Zn-Pb mineralization hosted in Lower Carboniferous shallow-water limestone of Late Courceyan age (Tournaisian): the Lower Pale Beds. They consist of two distinct lithologies. Mineralization is restricted to one lithology of lithoclastic and oolitic wackestones and packstones. The ore minerals are sphalerite, galena, and minor iron sulfides. Three different styles of mineralization are present: sulfide breccias, veins, and replacements. The sulfide breccias consist of fragments of banded sphalerite. The textural features of this banded sphalerite and its position in the diagenetic sequence indicate that it formed in the sedimentary or early-diagenetic environment. The brecciation, which post-dates the formation of this banded sphalerite, took place shortly after the deposition of the host rock, since uncemented carbonate grains were involved in the brecciation. Veins and replacements are epigenetic with respect to the host rock. However, since these veins and associated replacements occur below the sulfide breccias and display the same paragenesis, they are believed to have deposited simultaneously with the banded sphalerite.

The sediment petrographic features of the host rock, the thickness distribution of the Lower Pale Beds and the occurrence of feeder veins and brecciated sulfides indicate that both the deposition of the host rock and the mineralization are related to rapid differential subsidence.

Introduction

The Tatestown Zn-Pb prospect is located in the north of the Central Irish Plain, some 5 km northwest of the town of Navan (Fig. 1). It is considered as a satellite deposit to the major Navan Zn-Pb orebody (Andrew and Ashton 1985) which lies 3 km to the south. Both deposits are hosted in Lower Carboniferous (Courceyan) limestone which overlies unconformably the south flank of the Longford-Down Lower Palaeozoic Massif. The prospect was discovered in the 1970's by a drilling program of Irish Base Metals Ltd. (now Westland Exploration Ltd.).

[1]Katholieke Universiteit Leuven, Fysico-chemische Geologie, Celestijnenlaan 200C, B-3030 Heverlee, BELGIUM
[2]Westland Exploration Ltd., Piggott Street, LOUGHREA, Co. GALWAY., Ireland

Base Metal Sulfide Deposits
G.H. Friedrich, P.M. Herzig (Eds.)
© Springer-Verlag Berlin Heidelberg 1988

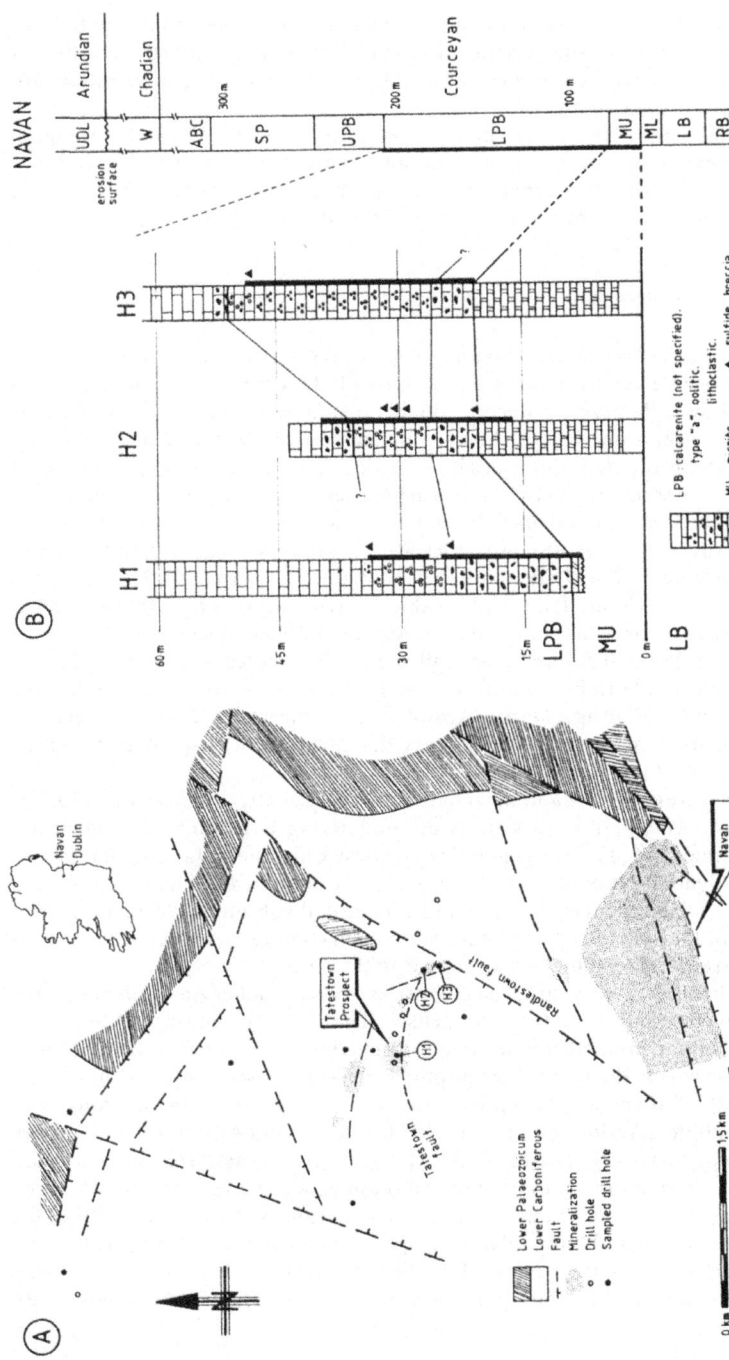

Fig. 1. A Location of the study area and simplified geological map. (After Andrew and Ashton 1985). **B** Lithostratigraphical correlation of the drill holes studied and stratigraphy at Navan (note the different vertical scale). For abbreviations see text

The aim of the study was to investigate the ore mineralogy and its relations to the host rock and the gangue minerals. For that purpose the chemistry of different minerals has also been determined. Implications for a metallogenetic model are derived from the data obtained.

Some 100 samples from three mineralized drill holes (H1, H2, and H3, Fig. 1) were investigated by means of microscope techniques (petrography and ore microscopy, cathodoluminescence) and microprobe analyses. They were compared with additional samples from surrounding barren drill holes.

General Geological Setting

The geology of the Navan area covering the stratigraphy as well as the structural geology has been given by Andrew and Ashton (1982, 1985) and is summarized in Fig. 1. The Lower Paleozoic is unconformably overlain by a Lower Carboniferous transgressive succession. The Carboniferous sedimentation started at Late Cour-ceyan times with the deposition of silici-clastics: the Red Beds (RB). They pass upwards into a series of increasingly carbonaceous sandstones, siltstones, and mudstones: the Laminated Beds (LB) and the Muddy Limestone (ML). These are overlain by more pure limestones representing a gradual deepening of the depo-sitional environment: the Micrite Unit (MU) and the Pale Beds (PB). The Pale Beds are followed by the Shaly Pales (SP): a thick varied succession of thinly bedded calc-argillites and crinoidal biomicrites indicative of further deepening. Above the Shaly Pales, ca. 250 m of monotonous argillaceous biomicrites, i.e., the Argillaceous Bioclastic Calcarenite (ABC) occur, followed by 200 m "Waulsortian" limestone (WL) of Lower Chadian age. During Arundian times the whole area was covered by the Upper Dark Limestone (UDL): a very thick (>1000 m) sequence of bedded agrillaceous limestone.

The structure of the Dinantian is characterized by gently dipping strata (20–30° to the WSW). These strata as well as the underlying Caledonian basement are strongly faulted (Fig. 1). Two generations of post-Caledonian faulting have been recognized. Syndepositional faults developed in Courceyan 4 to Chadian times as a consequence of sedimentary loading and differential subsidence (e.g., Tatestown Fault). Major faulting took place under Hercynian compressive forces and divided the area into several tectonic blocks (e.g., Randlestown Fault).

Mineralization in the study area comprises several smaller Zn-Pb deposits, the largest of which is the Tatestown prospect itself. They are all hosted in the Lower Pale Beds, some minor mineralization already occurring in the Micrite Unit. Miner-alization thus covers the same stratigraphic zone as the lower lenses of the Navan orebody. All of them are generally strata-bound, occurring in horizons with brecciated sulfides (Andrew and Poustie 1984; Andrew and Ashton 1985). They are aligned along two E-W striking faults (Fig. 1). The mineralized horizons are displaced by these faults and mineralization appears to be best developed in the hanging wall. Andrew and Poustie (1984), in a previous study, concluded that the Tatestown deposit is to be regarded as epigenetic and coeval with the Tatestown Fault development. On the basis of the inferred age of the fault, the mineralizing episode is thought to be Upper Chadian or early Arundian, synchronous with exhalative activity within the Boulder Conglomerate at Navan. This is in contrast to the mineralization at Navan, which had an extensive synsedimentary and syn-diagenetic component (Andrew and Ashton 1982; Kucha and Wieczorek 1984).

Petrography of the Host Rock

The drill holes in the Tatestown area transected the Pale Beds, the Micrite Unit, the Muddy Limestone and the Laminated Beds. In H1 the base of the Micrite Unit is missing due to faulting. The sampling covers the top of the Muddy Limestone, the Micrite Unit and the lowermost 60 m of the Lower Pale Beds. Mineralization occurs in this interval. At the base of the Micrite Unit there is a transition of the Muddy Limestone into the Micrite Unit, consisting of argillaceous biomicrite. Above the transition pure micrite with bird's eyes occur, passing upwards into bioclastic silty micrite.

Microscopic examination revealed the presence of two main types of limestone in the Lower Pale Beds (type "a" and type "b"). The Lower Pale Beds type "a" are characterized by the enormous heterogeneity of the allochemical components and the very poor sorting. These two features are in strong contrast to the other lithologies of the area. The major allochemical components are lithoclasts and ooids. Litho-clasts are fragments from the Micrite Unit ranging from 0.5 to 10 mm. Several types of ooids have been observed. Broken ooids occur throughout the type "a" section. Minor allochemical components are bioclasts and lumps. There is no evidence for an in situ fauna in the Lower Pale Beds. This confirms the observation of Andrew and Ashton (1982) at Navan. According to the dominant allochemical component, the Lower Pale Beds type "a" can be subdivided into dominantly lithoclastic sections and dominantly oolitic sections. The former are difficult to distinguish from the Micrite Unit by macroscopic examination only. These subdivisions can be cor-related stratigraphically (Fig. 1). Where ooids are the dominant allochemical component, the sediment is cemented mainly by sparry calcite to form packstones. Where MU-lithoclasts predominate, considerable quantities of unlithified lime mud containing some Zn (400–1000 ppm) seem to have deposited together with the lithoclasts thus forming wackestones. However, the same sparry cements are also present.

The Lower Pale Beds type "b" consist of sandy biosparite. They predominate in the upper part of the Lower Pale Beds. Since these are unmineralized in the Tatestown area, they were not investigated in detail.

The diagenesis of the host rock comprises several common early features such as micritization, geopetal infilling and the formation of fine-grained hypidiotopic dolomite at the expense of micrite. Four types of calcite cement are distinguished. Usually the first cement is a nonluminescent isopachous dogtooth cement. Its Mn and Fe content (Table 1) is always below the detection limit, indicating that the cement must have formed under oxidizing conditions. In micrite-rich sections this cement may be preceded by an isopachous fibrous calcite. At the top of some oolitic sections, a nonluminescent meniscus cement is present instead of the dogtooth cement. The remaining pore space is almost completely infilled by a blocky cement. It displays an orange to yellow cathodoluminescence zonation with etching zones. In

Table 1. Microprobe results of carbonates[a]

Analyzed phase	Number of analyses	Ca	Mg	Fe	Mn	Zn	Pb	Ba	$(Mg+Fe+Mn){:}Ca$ atomic ratio
Idiotopic dolomite	19	20.8–26.7 21.4	7.27–12.2 11.0	0.30–8.30 2.60	0.040–0.85 0.23	<1.85 0.53	/	<0.034	0.81–0.99 0.95
Baroque dolomite	8	21.1–24.2 22.1	9.59–12.1 10.6	0.43–4.58 2.38	0.13–0.26 0.19	<0.10 /	<2.77 /	<0.18 0.056	0.74–0.97 0.86
dogtooth cement	1	39.2	0.40	/	/	0.12	2.36	0.067	
Blocky cement	2	38.0–39.1	0.19–0.45	0.31	0.008–0.11	0.024–0.11	/	/	
Carbonate inclusions in banded sphalerite	3	36.8–39.1	0.07–0.17	0.23–0.39	0.10–0.16	1.60–3.77	/	<0.050	

[a] The range (minimum–maximum) of the data is given (in wt.-%) and also the median value if sufficient data are available. Element contents below the detection limit are presented by "/".

comparison with the dogtooth cement, the higher Mn and Fe contents imply calcite precipitation under oxidizing conditions.

Silicification is present as quartz overgrowths of detrital quartz and as authigenic quartz. Sometimes authigenic quartz contains inclusions of sulfides and seems to be associated with idiotopic dolomite. This idiotopic dolomite is closely associated with replacing sulfides and therefore may be regarded as a gangue mineral. It also contains inclusions of sulfides. It is usually zoned and has a brown to black luminescence. Stylolites and subvertical dissolution seams concentrating detrital quartz and feldspar, clay minerals, and organic matter, occur frequently.

Two main types of veins have been recognized. The first is related to the mineralization and contains sulfides, barite, baroque dolomite, and some calcite. The second type is associated with Hercynian faulting and consists of "tectonic calcite" with a bright orange luminescence. It is usually accompanied by recrystallization phenomena in the surrounding limestones. The sequence of the diagenesis and its relation to the mineralization will be given in a later section. A detailed discussion of the diagenesis on a regional scale is presented by Viaene et al. (1986).

Ore Mineralogy

Textures and Styles of Mineralization

The main ore minerals are sphalerite and galena. Minor quantities of iron sulfides are also present. In the three mineralized drill holes sulfides are restricted to the Lower Pale Beds type "a" lithology overlying the Micrite Unit (Fig. 1). Three different styles of mineralization have been recognized: sulfide breccias, veins, and replacements. In drill hole H1 most of the ore is found in veins, whereas sulfide breccias are more important in H2. H3 is poorly mineralized. Ore textures are illustrated in Fig. 2.

Sulfide Breccias. Sulfide breccias occur in all three drill holes but are best developed in H2. Five breccia horizons are present (Fig. 1). Fragments of banded sphalerite (Fig. 2a) are the main ore component of these breccias. Up to 80% by volume of the breccia may be occupied by fragments of banded sphalerite. The thickness of these bands ranges from 3 to 6 mm. The fragments are usually smaller than 3 cm. The banding is asymmetric. The fragments of banded sphalerite always have a substratum of Lower Pale Beds type "a" lithology upon which several layers of sphalerite are deposited. Each individual layer consists of subparallel elongated crystals of equal length standing perpendicular to the substratum. These banded sphalerites show a spectacular cathodoluminescence zonation of red, orange, yellow, and mainly blue colors. Within each crystal this zonation is parallel to the terminating crystal faces and points away from the substratum. This proves the growth direction to be perpendicular to and away from the substratum. Within the whole banded sphalerite the cathodoluminescence zonation defines a sphalerite microstratigraphy which is identical for all the fragments of banded sphalerite of any one sulfide breccia but is entirely different for each of the five breccia horizons. In

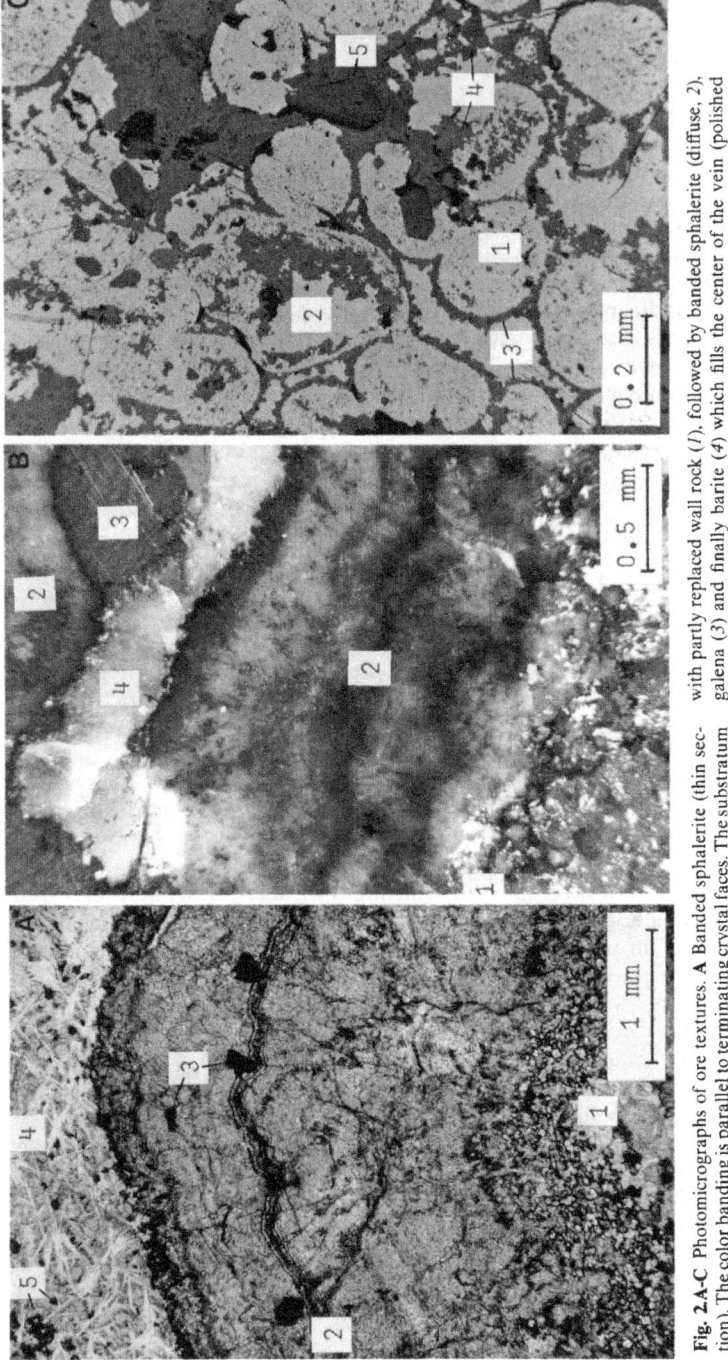

Fig. 2.A-C Photomicrographs of ore textures. **A** Banded sphalerite (thin section). The color banding is parallel to terminating crystal faces. The substratum (1) is partly replaced by sphalerite and idiotopic dolomite. Inclusions of Zn-bearing calcite are present in the dark band (2). Galena crystals (3) occur at two distinct levels in the banded sphalerite. The banded sphalerite is overgrown by a crust of acicular barite (4). Micrite sediment is present between the barite needles, as well as some framboidal pyrite overgrown with marcasite (5). **B** Vein with partly replaced wall rock (1), followed by banded sphalerite (diffuse, 2), galena (3) and finally barite (4) which fills the center of the vein (polished section). **C** Replacement of the host rock [ooids (1), lithoclasts (2), and calcite cements] by sphalerite. The contours of the allochems are marked by unreplaced dogtooth cement (3). Replacing sphalerite is associated with idiotopic dolomite (4). Some authigenic quartz (5) is present (polished section)

the outer sphalerite layers galena crystals are present as well as some minute carbonate inclusions (Fig. 2a). Microprobe results proved these inclusions to be Zn-bearing calcites (Table 1). Both have definite positions in the sphalerite microstratigraphy. The banded sphalerite is associated with pyrite framboids which occur in the substratum, as well as in or on top of the sphalerite. The top of the banded sphalerite is often overgrown by a crust of acicular barite (Fig. 2a). The space between the barite needles is infilled with micrite. All these features clearly formed before brecciation took place, since they are terminated abruptly at the edge of the fragments.

In the matrix between the banded sphalerite no contours of lithified limestone clasts are visible. The fragments of banded sphalerite are embedded in a mass of carbonate grains identical to those of the host rock (Lower Pale Beds type "a"). This suggests that loose carbonate grains, which were not yet subject to cementation, were involved in the brecciation and cemented afterwards. Further evidence for this is found in the fact that no brecciated cements were observed in the matrix between the banded sphalerite and in the distribution of the cements (although obscured by recrystallization and dolomitization). In the matrix of the breccia the dogtooth cement is not always present and crinoid ossicles may then be syntaxially overgrown by the later blocky cement. It thus seems that (due to brecciation) locally the precipitation of dogtooth cement did not take place and that the blocky cement was the first cement. In the substratum of the banded sphalerite the dogtooth cement is usually present and well identifiable, as well as the blocky cement, but it is uncertain whether the latter precipitated before or after brecciation. The remaining pore space has been infilled by later barite and baroque dolomite.

In a few samples there was a sharp vertical contact between the sulfide breccia and barren limestone of the Lower Pale Beds type "a". Along this contact a dissolution seam is present. In contrast to the carbonates in the breccia, recrystallization, dolomitization, and replacement are very weak or absent in this limestone and both dogtooth and blocky cement are present. As no stratification is visible in this limestone, it is impossible to determine whether it represents host rock or just a fragment in the breccia.

Iron sulfides occur in the breccia as pyrite framboids with a varying degree of recrystallization, overgrown by colloform pyrite, marcasite and eventually melnikovite. Pyrite framboids have also been observed in barren samples of the area but are more abundant in the sulfide breccias.

Veins. Veins and veinlets penetrated into the host rock along mechanical discontinuities and zones of weakness such as fractures and stylolites. The thickness of the veins (Fig. 2b) varies from 0.5 to 3 cm. The following minerals were deposited successively:

— Sphalerite grains with yellow internal reflections. Its luminescence is mainly dark blue, with light blue, green, and red zones. It is very closely associated with the idiotopic dolomite, often developing a poikilotopic texture with sphalerite.
— Banded sphalerite of fibrous crystals which differs from that in sulfide breccias in showing a rather diffuse and symmetrical banding with respect to the center of the vein.
— Coarse galena that is the most important vein sulfide.

- Iron sulfides: colloform and coarse pyrite, marcasite, and melnikovite occur frequently as overgrowth on the earlier phases.
- Gangue minerals: barite, baroque dolomite, and minor calcite.

The barite sometimes replaces banded sphalerite. According to the presence or absence of these phases four types of veins were distinguished. They are represented in Fig. 3. The order of the crystallization remains the same in the four types.

Replacements. Replacement of limestone, carbonate grains as well as cements, by sphalerite and occasionally by pyrite or galena is always accompanied by the idiotopic dolomite (Fig. 2c). Both replacement and dolomitization are selective, the dogtooth cement being the most resistant. Intense replacement is always related to veining. The first sphalerite in veins which is also associated with the idiotopic dolomite differs in no way from the replacing sphalerite and idiotopic dolomite. They thus seem to represent the same mineralizing phase.

The substratum of the banded sphalerite in sulfide breccias is also characterized by that replacing sphalerite and idiotopic dolomite association. The fact that the replacement and the dolomitization increase towards the interface banded sphalerite-idiotopic dolomite suggests that they took place before precipitation of the banded sphalerite. In some samples it is suggested that the sphalerite partly acted as a cement.

Ore Geochemistry

Sulfides were analyzed by microprobe for Fe, Mn, Zn, Pb, Cd, Cu, and As. The results are summarized in Table 2.

Sphalerite. The chemistry of the banded sphalerite is marked by a higher Cd content and a slightly lower Fe content than vein sphalerite and replacing sphalerite. In order to correlate its color banding and cathodoluminescence zonation with minor element variation, 118 point analyses were used to construct a profile (Fig. 4).

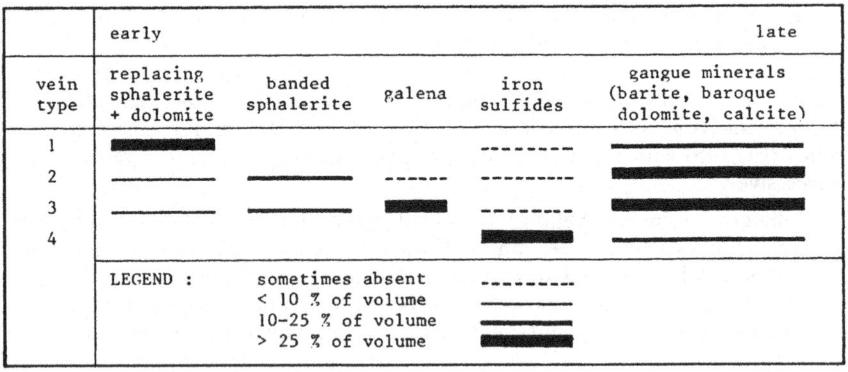

Fig. 3. Succession of sulfide deposition in different types of veins

Table 2. Microprobe results of sulfides[a]

Analyzed phase	Number of analyses	Cd	Fe	Mn	As	Cu	Pb
Banded sphalerite	118	0.006–1.86 / 0.78	<0.27 / 0.094	<0.11 / /	<0.076 / /	<0.41 / 0.017	<0.43 / /
Vein sphalerite	8	0.11–1.02 / 0.42	0.078–0.29 / 0.13	<0.085 / 0.017	<0.085 / 0.006	<0.39 / 0.079	<0.16 / /
Replacing sphalerite	7	0.12–0.72 / 0.36	<0.32 / 0.13	<0.13 / 0.029	/	<0.24 / 0.039	/

Analyzed phase	Number of analyses	Cd	Zn	Mn	As	Cu	Pb
Pyrite framboids	7	<0.12 / 0.070	<0.72 / 0.077	<0.012 / 0.002	<0.47 / 0.25	<0.16 / 0.081	/
Marcasite	6	<0.16 / .030	<0.23 / 0.15	<0.54 / 0.045	<0.060 / 0.010	<0.17 / 0.064	/
Vein pyrite	3	/	0.043–0.10	<0.046	0.66–2.87	<0.18	<0.081
Replacing pyrite	3	0.053–0.093	0.036–0.31	0.033–0.079	0.48–4.02	0.021–0.21	/

[a] The range (minimum-maximum) of the data is given (in wt.-%) and also the median value if sufficient data are available. Element contents below the detection limit are represented by "/".

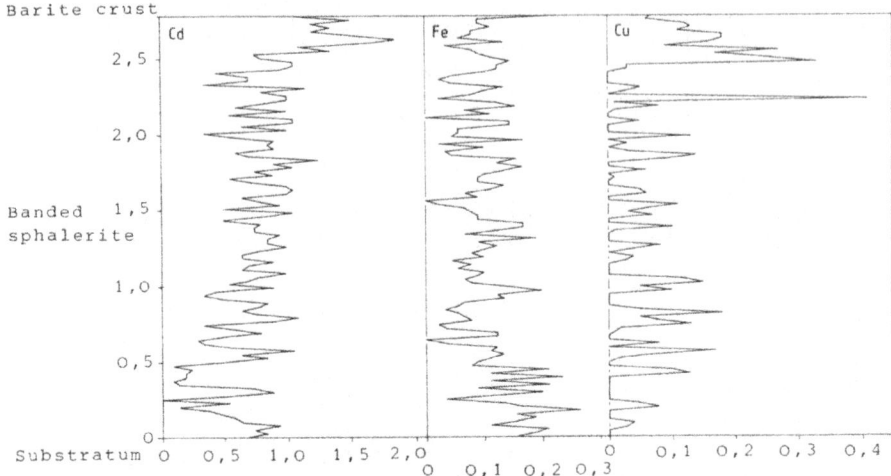

Fig. 4. Profile of point analyses (Cd, Fe, and Cu) of banded sphalerite (see Fig. 2A). Cd increases towards the top of the banded sphalerite. An abrupt increase of Cu is present in the outer zone of the banded sphalerite and is followed by a gradual decrease towards the top. The higher Fe content in the base reflects the presence of replacing sphalerite in the substratum. *Vertical scale* thickness in mm; *horizontal scale* wt.-% of Cd, Fe, Cu

Correlation of minor element variation with the color banding, however, is not straightforward from these data. The best correlation is obtained for combined Pb and Cu, since peaks of these elements usually coincide with darker zones. Fe seems to be too low to influence color banding. The correlation of the observed cath-odoluminescence zonation with minor element variation is even more difficult, since the individual zones are much smaller than for the color banding. According to Goni and Rémond (1969) the presence of Cu, Cd, and Mn is thought to be responsible for green, yellow, and orange-yellow luminescence respectively, while Fe acts as a quencher. The fact that up to 0.4 wt.-% of Cu does not cause green luminescence seems to be in contradiction with their observations. The observed global trends are also illustrated in Fig. 4.

As expected from their spatial relationships, vein and replacing sphalerite have comparable minor element contents. Replacing sphalerite generally tends to be more pure, although its Mn and Fe content is slightly higher. This Mn and Fe may be inherited from the replaced carbonates.

Iron Sulfides. The variations in minor element content of iron sulfides are less pronounced than in the sphalerites. However, early iron sulfides again differ from the later ones. The former have a higher Zn and a lower As content. The fact that framboids associated with banded sphalerites contain more Zn and As than do barren samples proves this spatial relationship to be also a time relationship. As also noted for sphalerites, replacing and vein pyrite have comparable minor element contents. Their low Zn content in comparison with the framboids is explained by the paragenesis in veins: iron sulfides are the last sulfides in the veins and are not associated in time with sphalerite. The minor element content of marcasite is low and does not show any variation.

Gangue Minerals. The idiotopic dolomite always has an excess of Ca. The atomic ratio (Mg + Fe + Mn):Ca should ideally equal unity, but may be as low as 0.75 for natural dolomites (Reeder 1983). For the analyzed dolomites this ratio ranges from 0.81 to 0.99, the median being 0.95. For three samples, however, it is abnormally low: 0.54–0.58. An explanation for this is not evident. The Fe content of the idiotopic dolomite varies from core to margin; the inner zones are Fe-poor (< 1 wt.-%) and the outer zones are richer (up to 8 wt.-%). The Zn content of this dolomite averages 0.53 wt.-% and shows no relation with the departure of the (Mg + Fe + Mn):Ca ratio from unity. The Ba and Pb content are mostly below detection limit.

In baroque dolomites the Ca-excess is higher as evidenced by the lower range and median for the (Mg + Fe + Mn):Ca ratio (Table 1). The Fe content exhibits the same evolution as does the idiotopic dolomite, but never exceeds 5 wt.-%. Baroque dolomite usually contains Ba (up to 0.2 wt.-%). This may be explained by its association with barite. Although not spatially associated with galena two samples contained about 2.5 wt.-% Pb.

Discussion

Lithostratigraphic Correlation

From the sediment petrographic study of all samples, a lithostratigraphic correlation with the base of the Micrite Unit as a datum line was constructed (Fig. 1). In the drill holes H2 and H3 the Micrite Unit is about 20 m thick. In H1 only the top of the Micrite Unit is transected. It becomes increasingly dolomitized towards the Tatestown Fault, which marks the end of the drill hole. On the basis of the correlation of a siltstone horizon in surrounding drill holes a thickness of about 10 m is estimated. In the three drill holes, the micrite is overlain by Lower Pale Beds type "a" lithology. The base is dominantly lithoclastic. A gradual transition into the oolitic horizons above is present. The decrease of lithoclasts is firstly accompanied by an increase of superficial ooids before ooids become dominant. There is also a gradual decrease in grain size towards the oolitic section. In H2 and H3 the oolitic horizon is overlain by a second lithoclastic horizon. Here the contact is sharply defined by an abrupt change of the dominant allochemical component and an abrupt increase in grain diameter, some lithoclasts reaching 10 mm. The presence of a meniscus cement just below this contact implies vadose cementation, whereas the other early cements such as the fibrous cement and the dogtooth cement are probably marine phreatic in origin (Cantrell and Walker 1985). The succession of a lithoclastic horizon by an oolitic one thus represents an upwards-shallowing sequence. All sulfide breccias are hosted in this type "a" section, but they seem to occur at different stratigraphic levels in the three drill holes. This is in agreement with their different sphalerite microstratigraphy. It indicates that the depositions of the banded sphalerite were local phenomena.

The thickness of the Lower Pale Beds increases remarkably towards the Navan deposit (Sevastopulo 1979; Andrew and Ashton 1985). Taking into account the surrounding drill holes the same is true for the type "a" lithology. However, this lithology is confined to a smaller area. It is absent 3 km to the NW of Tatestown and increases rapidly in thickness towards Tatestown, reaching at least 30 m in H3. There is no evidence that this trend is not continued towards Navan.

The deposition of the Lower Pale Beds is interpreted in terms of rapid differential subsidence centered around Navan. Evidence for this is found in the thickness distribution of the Lower Pale Beds and in the sediment petrographical features of the type "a" lithology which indicate a high energy depositional environment and instable conditions. Episodes of active subsidence caused reworking and redeposition of clasts from the Micrite Unit. On top of the accumulated lithoclastic sections ooids were deposited. The sedimentation rate compensated the subsidence resulting in upwards shallowing inter- to supratidal carbonate sequences.

The Relative Time Position of the Mineralization

Based on geometrical criteria, the sequence of diagenetic phases and of the ore minerals has been derived and is shown in Fig. 5. The first occurrence of sulfides, i.e., the banded sphalerite, is believed to have precipitated in the sedimentary or early-diagenetic environment. This interpretation is justified by:

— the occurrence and growth of the banded sphalerite on a substratum of type "a" lithology,

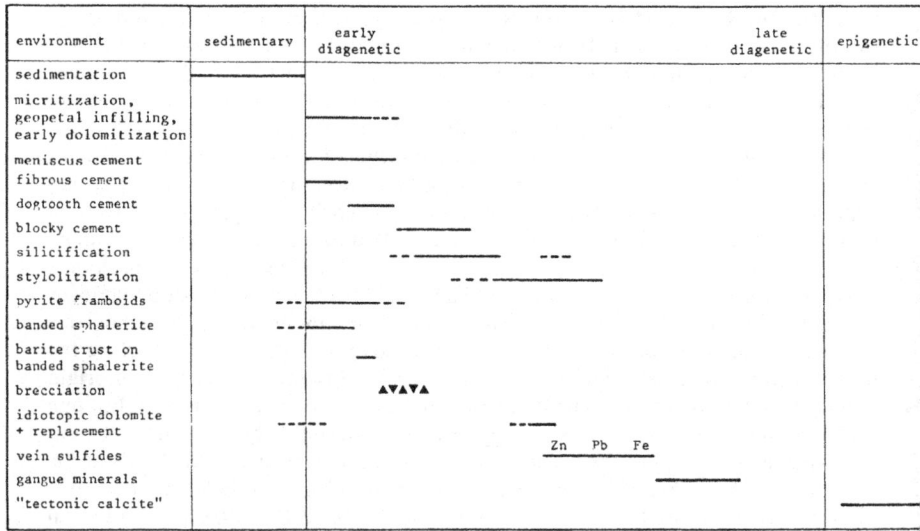

Fig. 5. Diagenetic sequence of the Tatestown prospect

- the association in space and time of banded sphalerite with pyrite framboids,
- the presence of micrite between the barite crystals on the banded sphalerite,
- the formation of banded sphalerite is postdated by the brecciation, which was a very early event; loose carbonate grains which had not been subjected to cementation were involved in the brecciation and have been subsequently cemented,
- the anomalous Zn and Pb content of the micrite sediment and the early cements.

Two objections can be made to this interpretation. The presence of vertical contacts between sulfide breccia and lithified limestone is difficult to interpret in terms of a pre-lithification brecciation. These contacts can be explained by the dissolution seam, which points to a secondary feature. The other objection is the identical paragenesis in sulfide breccias and veins. In both cases the ore fluids deposited successively, replacing sphalerite and idiotopic dolomite, banded sphalerite, galena, iron sulfides, barite, and baroque dolomite. Since almost no veins occur stratigraphically above the sulfide breccia horizons, this objection may prove to be an argument in favor of a feeder system in the deposition of sulfides in the sedimentary or early diagenetic environment. The presence of feeder systems has already been suggested by Andrew and Ashton (1985). The second occurrence of sulfides, veins, and replacements is clearly epigenetic with respect to the host rock. However, this does not exclude a simultaneous deposition of both occurrences, since the epigenetic sulfides represent the feeder system.

Metallogenetic Implications

According to this study, a metallogenetic model for the Zn-Pb mineralizations in the Tatestown area must take into account:

- The timing of the mineralization. As discussed in the previous section mineralization must have taken place during the deposition of the Lower Pale Beds type "a".
- The styles of mineralizations, i.e., sulfide breccias, veins, and replacements. The presence of several sulfide breccias at different stratigraphic levels requires episodic mineralization and brecciation.
- The relationship between the mineralization and the host rock. The mineralization is confined to a limestone with a high initial porosity (Lower Pale Beds type "a") deposited under conditions of rapid differential subsidence centered around Navan and of a high sedimentation rate. The importance of rapid subsidence for the Irish base-metal mineralizations was already stressed by Boyce et al. (1983). This rapid and differential subsidence may offer an explanation for the brecciation of the synsedimentary and early-diagenetic sulfides (banded sphalerite) and for the occurrence of the mineralizations along syndepositional faults. If brecciation is related to such faulting, the ore, of course, will preferentially accumulate in the down-thrown side. These faults may have been reactivated after the mineralization, thus causing displacement of the ore horizons. The ore-bearing fluids were probably released from the underground along these faults and migrated upwards, precipitating epigenetic (with respect to

the host rock) sulfides in the feeder system and synsedimentary or early-diagenetic sulfides at or near to the sediment-water interface.

The present study offers several arguments from different geological aspects for the mineralization at Tatestown being synchronous with the deposition of the host rock. From the study it is also clear that for mineral deposits of this type the examination of the sediment petrographic and diagenetic features of the host rock is a necessity for tackling the genesis of the deposit.

Acknowledgments. This study has been supported by a R & D programme of the Ministerie van Wetenschapsbeleid of Belgium and of the European Economic Community. The authors would like to thank the management of Westland Exploration Ltd. for permission to undertake research on their property and to publish the paper. We were fortunate to have stimulating discussions with Dr. J. Boissonnas, Dr. H. Kucha, Ph. Muchez, Dr. Swennen and Dr. J. Decleer. We thank J. Wautier from C.A.M.S.T. (Centre d'Analyse par Microsonde pour les Sciences de la Terre) for the microprobe analyses. Professor Dr. G. King improved the English text. Technical assistance has been given by G. Vanden Eynde and C. Moldenaers.

References

Andrew CJ, Ashton JH (1982) Mineral textures; metal zoning and ore environment of the Navan orebody, Co. Meath, Ireland. In: Brown AG, Pyne J (eds) Mineral exploration in Ireland: progress and developments 1971-1981 (Wexford Conference 1981) IAEG, Dublin, pp 35-46
Andrew CJ, Ashton JH (1985) The regional setting, geology, and metal distribution patterns of the Navan orebody, Ireland. Trans Inst Miner Metall B94:66-93
Andrew CJ, Poustie A (1984) Syndiagenetic or epigenetic? The evidence from the Tatestown prospect, Co. Meath. In: Geology and genesis of mineral deposits in Ireland (abstracts volume), IAEG, Dublin
Boyce AJ, Anderton R, Russell MJ (1983) Rapid subsidence and early Carboniferous base-metal mineralization in Ireland. Trans Inst Miner Metall B92:55-56
Cantrell DL, Walker KR (1985) Depositional and diagenetic patterns, ancient oolite Middle Ordovician, Eastern Tennessee. Jo Sediment Petrol 55:518-531
Goni J, Rémond G (1969) Localization and distribution of impurities in blende by cathodoluminscence. Miner Mag 37:152-155
Kucha H, Wieczorek A (1984) Sulfide-carbonate relationship in the Navan (Tara) Zn-Pb deposit, Ireland. Miner Dep 19:208-216
Reeder RJ (1983) Crystal chemistry of the rhombohedral carbonates. In: Reeder RJ (ed) Carbonates: mineralogy and chemistry, (reviews in mineralogy, vol 11) Miner Soc of Am Washington, p 32
Sevastopulo GD (1979) The stratigraphic setting of base-metal deposits in Ireland. In: prospecting in areas of glacial terrain, IMM, London, pp 8-15
Viaene WA, Van Oyen P, Clifford J (1986) Sediment petrography of lower Carboniferous sediments in the Navan area (Ireland) and its relationship to lithogeochemistry and base metal mineralizations. Unpublished EEC-report

**Part II Volcanic- and Volcanic-Sediment-Hosted Base
Metal Sulfide Deposits**

Geochemistry and Genesis of Sulfide Ore Deposits in the Volcano-Sedimentary Sequences of the Western Grauwackenzone (Eastern Alps, Austria)

M. TARKIAN and C.D. GARBE[1]

Abstract

The Ordovician/Silurian volcano-sedimentary sequence of the Wildschönauer Schiefer series in the western part of the Grauwackenzone hosts many strata-bound Fe-Cu(Zn-Pb) ore deposits. The most important ones occur between Mittersill and Zell am See and their geochemistry is described in this paper.

Fine-grained clastic sediments locally contain thick doleritic sills and less abundant pillow lavas. These have been affected by hydrothermal alteration and contain disseminated sulfides. The whole sequence has been influenced by a low-grade metamorphism. Primary sedimentary textures are preserved in carbon-rich black shales. In these black shales a positive correlation between sulfide and organic carbon content has been found. Pyrolitic carbon occurs in the mineralized sediments. It is thought to be produced by thermal effects of doleritic sill intrusion. These effects probably also caused the expulsion of pore water from unconsolidated sediments and played an important role in the mobilization of ore-bearing fluids. Petrology and geochemistry of the volcanic rocks and their geological setting suggests a continental rifting environment. Hydrothermal convection systems were produced by the thermal anomalies and caused metallogenesis by controlling the movement of seawater which became enriched in heavy metals extracted from sub-seafloor sediments and volcanics.

Introduction

Numerous strata-bound sulfide deposits are known in the Lower Paleozoic vol-cano-sedimentary complexes of the Grauwackenzone (GZ). They usually occur in low-grade metamorphic rocks (Friedrich 1953; Vohryzka 1968; Unger 1970–1973; Schulz 1983) and in most cases there is a significant relation between sulfides and greenschist facies-grade metabasics both spatially and temporally.

Synsedimentary exhalative processes have been suggested for the formation of some of the Eastern GZ deposits. These processes are assumed to be related to the volcanism associated with an intracontinental rift system (Schäffer and Tarkian 1984; Schlüter et al. 1984).

[1]Mineralogisch-Petrographisches Institut der Universität, Grindelallee 48, D-2000 Hamburg 13, FRG

Base Metal Sulfide Deposits
G.H. Friedrich, P.M. Herzig (Eds.)
© Springer-Verlag Berlin Heidelberg 1988

In the area studied in the Western GZ (Fig. 1), there are numerous strata-bound Fe-Cu(Zn-Pb) sulfide and siderite occurrences within the Wildschönauer Schiefer (WS) series. In contrast to metabasic rocks in the Eastern GZ, the volcanic rocks in this area have been influenced only by low-grade metamorphism. This results in better preservation of the original geochemistry allowing interpretation of magma type and its tectonic setting, both important parameters in controlling the ore genesis. A genetic model for the area has been developed based on ore textures, mineralogy, and igneous and sedimentary rock geochemistry.

Geological Setting

The Wildschönauer Schiefer (WS) are approximately 1000 m thick and have been subdivided by Mostler (1968, 1970) into Lower WS and Upper WS series. The age of the Lower WS series is still uncertain (Schönlaub 1979). The series consists of shales with thin intercalations of siltstones, sandstones, and, less abundantly, conglomeratic layers. Carbonatic rocks are developed locally. Numerous doleritic sills, less abundantly doleritic dykes and pillow lavas intrude this sedimentary sequence. The volcanic rocks are most extensive to the east and south of Kitzbühel (mainly tuffs and alkali dolerites) and near Viehhofen/Zell a.S. (tholeiitic sills and pillow lavas).

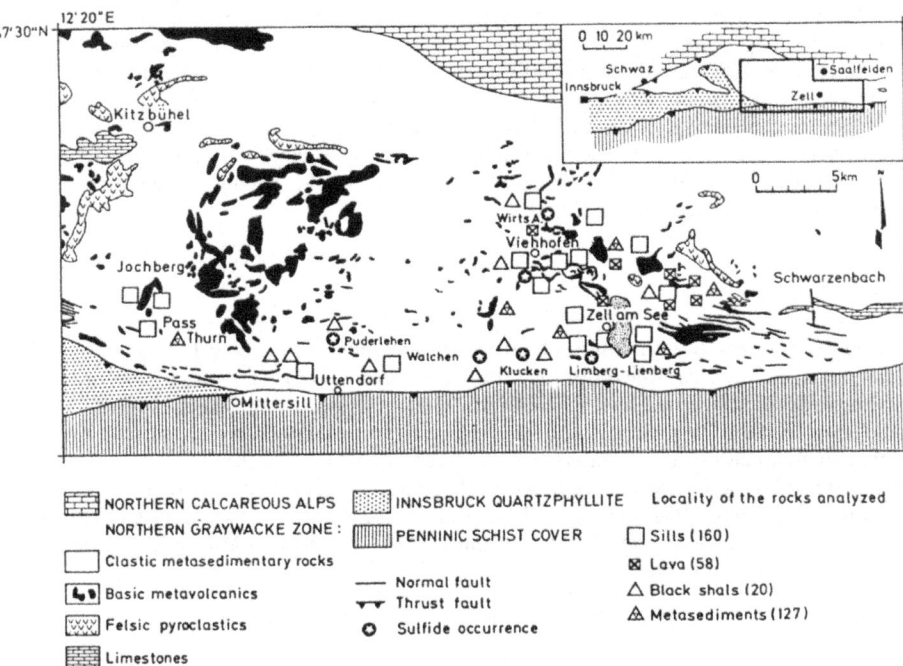

NORTHERN CALCAREOUS ALPS   INNSBRUCK QUARTZPHYLLITE   Locality of the rocks analyzed
NORTHERN GRAYWACKE ZONE: PENNINIC SCHIST COVER   ☐ Sills (160)
☐ Clastic metasedimentary rocks   ⊠ Lava (58)
◤◢ Basic metavolcanics   —— Normal fault   △ Black shals (20)
Felsic pyroclastics   ▼▼ Thrust fault   ⬭ Metasediments (127)
Limestones   ◯ Sulfide occurrence

Fig. 1. Geological situation of the investigated area with location of sulfide occurrences

The Lower WS series is overlain by the "Porphyroid" series of Upper Ordovician/Lower Silurian age. It consists of subaerially formed ignimbrites and tuffs (Mostler 1970) and forms part of an alkali rhyolite province within the GZ (Heinisch 1981).

Increasing subsidence and synsedimentary fault tectonics at the beginning of the Silurian resulted in transgression and formation of contrasting facies types. Varying lithologies, including bituminous shales and shallow-water carbonates, characterize the Upper WS series overlying the Porphyroid series (Mostler 1968).

During the Variscan and Alpidian, the WS sequence was affected by very-low-grade to low-grade regional metamorphism (Schramm 1980).

All the sulfide mineralization studied here occurs within the Lower WS series. General strike is E/W and dip varies from 40° to 80° both to the N and to the S, indicating regional synclines and anticlines.

Sediment Lithology

The main part of the sequence consists of a rhythmic layering of dark-grey to black phyllites and black shales with intercalated silt- and sandstone showing good layering and lamination. Graded bedding and cross bedding have been observed, particularly in silt- and sandstones. The most abundant textural features are load casts and flame structures. Convolute bedding and other slump structures also occur. Locally, dolomitic limestones and dolomites occur (Stimmelhöhe, Uttendorf, Puderlehen Alpe, and near Maishofen). The metapelites consist mainly of fine-grained quartz, muscovite, and a variable content of chlorite and albite. Less abundant are pyrophyllite, paragonite, and stilpnomelane. Zircon, tourmaline, apatite, and leucoxene are accessories. Graphite is found as flakes concentrated on foliation planes or finely dispersed. With increasing graphite content there is a greater abundance of dispersed framboidal pyrite. Phyllites and black shales locally contain black chert layers and also pyrite concretions.

Intercalated in this predominantly pelitic sequence are sandstone beds of up to 20 m thickness showing minor variation in matrix content, grain size, and mineral components. With increasing feldspar, chlorite, and leucoxene content they grade into volcanoclastics. The main portion of the detrital components consists of poorly rounded quartz and two generations of albite. One plagioclase generation is generally albitized with saussuritized cores indicating volcanogenic origin. The other generation consists of subhedral clear albite crystals of authigenic origin. The matrix is composed of chlorite, phengitic muscovite, quartz, and albite, sphene, rutile, graphite, and calcite.

The higher content of plagioclase in these sandstones is accompanied by an increased content of carbonate, sphene and Fe-rich ripidolite in the matrix trending towards a tuffitic composition. The volcanogenic components are assumed to originate from contemporary basic volcanism.

The sediment lithology, particularly the transition from euxinic sapropelitic facies to coarse clastic and carbonate sedimentation, points to a segmented sedimentary basin.

Petrology of the Volcanics

The shale sequence in the Viehhofen and Zell am See areas has thick intercalations of basic lavas and doleritic sills. The lavas often show pillow structure. Frequently they are rich in vesicles filled with quartz, calcite, muscovite, epidote, and sulfides. The abundance of vesicles in the pillows is not indicative of eruption in a shallow water environment (Dudás 1983), neither are brecciated hyaloclastic textures, which are abundant.

The doleritic sills have a known maximum thickness of 50 m. Significant silicification of the sedimentary host rocks and the occurrence of ptygmatic quartz or quartz/carbonate veins are evidence of sill intrusion into unconsolidated sediments. As a result of fractionation, some differentiated sills show transitions between augite-hornblende diorites and quartz granophyres.

Sills and lavas have very similar mineralogy. The most basic rocks are entirely or in part spilitized dolerites and basalts, respectively. They show phenocrysts of glomerophyric plagioclase together with variable amounts of clinopyroxene. Subophitic intergrowths of plagioclase with clinopyroxene are common. Pseudomorphs of chlorite and serpentine after olivine and orthopyroxene (?) are rare. Primary accessory minerals are ilmenite and magnetite. Clinopyroxene, chlorite, albite, epidote, actinolite, leucoxene, minor quartz, calcite, and sulfides of Fe-Cu-Pb-Zn form the recrystallized matrix of lavas and sills. Usually, primary igneous textures are well preserved.

Twenty eight microprobe analyses of clinopyroxenes from sills and pillow lavas of Viehhofen and Zell am See areas (Table 1) reveal common augites poor in Na and Ti. The pyroxene composition and its relation to magma type and tectonic setting will be discussed later.

Microprobe analyses of plagioclase show very low anorthite content (An 0.2 to 4%), which is assumed to be caused by submarine spilitization and by regional metamorphism.

Variable grades of hydrothermal alteration affect the pillow lavas. Pyroxenes have been partly replaced by chlorite and epidote; plagioclases show saussuritization. In addition, all volcanic rocks have been affected by a weak regional

Table 1. Representative microprobe analyses of clinopyroxenes from "Eastern Zone" sills and lavas (in wt.%)

	Sills		Lavas		
	1	2	3	4	5
SiO_2	51.83	52.23	52.53	54.74	53.64
TiO_2	1.01	1.19	0.86	0.45	0.45
Al_2O_3	2.97	2.51	3.68	1.43	1.68
FeO	9.16	9.16	6.28	6.64	8.01
MnO	0.18	0.20	0.15	0.17	0.16
MgO	15.53	15.13	16.69	18.95	16.76
CaO	19.53	19.74	19.77	17.88	20.24
Na_2O	0.29	0.31	0.25	0.18	0.21
Total	100.50	100.48	100.30	100.45	101.16

metamorphism of pumpellyite-prehnite to initial greenschist facies. The most common mineral paragenesis is: actinolite-chlorite-epidote/clinozoisite or pumpellyite-chlorite in addition to albite, quartz, and sphene. Pumpellyite has been observed optically and was verified by microprobe analyses. It is reported here for the first time from the metabasics of Western GZ. The occurrence of pumpellyite adjacent to clinozoisite/zoisite indicates low-grade to very-low-grade metamorphic conditions (300°–400°C).

Doleritic sills from the Uttendorf, Pass Thurn and Jochberg areas (Fig. 1) have a higher alkali content in comparison with those volcanic rocks described from the Viehhofen and Zell a.S. area. In these plagioclase-rich rocks secondary greenish biotite is abundant. Potassic feldspar, actinolite, epidote, chlorite, muscovite, calcite, sphene, and apatite are also present. Geochemically these alkali dolerites can be easily distinguished from the other volcanics described above (see also later).

Ore Structure and Mineralogy

In terms of ore texture and environment of formation four different ore types can be distinguished:

A: Banded pyritic ore in black shales.
B: Massive copper ores in metapelites.
C: Disseminated sulfides in hydrothermally altered volcanics.
D: Secondary sulfide enrichments.

Mineralization in Black Shales (Type A)

Banded pyritic ores occur in several zones throughout the black shales (Fig. 2). Owing to the very low grade of metamorphism, primary sedimentary structures are well preserved and syngenetic formation of the sulfides is evident. Mineralized black shales contain 5 vol.% sulfides on average. The ore mineralogy is rather uniform: Pyrite and pyrrhotite are the main minerals. In addition, chalcopyrite occurs in variable amounts. Less abundant are sphalerite, galena, and tetrahedrite. Pyrites often have framboidal texture which may coalesce to form idiomorphic aggregates.

Occasional enrichment of reworked fragmental pyrite in cross bedded units is evidence for previous existence of sulfides in the area of sedimentation.

Microprobe analyses reveal Ni- and Co-free pyrites corresponding to those from Eastern GZ (Stumpfl and Tarkian 1979; Schäffer and Tarkian 1984; Schlüter et al. 1984), which are assumed to be of syngenetic formation. Ni is either very low or absent in the pyrrhotites analyzed, which often occur as disseminated euhedral crystals. They are assumed to have also originated syngenetically, since no replacement or alteration fabrics between pyrite and pyrrhotite have been observed. Possibly, pyrrhotite together with chalcopyrite, sphalerite, and galena predate pyrite. The former appear as tiny inclusions in pyrite and in the interstices of framboidal pyrite and also as particulate grains. Coexisting pyrite and pyrrhotite are fairly common in sulfide precipitates of hydrothermal origin reflecting varying sulfur fugacities (e.g., Goldfarb et al. 1983).

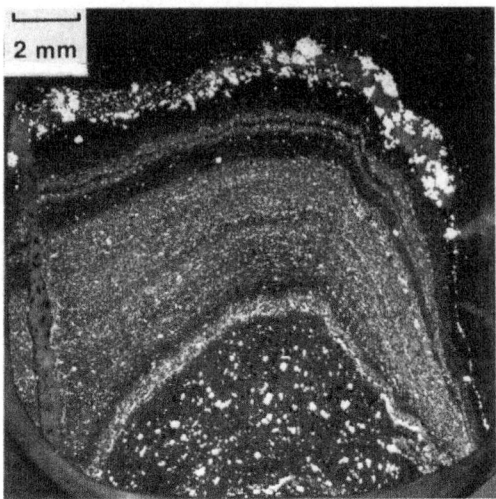

Fig. 2. Rhythmically banded pyritic ore in black shale. Polished specimen

All the black shales are characterized by high graphite content (1–5 vol.%) corresponding to organic carbon between 0.52 and 5.68 wt.% (Table 2). There is a positive correlation between sulfide and graphite content in the mineralized black shales. Therefore, a reducing environment with a variable concentration of sulfur, probably caused by biogenetic activity, appears to have been the main controlling factor during primary sulfide formation.

Unmineralized black shales show significant high Ba contents which are assumed to be of hydrothermal origin. Compared to the black shales described in the literature, Pb contents are rather high, while Cu and Zn are very low in some regions (Table 2). These low levels may be attributed to leaching processes during convection of high temperature solutions.

In some silicified sediments pyrolitic carbon has been observed appearing either in quartz veinlets together with sulfides or as disseminated grains. Optically it may be recognized as spherulites which show the typical Brewster cross (Fig. 3). From experiments it is known that these spherulites form at temperatures of >500°C (Stach 1975). The occurrence of pyrolitic carbon is attributed to thermal cracking of hydrocarbons caused by intrusion of doleritic sills into sediments rich in organic matter (Kisch and Taylor 1966; Simoneit and Lonsdale 1982).

Massive Copper Ores (Type B)

This ore type mainly occurs in the Viehhofen and Klucken areas within phyllitic shales. The ore bodies are small and have not been observed thicker than 1 m, but the Cu contents are rather high (Cu >15 wt.%). Chalcopyrite and pyrite are the main

Table 2. Comparison of black shales from mineralized investigation areas (XRF-analyses; major elements in wt.%, trace elements in ppm)

	1	2	3	4	5	6	7
SiO_2	44.61	64.36	50.06	64.70			
Al_2O_3	24.76	14.75	20.33	9.17			
FeO*	8.42	1.62	9.74	9.02			
MnO	0.07	0.02	0.06	0.03			
MgO	4.69	1.24	3.16	2.88			
CaO	0.22	0.26	0.01	0.13			
Na_2O	0.98	2.24	1.61	0.96			
K_2O	6.17	3.86	4.60	2.02			
TiO_2	1.19	1.04	0.84	0.61			
P_2O_5	0.17	0.19	0.03	0.09			
$H2O^+$	5.05	4.16	6.81	5.93			
$H2O^-$	0.18	0.63	0.25	0.93			
CO_2	n.d.	0.01	–	–			
C_{org}	2.33	5.68	0.52	2.97			
N	0.09	0.10	0.05	0.01			
S	0.72	0.03	2.16	0.94			
Total	99.56	100.19	100.25	100.38			
Ba	1451	858	955	713	955	(557–1751)	580	300
Cr	141	82	98	66	98	(66– 141)	90	100
Cu	17	<10	22	67	18	(3– 79)	45	70
Mn	522	168	454	264	453	(30– 627)	850	150
Ni	74	26	63	106	63	(5– 87)	68	50
Pb	40	9	136	33	27	(3– 136)	20	20
Rb	232	151	163	70	–		–	–
Sr	58	37	44	22	–		–	–
V	194	164	119	206	112	(46– 448)	130	150
Y	37	33	19	10	–		–	–
Zn	73	27	93	32	71	(13– 212)	95	300
Zr	206	163	123	105	–		–	–

1 = Puderlehen Alpe, N Uttendorf; 2 = SE Viehhofen; 3 = Walchen; 4 = Klucken; 5 = median of 14 black shales from the investigation area; 6 = "shales" (Turekian and Wedepohl 1961); 7 = median of 20 sets of 799 samples (Vine and Tourtelot 1970); * = total iron recalculated as FeO; n.d. = not detected; – = not determined

ore-forming minerals with quartz and siderite as gangue materials. Less frequent are pyrrhotite, sphalerite, tetrahedrite, galena, bornite, and chalcocite. These ore bodies are considered to be syngenetic. Granoblastic intergranular textures in the sulfides, carbonates, and quartz and the deformation fabrics reflect a high degree of recrystallization. Metamorphic remobilization created secondary ore enrichment in vein-like structures.

Occasionally lamellae of machinawite in chalcopyrite have been observed. If these are considered to be an exsolution from high-temperature chalcopyrite, they would reflect a temperature of $>250°C$. Of course this temperature would have been exceeded during regional metamorphism.

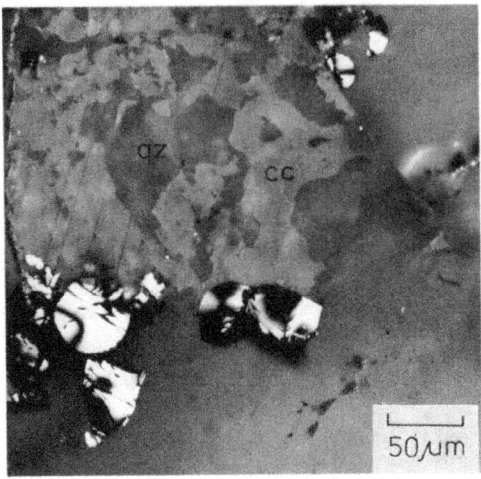

Fig. 3. Pyrolitic carbon from quartz filled veinlet in black shale. Crossed pols. *qz* quartz; *cc* calcite

Disseminated Sulfides in Volcanics (Type C)

All the sills and lavas studied contain disseminated sulfides. Significant sulfide enrichments were found in high-grade hydrothermally altered volcanics. Pyrrhotite is the main mineral. In addition less abundant chalcopyrite, Fe-bearing sphalerite, galena, pyrite, and subordinate marcasite occur. Pillows show disseminated as well as characteristic stringer ores. Cooling cracks and interstices of pillows are filled by recrystallized quartz, calcite, epidote/clinozoisite, chlorite, muscovite, pyrite, pyrrhotite, chalcopyrite, rarely sphalerite and galena. Vesicles of pillows show a similar paragenesis, but pyrite is lacking. Brecciated sulfides have been observed in the interstices of the pillows. All textural features point to a high energy exhalation of boiling solutions onto the seafloor.

Generally all the volcanics are poor in Ni. The Ni content of doleritic sills is 41 ppm on average, whereas pillow lavas show 82 ppm Ni (median of 160 and 58 whole-rock XRF-analyses, respectively). Accordingly, no Ni minerals have been found. Pyrrhotite is very poor in Ni corresponding to that in black shales (Table 3).

The sulfide paragenesis of all three ore types described is very similar. Therefore, it is deduced that the disseminated sulfides in volcanics and subvolcanics were produced by the same hydrothermal activity which was also responsible for the formation of massive copper ores.

Secondary Sulfide Enrichments (Type D)

Sulfide enrichment caused by regional metamorphic remobilization mostly occurs in zones of tectonic stress. Quartz lenses, veinlets, and narrow joints in metapelites contain chlorite, muscovite, and calcite with granoblastic, percrystalline structure

Table 3. Microprobe analyses of sulfides (in wt.%)

	Pyrrhotite				Tetrahedrite		Sphalerite	
	1 (A)	2 (B)	3 (B)	4 (D)	5 (D)	6 (C)	7 (C)	8 (C)
Fe	60.4	60.9	59.8	60.0	5.7	5.6	6.3	7.0
Co	0.06	0.02	0.25	0.16	–	–	–	–
Ni	0.03	0.10	0.13	0.03	–	–	–	–
Zn	–	–	–	–	1.2	1.4	59.6	59.3
Cu	–	–	–	–	38.8	38.6	–	–
Sb	–	–	–	–	27.3	28.3	–	–
As	–	–	–	–	1.4	0.8	–	–
S	39.1	38.9	39.5	40.0	26.0	25.8	33.0	33.2
Total	99.59	99.92	99.68	100.19	100.4	100.5	98.9	99.5

A = banded pyritic ore in black shale; B = stringer ore in pillow lava; C = disseminated sulfides in hydrothermally altered volcanics; D = secondary sulfide enrichments; – = not detected;

Formula: 5: $Cu_{9.92}Fe_{1.66}Zn_{0.29})_{11.87}(Sb_{3.64}As_{0.30})_{3.94}S_{13.19}$
              6: $Cu_{9.99}Fe_{1.64}Zn_{0.33})_{11.96}(Sb_{3.77}As_{0.16})_{3.93}S_{13.11}$

together with chalcopyrite, pyrrhotite, Ag-free tetrahedrite (Table 3), and galena. Traces of bornite and native gold were also found.

Magma Type and Tectonic Setting of the Volcanic Rocks

230 samples of volcanic rocks have been analyzed for major and trace elements using XRF (Tables 4 and 5). The considerable correspondence between geochemical data for doleritic sills and pillow lavas points towards a close genetic relationship in terms of comagmatic origin. In contrast, alkali dolerites from Uttendorf, Pass Thurn, and Jochberg areas show a different magmatic development. The first indication that the former are typical subalkali basalts is given by Na- and Ti-poor clinopyroxene mineral chemistry (Fig. 4). In the "F1-F2" discrimination diagram (Fig. 5) the augites occupy the ocean floor basalt (OFB) discrimination field which is overlapping with volcanic arc basalts (VAB) and within-plate basalts (WPB). However, because of dependence of pyroxene chemistry on the magma cooling process (Gibb 1973; Offler 1979; Mevel and Velde 1976), this trend has to be verified by examination of trace element data from the rocks concerned.

No major changes in whole rock "immobile" trace element chemistry have been observed. Therefore, some trace elements (Floyd and Winchester 1978; Pearce and Norry 1979; Pearce 1982) are used for petrogenetic studies.

The tholeiitic character of doleritic sills and pillow lavas from the Viehhofen and Zell am See areas is displayed in the TiO_2 versus Zr/P_2O_5 diagram (Fig. 6). All samples plot into the discrimination field of modern oceanic tholeiites. In the following discussion they are abbreviated to "*volcanics of the eastern zone*". In contrast, alkali dolerites from Uttendorf, Pass Thurn, and Jochberg areas show alkaline affinity. They are abbreviated to "*volcanics of the western zone*" (see Fig. 1). The Ti/Cr-Ni ratios of the tholeiitic "volcanics of the eastern zone" indicate ocean

Table 4. Representative XRF-analyses of volcanics from Viehhofen/Zell am See Area (major elements in wt.%, trace elements in ppm)

	Doleritic sills				Pillow lavas			
	1	2	3	4	5	6	7	8
SiO_2	46.74	48.21	48.70	53.80	47.19	48.13	47.21	48.29
Al_2O_3	16.71	15.87	14.17	14.21	15.20	14.08	15.00	13.61
Fe_2O_3*	10.28	11.53	12.70	12.54	11.33	11.17	12.27	13.08
MnO	0.15	0.19	0.18	0.25	0.17	0.17	0.16	0.19
MgO	7.15	6.06	7.24	3.23	7.24	6.85	6.74	7.11
CaO	11.55	9.29	8.37	2.86	10.06	11.26	9.60	8.23
Na_2O	2.61	3.84	2.77	5.82	3.13	3.21	3.40	3.62
K_2O	0.17	0.11	0.19	0.56	0.08	0.07	0.09	0.02
TiO_2	1.36	1.69	1.87	2.59	1.62	1.52	1.64	1.79
P_2O_5	0.10	0.14	0.14	0.37	0.14	0.12	0.14	0.18
SO_3	0.10	0.01	0.18	0.97	0.03	0.17	0.24	0.29
L.O.I.	3.01	2.77	2.96	2.83	2.92	3.04	2.95	2.29
Total	99.94	99.71	99.47	100.03	99.11	99.79	99.44	98.67
Ba	277	355	424	992	193	292	339	95
Ce	<10	<10	17	79	21	<10	<10	44
Cr	238	138	183	41	253	249	267	295
Cu	91	16	56	64	92	81	73	62
Ga	19	17	17	–	19	16	14	15
La	<10	<10	13	39	<10	<10	<10	<10
Mn	1180	1474	1419	1936	1309	1325	1237	1476
Nb	13	27	17	23	6	25	24	24
Nd	<10	<10	21	59	11	<10	<10	25
Ni	71	51	56	18	70	76	84	110
Pb	16	7	9	–	27	<5	<5	<5
Rb	<5	<5	<5	20	<5	<5	<5	<5
Sc	25	37	47	24	–	43	40	–
Sr	497	549	421	327	247	124	159	121
V	278	282	337	252	320	278	323	268
Y	23	33	37	49	25	36	33	38
Zn	93	151	77	98	75	62	80	184
Zr	75	102	127	322	102	114	116	130

1 and 2 = E Viehhofen; 3 = Glemmtal, ENE Viehhofen; 4 = S Viehhofen; 5 and 8 = Michelgraben, NE Zell a.S.; 6 and 7 = Ratzensteinhöhe, N Zell a.S.; L.O.I. = loss on ignition; * = total iron as Fe_2O_3; – = not determined

Table 5. Representative XRF-analyses of sills from Uttendorf and Pass Thurn areas (major elements in wt.%, trace elements in ppm)

	1	2	3	4	5	6
SiO$_2$	45.31	46.12	47.37	53.25	45.19	48.94
Al$_2$O$_3$	13.64	12.93	13.93	18.25	12.89	15.58
Fe$_2$O$_3$*	12.94	14.56	12.36	9.58	13.36	11.89
MnO	0.22	0.19	0.17	0.11	0.17	0.15
MgO	4.98	5.65	5.71	2.51	6.89	4.44
CaO	9.91	8.43	9.01	3.51	9.18	5.80
Na$_2$O	3.12	3.12	2.97	6.49	3.24	4.41
K$_2$O	1.19	1.07	1.07	1.26	0.50	1.49
TiO$_2$	3.03	3.95	2.67	1.93	4.24	3.25
P$_2$O$_5$	0.43	0.34	0.38	0.42	0.39	0.50
SO$_3$	0.04	0.06	0.06	0.04	0.01	0.10
L.O.I.	4.07	2.04	2.88	1.97	2.52	2.16
Total	98.88	98.46	98.58	99.32	98.58	98.71
Ba	814	895	741	614	1075	946
Ce	110	87	121	125	89	61
Cr	9	49	176	85	201	205
Cu	<5	<5	<5	<5	139	19
La	39	39	38	41	<10	35
Mn	1704	1471	1317	852	1300	1142
Nb	30	30	35	20	35	36
Nd	50	49	58	50	38	32
Ni	<5	31	84	50	107	14
Pb	<5	<5	<5	<5	<5	<5
Rb	19	19	19	27	11	33
Sc	<10	20	17	27	–	–
Sr	577	376	868	400	629	953
V	228	308	255	130	1132	722
Y	29	27	37	29	39	38
Zn	100	128	92	42	93	117
Zr	311	247	327	178	254	279

1 to 4 = Manlitzbach, W Uttendorf; 5 and 6 = Pass Thurn; * total iron as Fe$_2$O$_3$; – = not determined; L.O.I. = loss on ignition

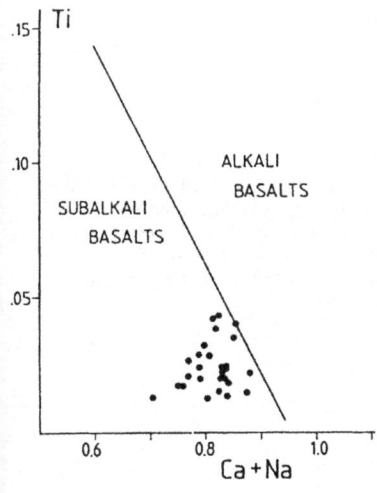

Fig. 4. Clinopyroxenes in sills and lavas from Viehhofen and Zell am See areas in a Ti vs. Ca+Na- plot. (After Leterrier et al. 1982). The elements are recalculated as cationic values from the structural formula

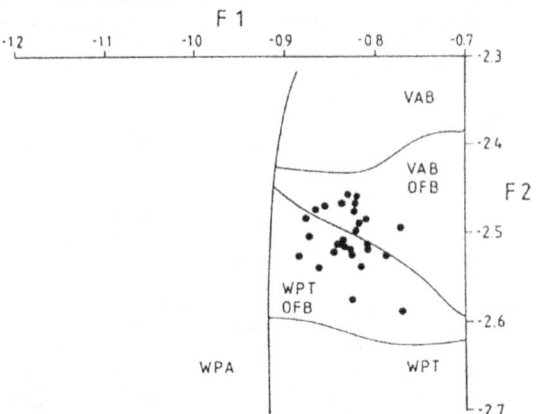

Fig. 5. Clinopyroxenes in sills and lavas from Viehhofen and Zell am See areas in F1 vs. F2 diagram. (After Nisbet and Pearce 1977). The analyzed augites occupy the OFB discrimination field. *WPT* within-plate tholeiites; *WPA* within-plate alkali basalts; *VAB* volcanic arc basalts; *OFB* ocean floor basalts

$$F1 = -0.0120*SiO_2 \quad -0.0807*TiO_2 \quad +0.0026*Al_2O_3 \quad -0.0012*FeO^*$$
$$-0.0026*MnO \quad -0.0087*MgO \quad -0.0128*CaO \quad -0.0419*Na_2O,$$
$$F2 = -0.0469*SiO_2 \quad -0.0818*TiO_2 \quad -0.0212*Al_2O_3 \quad -0.0041*FeO^*$$
$$-0.1435*MnO \quad -0.0029*MgO \quad -0.0085*CaO \quad -0.0160*Na_2O$$

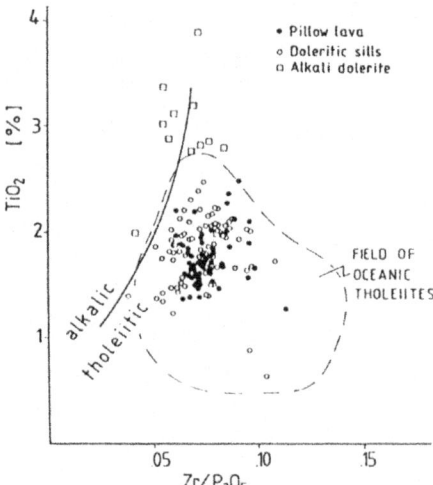

Fig. 6. Plot of TiO_2 vs. Zr/P_2O_5. (After Winchester and Floyd 1976). "Eastern Zone" volcanics plot into the field of modern oceanic tholeiites whereas "Western Zone" alkali dolerites are separated due to their elevated TiO_2 content

floor affinity (Fig. 7); an island arc origin can be excluded. Ti-Zr-Y and Zr/Y-Zr diagrams indicate a transitional tectonic setting of these volcanics. They show trends between OFB and WPB (Fig. 8) and mid-ocean ridge basalts (MORB) and WPB, respectively (Fig. 9). In contrast "volcanics of the western zone" are well defined as WPB. The same trend is picked out in the discrimination diagram TiO_2-Zr (Fig. 10). Evidence for volcanic arc basalts (VAB) is lacking and therefore subduction-related volcanism can be excluded for the entire area studied.

Comparison of the trace element data with those from modern mid-ocean ridge basalts (N-type MORB after Pearce 1982) shows selective enrichment of Sr, K, Rb, Ba, Nb, and Ce. Ba in particular is strongly enriched (Fig. 11). Altogether, "humped" geochemical patterns from the tholeiitic "volcanics of the eastern zone" (pillow lavas

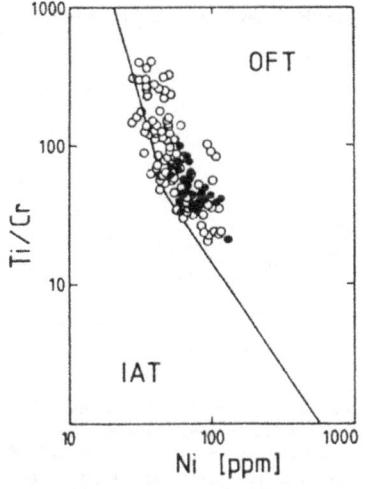

Fig. 7. Tholeiitic "Eastern Zone" volcanics in the Ti/Cr vs. Ni plot. (After Beccaluva et al. 1979). *OFT* ocean floor tholeiites; *IAT* island arc tholeiites

• Pillow lava ○ Doleritic sills

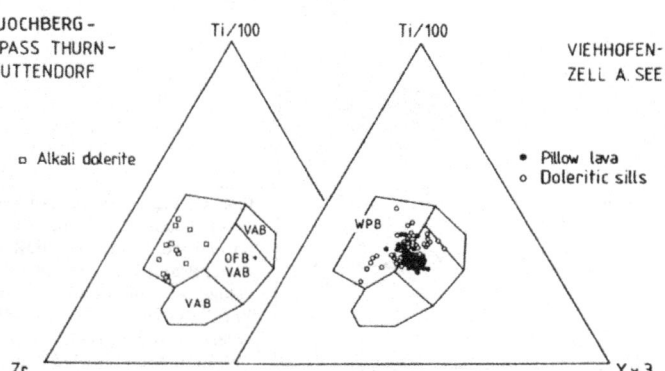

Fig. 8. "Western Zone" volcanics (Jochberg-Pass Thurn-Uttendorf) discriminated as WPB in the Ti-Zr-3Y diagram. (After Pearce 1975). "Eastern Zone" volcanics (Viehhofen-Zell am See) are transitional OFB/WPB. *WPB* within-plate basalts; *OFB* ocean floor basalts; *VAB* volcanic arc basalts

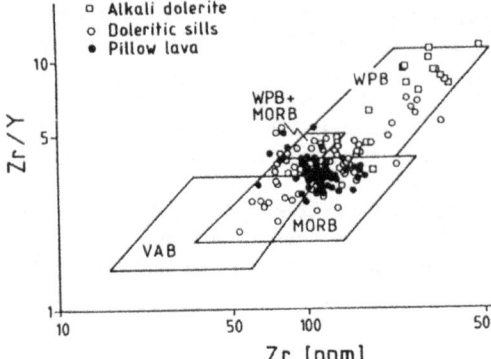

Fig. 9. Plot of Zr/Y vs. Zr. (After Pearce and Norry 1979). Due to elevated Zr, "Western Zone" alkali dolerites (WPB) separate from the "Eastern Zone" tholeiites which resemble modern MORB. *WPB* within-plate basalts; *VAB* volcanic arc basalts; *MORB* mid-ocean ridge basalts

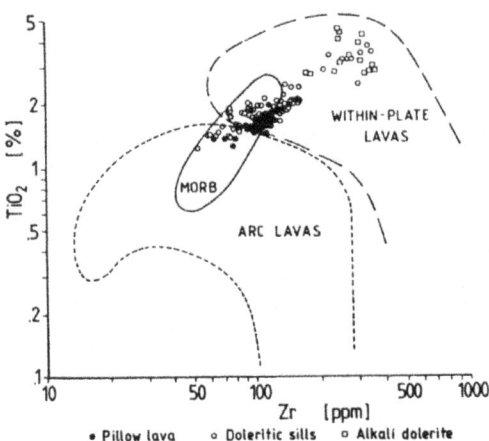

Fig. 10. "Eastern Zone" volcanics show affinities to modern MORB. "Western Zone" dolerites are characterized by high TiO₂ and Zr concentrations (discrimination fields after Gale and Pearce 1982). *MORB* mid ocean ridge basalts

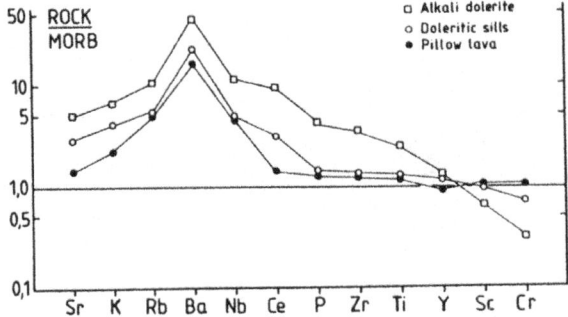

Fig. 11. Geochemical patterns of the studied volcanic rocks normalized to N-type MORB. (After Pearce 1982). The plotted values are medians of "Eastern Zone" pillow lavas (n = 58), doleritic sills (n = 123) and "Western Zone" alkali dolerites (n = 9)

and sills) support the assumption of a transitional tectonic position between WPB and OFB. Alkali dolerites from the "western zone" are in addition characterized by significant enrichment of Nb, Ce, Zr, and Ti emphasizing their within-plate setting again. The geochemical patterns correspond to those for tholeiitic/alkalic or alkalic WPB illustrated by Pearce (1982).

The uniform spatial and temporal distribution of alkalic WPB as well as the transitional tholeiitic OFB/WPB character of the volcanic rocks studied here may be explained by the assumption of a progressing from an initial to an advanced zone of rifting for the tectonic setting of these volcanics.

Colins et al. (1980) studied volcanic rocks from the Western GZ. They established a MORB affinity with transitions to WPB. In the course of their discussion they could not preclude an oceanic island setting, though. Within the WPB group, a more definite geochemical discrimination has not been possible (Pearce and Cann 1973; Shervais 1982). However, as an argument against their assumption of an oceanic island environment (within-plate-oceanic crust) it must be noted that rock types representing true oceanic crust are not known from the GZ. On the contrary, the magma was erupted on or through continental crust (Daurer and Schönlaub 1978), which was attenuated by rifting. For the Upper Ordovician acidic "porphyroids" in the GZ (e.g., Kitzbühel area and W of Schwarzenbach, see Fig. 1), Heinisch (1981) concluded that this volcanism cannot be explained by subduction processes.

Summarizing, Paleozoic volcanism in the GZ yielded bimodal products and transitional magma types, which are common in continental rift zones and are in accord with an advanced rifting stage. Such processes could have taken place in an epicontinental marginal basin. The Gulf of California is a more recent example of such extensional tectonism at continental margins (Karig 1971). Injections of doleritic to granophyric sills are symptoms of crustal instability on continental margins (Sawkins 1976).

An alternate model is that of the strike-slip orogenic belt along which lithosphere is essentially conserved and neither accreted nor consumed (Reading 1980). During periods and within zones of transtension (divergent strike-slip), small basins develop and, as extension proceeds, crustal thinning may allow magma to rise and to be emplaced. Strike-slip basins are elongated parallel to the strike-slip system. They are commonly very deep in relation to their width and characterized by lateral contrasting facies variations. The best example today is the Gulf of California.

Conclusions on the Metallogenesis

During the Lower Paleozoic era no true oceanic environment existed in the western part of GZ. As a result of rifting extensional trough faults were formed — probably close to a passive continental margin. Crustal stretching and subsidence led to the development of rift basins. Thermal anomalies caused by basic minor intrusions and growth faulting controlled convection of seawater. Accordingly, the initiated hydrothermal systems were enriched in base metals derived from leaching of sub-seafloor sediments and partly also from hydrothermal alteration of the volcanics themselves.

The Guaymas Basin south of Gulf of California offers an example of an embryonic oceanic basin. There a high heat flow exists which is induced by intrusion of basaltic sills into highly porous sediments (Einsele 1986). Thermal alteration and silicification of surrounding rocks, migration of organic material, filtration of heated pore water, and mobilization of ore metals as a consequence are the most significant effects caused by this abnormal heat flow.

Analogous processes may be assumed for the area under investigation. Particularly the formation of numerous small sulfide deposits, located in several stratigraphic horizons and formed at different times, could convincingly be explained by shallow level sill intrusion (Fig. 12). Owing to the rapid cooling of a sill, hydrothermal activity is short-lived and can only be revived by intrusion of a younger sill. The mobilized ore solution would have had temperatures not exceeding 200°C (Rona 1984). Doleritic intrusions into unconsolidated sediments resulted in extensive silicification. Raised thermal gradients remobilized the organic matter in the form of gaseous hydrocarbons which partly condensed as pyrolitic carbon.

Long-lasting thermal anomalies were caused by deeper-seated magma chambers which acted as a heat source and played a significant role in driving a hydrothermal convection system. A close relationship between the intensity of magmatic activity and ore formation is obvious in the area between Viehhofen and Zell am See, where some massive Cu-ores have been observed. A small portion of the metals — especially Cu — may already have been precipitated as a sub-seafloor formation by the ascent of hot solutions. The occurrence of small massive Cu-ore bodies at the footwall of a doleritic sill east of Viehhofen is evidence for such an ore-forming process. However, according to the most abundant structural features, mineralization mainly took place by venting of ore solutions onto the seafloor, forming syngenetic sulfides within black shales.

Sulfide deposits from Eastern and Western GZ which have been investigated geochemically show comparable mineral parageneses and host rock composition. Spatial and temporal relations between mineralization and basic volcanism have been established in the areas studied (Table 6). Detailed geochemical data for some areas within the GZ are still lacking. However, the similar tectonic setting of the volcanism (OFB-WPB or WPB) and other corresponding geological features suggest that continental rifting and mineralization analogous to the Western GZ may also have taken place in other regions of the GZ during Lower Paleozoic time.

Acknowledgments. We gratefully acknowledge financial support of the research project (Ta 66/9-1) given by the Deutsche Forschungsgemeinschaft. Thanks are due to Mrs. B. Cornelisen and H.J. Bernhardt for help with microprobe analyses.

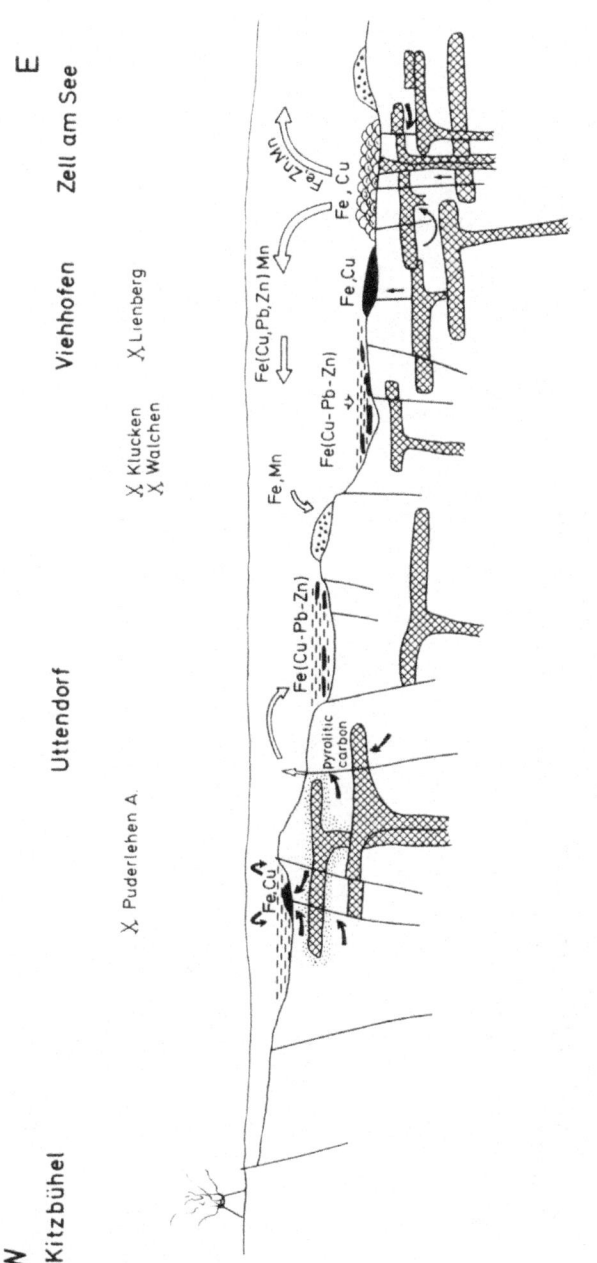

Fig. 12. Diagramatic scheme showing the most significant geological features of sulfide occurrences in the investigated area. See text for explanation

Table 6. General features of the sulfide mineralization from the Eastern and Western Grauwackenzone

Deposit	Ore	Sulfides	Country Rock	Volcanism	Reference
Kalwang	Fe-Cu (Zn, Pb)	py, cpy, po	qz-/ser-phyllites, dolomite marble	thol. WPB → OFB mafic tuffs, intrusives	Schäffer and Tarkian (1984)
Öblarn i. Walchental	Fe-Cu (Zn, Pb)	py, cpy, po	qz-/ser-/graphite-phyllites	alk. WPB mafic tuffites	Schlüter et al. (1984)
Western NGZ			ser-/graphite phyllites turbidites	thol. OFB → WPB pillow lava, intrusives	Colins et al. (1980)
Western NGZ (E-Zone)	Fe-Cu (Pb, Zn)	po, py, cpy	ser-/graphite phyllites black shales	thol. OFB → WPB	This paper
Western NGZ (W-Zone)	Fe-Cu (Pb, Zn)	po, py, cpy	phyllites, black shales	alk. WPB	This paper
Kitzbühel	Fe-Cu	cpy, py	ser-phyllites, quartzite	alk. WPB? mafic tuffs, intrusives	Schulz (1972)

References

Beccaluva L, Ohnenstetter D, Ohnenstetter M (1979) Geochemical discrimination between ocean-floor and island-arc tholeiites – application to some ophiolites. Can J Earth Sci 16:1874–1882

Colins E, Hoschek G, Mostler H (1980) Geologische Entwicklung und Metamorphose im Westabschnitt der Nördlichen Grauwackenzone unter besonderer Berücksichtigung der Metabasite. Mitt österr Geol Ges 71/72:343–378

Daurer A, Schönlaub HP (1978) Anmerkungen zur Basis der Nördlichen Grauwackenzone. Mitt österr Geol Ges 69:77–88

Dudás Fö (1983) The effect of volatile content on the vesiculation of submarine basalts. In: Ohmoto H, Skinner BJ (eds) The Kuroko and related vulcanogenic massive sulfide deposits. Econ Geol Monogr 5:134–141

Einsele G (1986) Interaction between sediments and basalt injections in young Gulf of California-type spreading centers. Geol Rundsch 75:197–208

Floyd PA, Winchester JA (1978) Identification and discrimination of altered and metamorphosed volcanic rocks using immobile elements. Chem Geol 21:291–306

Friedrich OM (1953) Erläuterungen zur Erzlagerstättenkarte der Ostalpen. Radex Rundsch, Radenthein, pp 371–407

Gale GH, Pearce JA (1982) Geochemical patterns in Norwegian greenstones. Can J Earth Sci 19:385–397

Gibb FG (1973) The zoned clinopyroxenes of the Shiant Isles sill. J Pet 14:203–230

Goldfarb MS, Converse MR, Holland HD, Edmond JM (1983) The genesis of hot spring deposits on the East Pacific Rise, 21° N. In: Ohmoto H, Skinner BJ (eds) The Kuroko and related vulcanogenic massive sulfide deposits. Econ Geol Monogr 5:184–197

Heinisch H (1981) Zum ordovizischen "Porphyroid"-Vulkanismus der Ost- und Südalpen. Stratigraphie, Petrographie, Geochemie. Jahrb Geol Bundesanse 124:1–109

Karig DE (1971) Origin and development of marginal basins in the western Pacific. J Geophys Res 76:2542–2561

Kisch HJ, Taylor GH (1966) Metamorphism and alteration near an intrusive coal contact. Econ Geol 61:343–361

Leterrier J, Maury RC, Thonon P, Girard D, Marchal M (1982) Clinopyroxene composition as a method of identification of the magmatic affinities of paleo-volcanic series. Earth Planet Sci Lett 59:139–154

Mevel C, Velde D (1976) Clinopyroxenes in mesozoic pillow lavas from the French Alps. Influence of cooling rate on compositional trends. Earth Planet Sci Lett 32:158–164

Mostler H (1968) Das Silur im Westabschnitt der Nördlichen Grauwackenzone. Mitt Ges Geol Bergbaustud 18:89–150

Mostler H (1970) Struktureller Wandel und Ursachen der Faziesdifferenzierung an der Ordoviz/Silur-Grenze in der Nördlichen Grauwackenzone (österreich). Festbd Geol Inst 300-Jahr-Feier Univ Innsbruck, pp 507–522

Nisbet EG, Pearce JA (1977) Clinopyroxene composition in mafic lavas from different tectonic settings. Contrib Miner Pet 63:149–160

Offler R (1979) Pyroxenes in altered volcanic rocks, Glenrock Station, NSW, Australia. Miner Mag 43:497–503

Pearce JA (1975) Basalt geochemistry used to investigate past tectonic environments on Cyprus. Tectonophysics 25:41–67

Pearce JA (1982) Trace element characteristics of lavas from destructive plate boundaries. In: Thorpe R (ed) Andesites. Wiley, New York, pp 525–547

Pearce JA, Cann JR (1973) Tectonic setting of basic volcanic rocks determined using trace element analyses. Earth Planet Sci Lett 19:290–300

Pearce JA, Norry MJ (1979) Petrogenetic implications of Ti, Zr, Y and Nb variations in volcanic rocks. Contrib Miner Pet 69:33–47

Reading HG (1980) Characteristics and recognition of strike-slip fault systems. Spec Publ Int Assoc Sediment 4:7–26

Rona PA (1984) Hydrothermal mineralization at seafloor spreading centers. Earth Sci Rev 20:1–104

Sawkins FJ (1976) Metal deposits related to intracontinental hot-spot and rifting environment. J Geol 80:1028–1041

Schäffer U, Tarkian M (1984) Die Genese der stratiformen Sulfidlagerstätte Kalwang (Steiermark), der Grünsteinserie und einer assoziierten silikatreichen Eisenformation. Tschermaks Miner Pet Mitt 33:169–186

Schlüter J, Tarkian M, Stumpfl EF (1984) Petrochemische Untersuchungen an der stratiformen Sulfidlagerstätte Walchen (Steiermark), Österreich. Tschermaks Miner Pet Mitt 33:287–296

Schönlaub HP (1979) Das Paläozoikum in Österreich. Abh Geol Bundesanst Wien 33:1–124

Schramm JM (1980) Bemerkungen zum Metamorphosegeschehen in klastischen Sedimentgesteinen im Salzburger Abschnitt der Grauwackenzone und der Nördlichen Kalkalpen. Mitt österr Geol Ges 71/72:379–384

Schulz O (1972) Horizontgesundeue altpaläozoische Kupfer Biesvenenzungen in des Nordtirolen Grauwacken-Zone, Österreich. Tschermaks Miner Pet Mitt 17:1–18

Schulz O (1983) Recent results and critical considerations of the Eastern Alpine metallogenesis. In: Schneider HJ (ed) Mineral deposits of the Alps and of the Alpine Epoch in Europe, Springer Berlin Heidelberg New York, pp 19–27

Shervais JW (1982) Ti-V plots and the petrogenesis of modern and ophiolitic lavas. Earth Planet Sci Lett 59:101–118

Simoneit BRT, Lonsdale PF (1982) Hydrothermal petroleum in mineralized mounds at the seabed of Guaymas Basin. Nature 295:198–202

Stach E (1975) Coal petrology. Borntraeger, Berlin

Stumpfl EF, Tarkian M (1979) Schichtgebundene Sulfidvererzung in den Schladminger Tauern. österr Akad Wiss Math Natwwiss Jahrg 79, 5:111–115

Turekian KK, Wedepohl KH (1961) Distribution of the elements in some major units of the earth's crust. Bull Geol Soc Am 72:172–202

Unger HJ (1970) Der Lagerstättenraum Zell am See. I: Gries bei Saalfelden; II: Fürther Graben; III: Limberg-Lienberg. Arch Lagerst Forsch Ostalpen 11:33–84

Unger HJ (1971) Walchen im Oberpinzgau, Salzburg, ein Kupfer- und Schwefelkies-Bergbau. Arch Lagerst Forsch Ostalpen 12:63–67

Unger HJ (1972) Der Lagerstättenraum Zell am See. IV: Klucken; V: Weikersbach; VI: Saal-Alm. Arch Lagerst Forsch Ostalpen 13:75–109

Unger HJ (1973) Der Lagerstättenraum Zell am See. VII: Viehhofen im Saalachtal. Arch Lagerst Forsch Ostalpen 14:15–53

Vine JD, Tourtelot EB (1970) Geochemistry of black shales – a summary report. Econ Geol 65:253–273

Vohryzka K (1968) Die Erzlagerstätten von Nordtirol und ihr Verhältnis zur alpinen Tektonik. Jahrb Geol Bundesanst 111:3–88

Winchester JA, Floyd PA (1976) Geochemical magma type discrimination. Application to altered and metamorphosed basic igneous rocks. Earth Planet Sci Lett 28:359–469

Base Metal Mineralization in the Evros Region, N.E. Greece

K.L. ASHWORTH[1], M.F. BILLETT[1], D. CONSTANTINIDES[2], A. DEMETRIADES[2], C. KATIRTZOGLOU[2] and C. MICHAEL[2]

Abstract

The stratiform and vein base metal sulfide mineralization of the Evros region has been emplaced during three major metallogenetic periods: the Pre-, Early- and Mid-Alpidic orogenic era. The Pre-Alpidic mineralization is associated with a metamorphosed ophiolitic mafic-ultramafic sequence (Rhodope Massif), the Early-Alpidic with tholeiitic metabasalt (Circum Rhodope Belt) and the Mid-Alpidic mineralization has its major development in Tertiary sedimentary and calc-alkaline igneous rocks.

These types of mineralization, depending on their geotectonic setting, are considered to be similar to that of the Limassol Forest Plutonic Complex (Rhodope Massif), to volcanic-exhalative and analogous to Cyprus volcanogenic massive sulfides (Circum Rhodope Belt), and to stratiform sediment hosted and veins of volcanic affiliation (Tertiary volcano-sedimentary basins).

Introduction

The Evros region is situated in north-eastern Greece and is bounded to the north and east by Bulgaria and Turkey (Fig. 1). It shows a history of some minor exploration and mining activity which started during the Turkish and Bulgarian occupation of Thrace in the late 19th and early 20th century. Over the last 10 years exploration activities by IGME (Institute of Geology and Mineral Exploration) has increased. The use of such exploration techniques as investigations of stream sediment, soil and rock geochemistry, geophysics, and drilling has led to the delineation of new target regions. A number of mineralized areas have now been located by IGME and although the most important of these are base metal prospects, the Evros region also contains gold, silver, chromite, and manganese mineralization (Constantinides et al. 1983).

[1]Department of Geology, University of Southampton, Southampton S09 5NH, England
[2]Institute of Geology and Mineral Exploration, 70 Messoghion Str, 115 27 Athens, Greece

Base Metal Sulfide Deposits
G.H. Friedrich, P.M. Herzig (Eds.)
© Springer-Verlag Berlin Heidelberg 1988

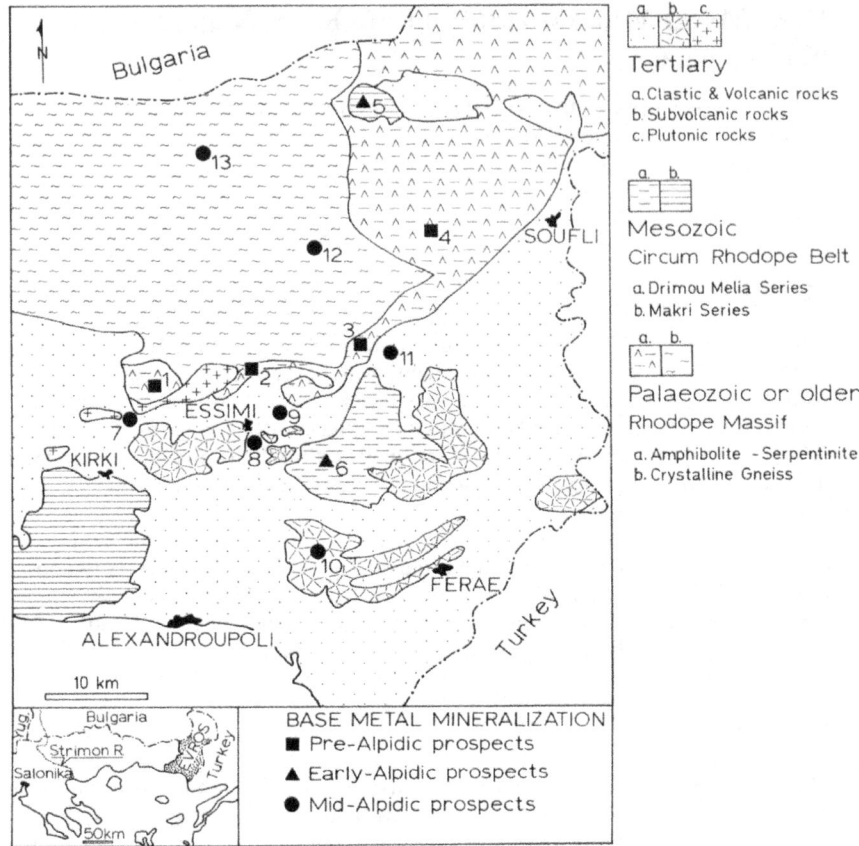

Fig. 1. Simplified geological map of the Evros region showing the location of the major base metal prospects. The names of the numbered prospects are given in Table 1. (After Constantinides et al. 1983)

Geology

The Evros region comprises three major structural units (Fig. 1): the Rhodope Massif with medium to high grade metamorphic rocks, the Mesozoic Circum Rhodope Belt consisting of weakly metamorphosed volcano-sedimentary and flysch formations, and the Tertiary volcano-sedimentary basins. Metallogenetically the Evros region is part of the East Rhodope metallogenetic district which extends northwards into Bulgaria (Jankovic 1979; Boncev 1980).

Rhodope Massif

The oldest rocks in the Evros region are the metamorphics of the Rhodope Massif. They comprise deformed and highly metamorphosed sedimentary and igneous rocks. Several phases of deformation have affected the massif during which metamorphism reached upper greenschist to lower amphibolite facies (Papanikolaou and Scarpelis 1980; Papanikolaou et al. 1982; Zachos and Dimadis 1983). The present-day stratigraphy of the Rhodope Massif in the Evros region consists of a lower unit of leucocratic orthogneiss, mica-schist, amphibolite and thin marble horizons, and an upper unit of ophiolitic amphibolite and serpentinite. The latter unit is important with respect to base metal mineralization, hence its geology is elaborated further.

The Amphibolite-Serpentinite Unit comprises a lower sequence of mafic amphibolite and banded quartz-amphibolite, a mid-sequence of amphibolite and marble and an upper of podiform serpentinite. Field relationships and the preservation of igneous layering and massive gabbroic textures in the coarse grained mafic amphibolite, the primary igneous minerals such as olivine, pyroxene and plagioclase in the centers of large podiform serpentinite masses, and the mafic units of 0.5–1.5 m thickness in the banded quartz-amphibolite which are interpreted as representing highly deformed dykes, all suggest that the Amphibolite-Serpentinite Unit may represent a deformed and metamorphosed mafic-ultramafic complex, which has been intruded by granitic material, presumed to be the plagiogranite of ophiolite complexes.

Circum Rhodope Belt

The Circum Rhodope Belt formations overlie unconformably the crystalline basement and comprise sub-greenschist facies volcano-sedimentary sequences. It is subdivided into two units, the Makri or Phyllite Series of Jurassic to Lower Cretaceous age and the Drimou Melia Series of Cretaceous age. The *Makri Series* mainly consist of clayey, sericitic, calcareous, and quartzitic phyllite, limestone, and greenstones. The *Drimou Melia Series* unconformably overlies the Makri Series and comprises shale, clayey-bituminous marl, quartzite, sandstone, conglomerates, pyroclastics, and mafic volcanic rocks (Papadopoulos 1980).

Tertiary Volcano-Sedimentary Basins

The Tertiary basins in the Evros region (e.g., Kirki-Essimi) are elongated with the major axis trending E-W and/or NE-SW, directions followed by deep crustal faults. Basin subsidence was initiated during Mid-Eocene (Papadopoulos 1980) with sedimentation followed by concurrent intense volcanic activity from Upper Eocene to Oligocene (Fytikas et al. 1979, 1984; Innocenti et al. 1984). Sedimentation and volcanism are both controlled by reactivated faults of mainly E-W, N-S and NE-SW directions.

The Tertiary formations belong to three stratigraphic eras: (a) Lutetian (Middle Eocene), (b) Priabonian (Upper Eocene), and (c) Oligocene. Each formation is characterized by an unconformity to the underlying rocks. A notable feature of the Lutetian is the complete absence of volcanic products and mineralization.

The *Lutetian Stage* (Middle Eocene) consists of a basal breccio-conglomerate, marl and nummulitic limestone. The *Priabonian Stage* (Upper Eocene) is made up of a basal breccio-conglomerate, shale, clayey sandstone, calcareous sandstone, and volcanic rocks such as andesite, dacite, tuff, and tuffite, which occur within the sedimentary strata. The volcano-sedimentary sequence is crosscut by subvolcanic rocks of intermediate composition. The *Oligocene Stage* includes mainly acid volcanic rocks, e.g., tuff, volcanic breccia, and rhyolite porphyry in the form of domes and dykes, but also granitoid stocks occur. Its sedimentary rocks have a restricted regional development.

Mineralization

The base metal mineralization of the Evros region has been classified into three metallogenetic epochs, the Pre-, Early- and Mid-Alpidic orogenic era, according to its emplacement relative to the different phases of activity during the Alpine orogeny (Constantinides et al. 1983). The most important base metal prospects of the region were classified using this scheme (Table 1, Fig. 1).

Table 1. Classification of the Evros region base metal prospects into metallogenetic epochs

Pre-Alpidic	Early-Alpidic	Mid-Alpidic	
1. Aberdeen	5. Mikro Dherio	7. St. Philip	11. Virini
2. Baiko	6. Elia	8. Mili	12. Tris
3. Pessani		9. Pr. Elias	Vrises
4. Yiannouli		10. Pefka	13. Ano Kambi

Note: The numbers assigned to each prospect represent their location in Fig. 1. The most important prospects are discussed in the text.

Pre-Alpidic Mineralization

The base metal mineralization in the metamorphic basement is closely associated with the Amphibolite-Serpentinite Unit, where a number of disseminated and massive pyrite-chalcopyrite ($\pm$ sphalerite $\pm$ galena) prospects have been located. The most important are those at Aberdeen, Baiko, and Pessani (Fig. 1). The styles of mineralization recognized in all three prospects are stratiform and vein type.

The *Aberdeen prospect* occurs in the lower part of the serpentinite unit close to its contact with the mafic amphibolite, and lies at the intersection of two N-S and E-W trending vertical fault zones. The stratiform pyrite-chalcopyrite mineralization follows the foliation and is confined to the mafic amphibolite horizon. The minor constituents are sphalerite, galena, marcasite, enargite, bornite, magnetite, rammelsbergite, bismuthinite, argentite, hessite, siegenite, and sulfosalts of Bi, Pb, Ag, and Te. Chromite is widespread in the host rocks. The effects of deformation and metamorphism on the base metal mineralization may be observed in crystalloblasts of pyrite and chalcopyrite. The epigenetic vein-type mineralization, on the other hand, occurs as fracture filling mainly along the N-S fault zone, and as minor veinlets crosscutting the stratiform mineralization. The major and minor ore mineral association is similar to that in the stratiform mineralization, but in the vein-type pyrite and chalcopyrite have been brecciated by tectonic movements and are cemented by quartz.

At the *Baiko prospect* two mineral associations, disseminated pyrite-chalcopyrite (± sphalerite ± galena) and disseminated magnetite-pyrite-chalcopyrite, have been recognized. The disseminated pyrite-chalcopyrite association occurs close to the contact between the mafic amphibolite and the serpentinite. Pyrite is present both as a fine-grained disseminated phase orientated parallel to the regional foliation, and as an euhedral phase occurring in lenses and veins. Chalcopyrite typically occurs in veinlets which crosscut the regional foliation. Disseminated magnetite-pyrite-chalcopyrite mineralization is found in a unit of amphibolite-marble. The mineralization is concentrated in metasomatic, epidote-rich halos at the contacts of marble horizons with amphibolite bands. The presence of carbonate horizons within the metabasite suggests that this unit represents a thin sequence of intercalated mafic volcanics and carbonates.

The stratiform mineralization at the *Pessani prospect* is subdivided into two types (Michael et al. 1984; Billett and Nesbitt 1986). The first is concentrated in metasomatic, epidote-rich halos at the contacts of marble horizons with amphibolite bands. The mineralization is mainly disseminated and rarely occurs in aggregates. The major mineral is pyrite. Chalcopyrite is usually a minor constituent but locally becomes more abundant. Actinolite, tremolite, garnet, epidote, and chlorite are closely associated with the mineralization. The second type occurs close to the contact of the banded quartz amphibolite with large serpentinite bodies. It comprises disseminations and small lenses (5–30 cm) of mainly fine-grained and crystalloblastic pyrite. Chalcopyrite and galena occur as minor constituents. The fine-grained pyrite appears rarely in relict spherical forms of probably colloidal origin. Most of the pyrite has been recrystallized in crystalloblasts which are often brecciated and cemented by chalcopyrite, whereas galena shows signs of plastic deformation. The epigenetic mineralization consists of small quartz veins with pyrite, galena, sphalerite and chalcopyrite, which occur in the amphibolite with a NW-SE trend.

Early-Alpidic Mineralization

The base metal mineralization of the Early-Alpidic epoch is associated with the mafic volcanics of the Circum Rhodope Belt at Mikro Dherio (Makri Series) and Elva (Drimou Melia Series) (Fig. 1). In both cases the mineralization occurs as disseminations and veinlets in a sequence of deformed, weakly metamorphosed and altered basaltic pillow lavas. Chlorite, epidote, and zeolites appear as alteration minerals.

Mid-Alpidic Mineralization

The Mid-Alpidic base metal sulfides mainly occur in rocks of the Tertiary volcano-sedimentary basins in two styles, stratiform and vein form. The latter is also present in the Rhodope Massif and Circum Rhodope Belt.

Vein Mineralization. The vein mineralization in the Evros region is widespread in contrast to the known statiform ore (Fig. 1). It occurs in fracture zones of NW-SE (Pr. Elias area) and/or N-S (St. Philip) general trend (Fig. 2). Within these fracture zones, apart from the vein mineralization, disseminated ores of restricted development, and rarely veinlets forming stockworks may be observed. The ore veins have a thickness of a few centimeters up to 3 m, whereas their length and vertical dimension vary from a few to about 100 m. The vein mineralization is irregularly distributed, both in the vertical and horizontal sense and even over short distances.

At *Prophetis Elias* pyrite-sphalerite-galena and chalcopyrite mineralization with minor arsenopyrite, marcasite, pyrrhotite, cosalite and bismuthinite has been described (Constantinides et al. 1983), associated with veins in the Priabonian sediments. The gangue minerals are calcite (with some dolomite), quartz, chlorite, and epidote.

The *St. Philip* vein base metal mineralization is presently the only one in the Evros region with known economic potential (Figs. 1, 2, 3). The 1986 estimated ore reserves are 1.2 million tons with an average grade of 9 wt.-% combined Pb + Zn and 50–150 ppm Ag. Two ore mineral associations have been recognized: (a) sphalerite, galena, and pyrite, and (b) chalcopyrite, pyrite, and As-, Sb-sulfosalts as

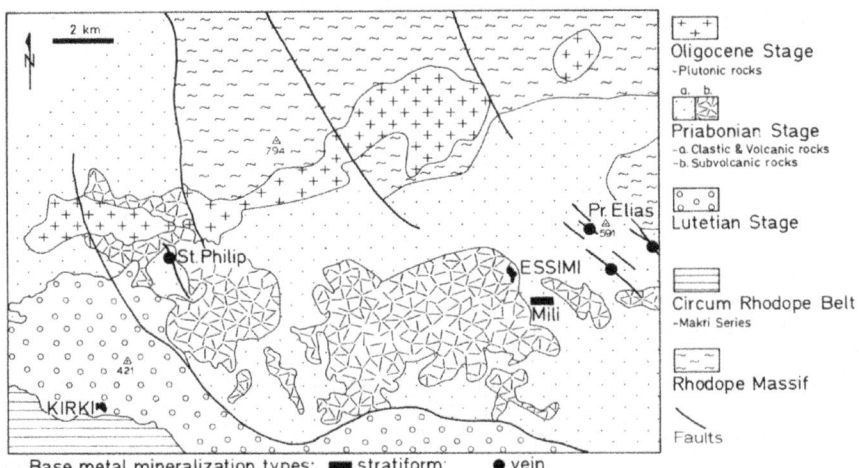

Base metal mineralization types: ▬ stratiform; ● vein.

Fig. 2. Simplified geological map of the Kirki-Essimi Tertiary volcano-sedimentary basin showing the location of the major stratiform and vein mineralization prospects. (After more detailed maps by P. Papadopoulos, K. Arikas, I. Romaides, E. Dimadis, C. Michael and C. Katirtzoglou)

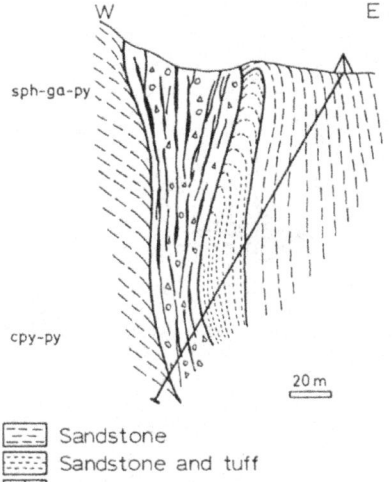

Sandstone
Sandstone and tuff
Mineralized fracture zone

Fig. 3. Diagrammatic geological section of the St. Philip mineralized fracture zone

well as Bi- and Sn-sulfides (e.g., bismuthinite, stannite, tennantite, luzonite, selig-mannite, jordanite, and tetrahedrite). Quartz, carbonates and baryte are gangue minerals. The first mineral association is commonly found vertically above the second one (Fig. 3).

The *Pefka* area is covered by basic and acid calc-alkaline volcanic rocks which were emplaced in a continental environment (lahar forms), with local indications of shallow sea conditions. The polymetallic sulfide mineralization of the Pefka pro-spect comprises two ore-mineral associations: (a) galena, sphalerite, and pyrite, and (b) pyrite, chalcopyrite, tetrahedrite, antimonite, tennantite, and gold. The gangue minerals include quartz, baryte, and carbonates. There is an apparent mineral zonation, the ore minerals of the first association are above those of the second one. Native gold has been observed in the form of grains of a few microns in diameter within quartz and rarely as inclusions in tenantite-tetrahedrite. The veins occur in silicified N-S and E-W-trending fracture zones within intensely altered volcanic rocks. The main ore concentrations are located at the intersections of cross-cutting faults.

Stratiform Mineralization. The stratiform base metal sulfides of the *Mili* prospect (Figs. 1, 2) are concentrated in certain horizons, all of which occur in the center of a clastic sedimentary sequence belonging to the Priabonian Stage (Fig. 4). The mineralization varies from semi-massive to disseminated ores, and passes later-ally to pyrite only. The ore minerals are generally finer-grained than those of the vein mineralization and include sphalerite, galena, and pyrite, with minor marcasite, bornite, bournotite, and arsenopyrite. The gangue minerals are calcite, ankerite, dolomite, quartz, chlorite, sericite, and baryte. The most dominant authigenic minerals in the mineralized area are ankerite and chlorite with large amounts of sericite and quartz. The only allogenic mineral is quartz in variable quantities.

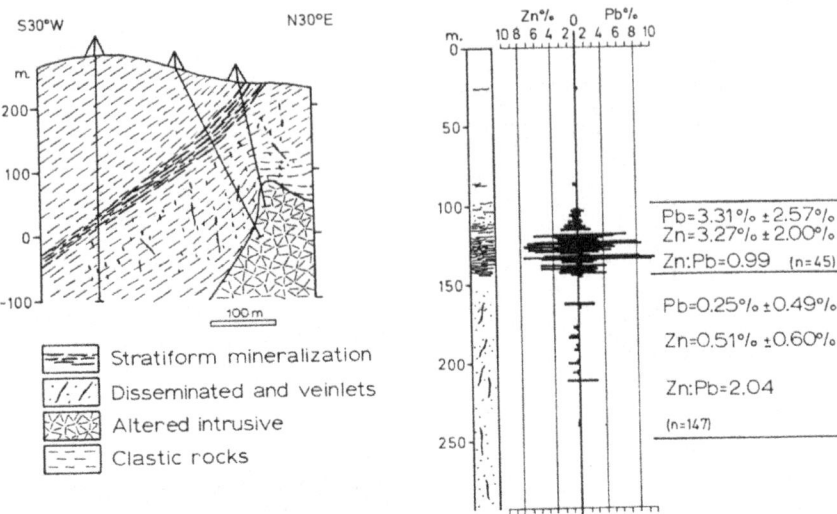

Fig. 4. Diagrammatic section of the Mili stratiform sediment hosted base metal mineralization (*left*) and diamond drill-hole section showing the mineralized zone and the Zn and Pb distribution (*right*)

The clastic sedimentary rocks underlying the mineralized sequence are silicified and contain large amounts of sericite, ankerite, and chlorite as well as veinlets and disseminations of base metal sulfides. Secondary silicification has produced epitaxic quartz grains. The overlying clastic rocks, on the other hand, contain considerable amounts of calcite and abundant ankerite, sericite, montmorillonite, and some albite and kaolinite. Carbonate appears to have been extensively recrystallized.

Porphyritic igneous rocks of felsic composition which are intruded into the clastic sedimentites are strongly carbonated and feldspars are converted to sericite, indicating the activity of hydrothermal fluids. The adjoining clastic sedimentary rocks are also intensely altered. Pyrite appears to have been crystallized later than silicate and carbonate minerals, but its widespread distribution and the presence of some pyrite-rich bands suggests that Fe and S were either original components of the host rocks or the hydrothermal fluids were enriched in these two constituents. Sphalerite (with minor chalcopyrite) and galena were the last minerals to crystallize and appear to be interstitial or to replace some feldspar and possibly quartz. The general appearance of quartz and particularly the irregular and ragged nature of the carbonate patches suggests that recrystallization took place in the presence of fluids.

Geotectonic Setting

The variation of Cr and Ni vs. total FeO/MgO in the mafic amphibolite and serpentinite is shown graphically in Fig. 5. On the Cr vs. total FeO/MgO diagram the mafic and ultramafic rocks exhibit a continuous fractionation trend of decreasing Cr content with increasing fractionation index. The Ni vs. total FeO/MgO plot,

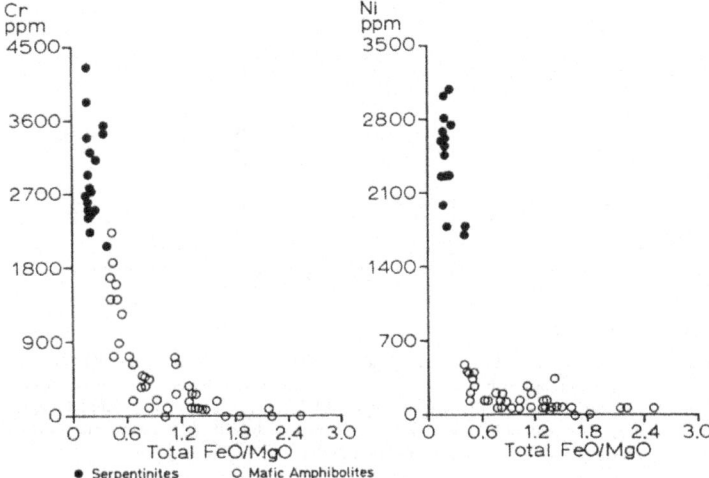

Fig. 5. Cr vs. total FeO/MgO and Ni vs. total FeO/MgO variation diagrams for the mafic amphibolite and serpentinite. (After Billett and Nesbitt 1986)

however, shows a distinct composition gap between Ni-rich serpentinite and relatively Ni-poor mafic amphibolite. According to Billett and Nesbitt (1986), we are dealing with a comagmatic suite of mafic and ultramafic rocks, possibly of ophiolitic affinity. Furthermore, the trace element data, particularly the gap in the Ni distribution, suggests that at least parts of the serpentinite unit are residual ultramafic rocks, i.e., tectonites in ophiolite terminology. A tectonic environment is envisaged similar to that of the Troodos Ophiolite Complex, which is an early back-arc at a constructive plate margin (Smith 1971; Pearce 1975; Shelton and Gass 1980). At some stage in the Lower Paleozoic or even Precambrian, the mafic-ultramafic sequence was obducted onto the sedimentary-igneous rocks of the Rhodope Massif. Both units were then affected by the thermodynamic metamorphism of at least the Caledonian, Hercynian and Alpine orogenic cycles (S. Zachos, pers. commun.).

Two discrimination plots Ti vs. Zr (Pearce and Cann 1973) and Zr/Y vs. Zr (Pearce and Norry 1979) have been used for the identification of the tectonic environment of the volcanic rocks (Fig. 6). The metabasalts of the Circum Rhodope Belt (Makri Series, Mikro Dherio prospect) fall in the low K-tholeiite field of the Ti vs. Zr diagram and in the island arc tholeiite of the Zr/Y vs. Zr plot. The interpretation given is that the metabasalts were formed at a constructive plate margin back-arc environment, similar to the Cyprus pillow lavas setting.

The Tertiary calc-alkaline volcanics from Essimi and Pefka areas plot as two distinct populations on the above two diagrams. Both sets fall in the orogenic calc-alkaline basalt field of the Ti vs. Zr diagram. On the Zr/Y vs. Zr diagram the Essimi volcanics plot in the mid-ocean ridge basalt (MORB) field, and those from Pefka at the boundary between MORB and within plate basalt (WPB). It can be argued that the Essimi calc-alkaline volcanics were emplaced in Priabonian times during the initial rifting and subduction of the Apulian microplate under the

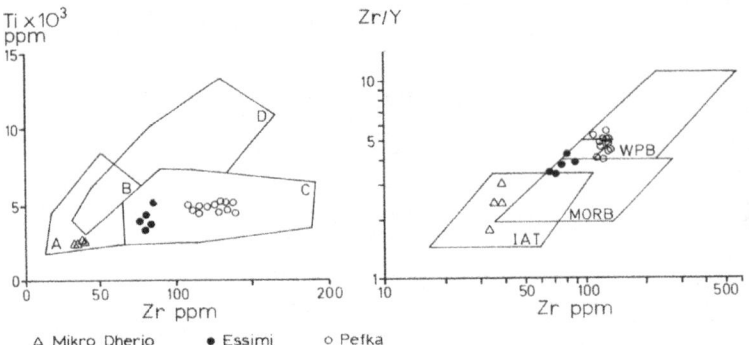

Fig. 6. Position of metabasalts from Mikro Dherio (Makri Series, Circum Rhodope Belt) and basic members of the calc-alkaline volcanics from Essimi (Priabonian) and Pefka (Oligocene) plotted on Ti vs. Zr (Pearce and Cann 1973) and Zr/Y vs. Zr (Pearce and Norry 1979) discrimination diagrams ($A + B$ low potash island arc tholeiite; $B + C$ calc-alkaline basalt; D ocean-floor basalt; IAT island arc tholeiitic basalt; $MORB$ mid-ocean ridge basalt; WPB within plate basalt)

Rhodope Massif, the southern margin of Eurasia (Fytikas et al. 1984). During the Oligocene, the emplacement of the majority of the more differentiated calc-alkaline volcanics, subvolcanic and plutonic rocks took place. The Pefka volcanics with a higher Zr content were most likely emplaced in a continental setting.

Discussion

The base metal sulfide mineralization of the Evros region occurs in three different geological settings: in the ophiolitic Amphibolite-Serpentinite Unit, in the tholeiitic lavas of the Circum Rhodope Belt and the Tertiary volcano-sedimentary basins, and was emplaced during three different metallogenetic epochs: the Pre-, Early- and Mid-Alpidic orogenic era.

The mineralization in the Amphibolite-Serpentinite Unit has different modes of origin. The amphibolite of the amphibolite-marble member is considered to be a submarine metabasalt. This inference is supported by field relationships and the presence of marble horizons. The base metal sulfides are believed to be of syn-genetic-exhalative origin, analogous to Cyprus-type volcanogenic massive sulfides. The occurrence of mineralization close to the contact between the mafic amphibolite and serpentinite suggests that the latter may have played an important role in the remobilization and concentraction of sulfides, possibly during the introduction of fluids into the system, which mobilized metals from the amphibolite. Alternatively, fluids may have been introduced along the ductile contacts between the ultramafic bodies and the amphibolite. The suggestion that mobilization of sulfides has taken place close to the contacts is supported by the presence of epigenetic pyrite-rich veins in the alteration zone between the two units. This type of mineralization associated with serpentinite masses is rather unusual because high base metal contents tend to be associated with mafic extrusive rather that with ultramafic plutonic rocks. An analogy can be drawn, however, to the deposits of the Limassol Forest Plutonic Complex, where lensoid, disseminated, and vein type Cu-Ni-Co-Fe sulfide deposits are sporadically distributed along the peripheral zone in the shattered serpentinite (Panayiotou 1980).

The base metal mineralization associated with the tholeiitic pillow lavas in the Circum Rhodope Belt (Makri and Drimou Melia Series) is considered to be of volcanic-exhalative origin and, consequently, there is probably a potential for Cyprus-type massive sulfides.

The calc-alkaline magmatic activity and mineralization associated with the Tertiary volcano-sedimentary basins are connected with the continental collision and subduction of the African plate under the Eurasian (Innocenti et al. 1984; Fytikas et al. 1984). In the Evros region this activity occurred from the Upper Eocene to Oligocene.

The following genetic concept is proposed for the Tertiary stratiform and vein mineralization (Fig. 7). Early tensional conditions related to an incipient rifting of the basement (Rhodope Massif and Circum Rhodope Belt) caused the creation of intra-continental sedimentary basins, filled by fine to medium-grained quartz, feldspar, and considerable carbonate sediments. Apparently the sediments derived from the weathering of the metamorphic basement rocks also contained Fe sulfides and small amounts of Zn, Pb, and other metallic elements. The latter is supported by geochemical analyses of unmineralized clastic sedimentary rocks, which give metal contents above normal values. The possible early introduction of metals by hydrothermal fluids due to the existence of high geothermal gradients cannot be precluded.

During the later stages of deposition and while the sediments were still saturated with brines, felsic intrusives were injected into the sedimentary sequence. The location of these intrusives was controlled, in part, by major deep-seated fracture zones in mainly E-W and NW-SE directions. The intrusions became strongly carbonated due to reaction with the wet carbonaceous sediments, and base metals such as Zn, Pb, and small amounts of Cu were released into the brines. The intrusives also formed heat centers, causing brine circulation through the sediments and leaching of additional Zn, Pb, and other metallic elements.

The more porous beds were suitable for the migration of solutions and metal deposition took place when temperature decreased, and/or where the physico-chemical environment was suitable, e.g., high porosity and change in carbonate content. The above mechanism does explain the emplacement of the stratiform mineralization, and particularly, its limited lateral extent, the proximity to intensely altered felsic intrusives and the irregular layering of the ore minerals.

After diagenesis of the sediments, other subvolcanic masses of andesite-dacite composition intruded during the late Priabonian (Upper Eocene). Finally, in the Oligocene, granodioritic plutons were emplaced followed by rhyolitic dykes, which intruded along reactivated fracture zones of mainly NW-SE direction. In a very close temporal and spatial relationship, the main phase of epigenetic vein mineralization took place. This acid intrusive and mineralizing phase appears to have affected earlier mineralization in the Amphibolite-Serpentinite Unit and Circum Rhodope

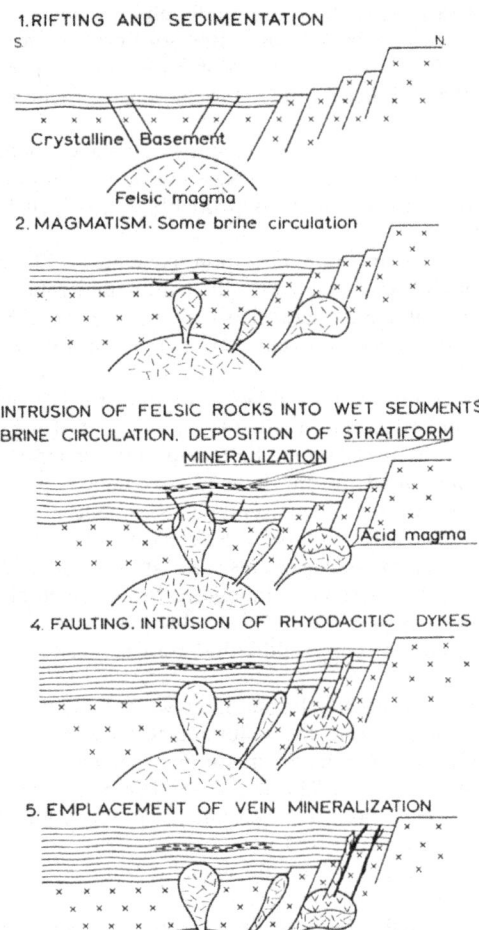

1.RIFTING AND SEDIMENTATION

2. MAGMATISM. Some brine circulation

3.INTRUSION OF FELSIC ROCKS INTO WET SEDIMENTS.
BRINE CIRCULATION. DEPOSITION OF STRATIFORM
MINERALIZATION

4. FAULTING. INTRUSION OF RHYODACITIC DYKES

5. EMPLACEMENT OF VEIN MINERALIZATION

Fig. 7. Conceptual diagrams showing
the history of development of the Tertiary Kirki-Essimi volcano-sedimentary basin and the emplacement of
stratiform and vein mineralization

Belt, e.g. the fracture-filling mineralization at Aberdeen, Baiko, and Pessani prospects is considered to be due to remobilization of the primary stratiform
mineralization.

The above metallogenetic model may be modified or changed when additional
information will be available. For example the role of the calc-alkaline granitoids,
which are the subject of current investigations, has to be evaluated.

Acknowledgments. Financial support by the E.E.C. R. & D. programme (Project No MSM 133 GR) and
IGME is gratefully acknowledged, as well as the permission to published the results.

References

Billett MF, Nesbitt RW (1986) Base metal mineralization associated with mafic and ultramafic rocks, E. Rhodope Massif, Greece. Trans Inst Miner Metall 95:B37–B45

Boncev E (1980) The trans-Balkan strip of post-Lutetian tectono-magmatic and metallogenic mineralization. Geol Balc 10(4):3–22

Constantinides D, Katirtzoglou C, Michael C, Demetriades A, Angelopoulos A, Constantinidou E (1983) Metallogenic map of Evros county. Athens, Inst Geol Miner Exp (int. rep.), pp 136

Fytikas M, Giuliani O, Innocenti F, Manetti P, Mazzuoli R, Peccerillo A, Villari L (1979) Neogene volcanism of the northern and central Aegean region. Ann Geol Pays Hell 30:106–129

Fytikas M, Innocenti F, Manetti P, Mazzuoli R, Peccerillo A, Villari L (1984) Tertiary to Quaternary evolution of volcanism in the Aegean region. In: Dixon JE, Robertson AHF (eds) The Geological Evolution of the Eastern Mediterranean. Geol Soc (Lond) Spec Publ 17:687–699

Innocenti F, Kolios N, Manetti P, Mazzuoli R, Peccerillo A, Rita F, Villari L (1984) Evolution and geodynamic significance of the Tertiary orogenic volcanism in northeastern Greece. Bull Volcanol 47(1):25–37

Jankovic S (1979) Report on the mixed sulphide mineralization: a consultant's report. United Nations Proj GRE/77/007:44

Michael C, Demetriades A, Mastroyiannidou K, Angelopoulos A (1984) Mineral exploration in the Virini-Pessani area, Lefkimi, Evros County. Athens, Inst Geol Miner Exp (int. rep.), 68

Panayiotou A (1980) Cu-Ni-Co-Fe sulphide mineralization, Limassol Forest, Cyprus, In: Panayiotou A (ed) Ophiolites. Proceedings International Ophiolite Symposium Cyprus 1979. Cyprus Geol Surv Dept, pp 102–116

Papadopoulos P (1980) Geological map of Ferrae (1:50.000) Athens, Inst Geol Miner Exp

Papanikolaou DJ, Scarpelis N (1980) Geotraverse southern Rhodope Crete (Preliminary results). In: Sassi FP (ed) IGCP Project No 5. Newsletter 2:41–48

Papanikolaou DJ, Sassi FP, Scarpelis N (1982) Outlines of pre-Alpine metamorphism in Greece. In: Sassi FP (ed) IGCP Project No 5. Newsletter 4:56–62

Pearce JA (1975) Basalt geochemistry used to investigate past tectonic environments on Cyprus. Tectonophysics 25:41–67

Pearce JA, Cann JR (1973) Tectonic setting of basic volcanic rocks determined using trace element analyses. Earth Planet Sci Lett 19:290–300

Pearce JA, Norry MJ (1979) Petrogenetic implications of Ti, Zr, Y, and Nb variations in volcanic rocks. Contrib Miner Pet 69:33–47

Shelton AW, Gass IG (1980) Rotation of the Cyprus microplate. In: Panayiotou A (ed) Ophiolites, Proc Int Ophiolite Symp Cyprus 1979. Cyprus Geol Surv Dept: 61–65

Smith AG (1971) Alpine deformation and the oceanic areas: the Tethys, Mediterranean and Atlantic. Geol Soc Am Bull 82:2039–2070

Zachos S, Dimadis E (1983) The geotectonic position of the Skaloti-Echinos granite and its relationship to the metamorphic formations of Greek Western and Central Rhodope. Geol Balc 13(5):17–24

A Mineralogical, Geochemical and Thermal Profile Through the Agrokipia "B" Hydrothermal Sulfide Deposit, Troodos Ophiolite Complex, Cyprus

P.M. HERZIG[1]

Abstract

The Agrokipia "B" deposit represents a typical example of sulfide mineralization formed by sub-seafloor mixing of ascending hydrothermal fluids of ca. 350°C with downwelling seawater in Cretaceous pillow lavas. Based on borehole data of secondary mineral distribution, four major alteration zones were distinguished: (1) uppermost low-grade zone (0-154 m), (2) silicified and argillized stockwork zone (154-300 m), (3) underlying propylitic transition zone (300-400 m) and (4) lowermost "lower greenschist facies" zone (400-689 m). The stockwork zone consists of pyrite-quartz veins and disseminated pyrite locally associated with sphalerite + chalcopyrite +/− pyrrhotite hosted in a matrix of chlorite + sericite + illite. Epidote only occurs below about 400 m depth in association with chlorite + albite + sphene and fracture-filling quartz + minor sulfides and is abundant within the Sheeted Dike Complex (>530 m). Mg-chlorite is the dominant phyllosilicate in and beneath the stockwork zone.

Sulfide mineralogy displays a complex intergrowth pattern. Various replacement relationships indicate multiple stages of sulfide formation during fluid-seawater mixing, causing local variations in fluid temperature, oxygen fugacity, and metal concentration. While chalcopyrite is quite homogeneous in chemistry, erratically high Co content in pyrite and elevated Cd concentration in sphalerite were found below 600 m depth. Elevated As values are characteristic for pyrite throughout. Sphalerite shows a distinct bimodal downhole distribution in the Fe content reaching 9.7 mol% FeS above and 19.7 mol% FeS below 270 m depth, which apparently is temperature-related.

Bulk rock chemical analyses show that the greenschist facies and propylitic altered basalts are K- and Ca-depleted and Na-enriched relative to fresh lava composition. These trends are explained by albitization of plagioclase feldspar during hydrothermal activity. The sulfide stockwork is enriched in Fe, Zn, Cu, Si, As, Ba, and K, the latter incorporated into sericite and illite. Barium within the stockwork substitutes for K in the sericite-illite host. Kalium and Ba probably originate from both basement leaching and seawater downwelling during the early and final stages of sulfide formation. Calcium, Na, and Sr are distinctly depleted due to almost complete decomposition of primary phases. The uppermost pre-miner-

[1]Institut für Mineralogie und Lagerstättenlehre der RWTH Aachen, Wüllnerstraße 2, D-5100 Aachen, FRG

Base Metal Sulfide Deposits
G.H. Friedrich, P.M. Herzig (Eds.)
© Springer-Verlag Berlin Heidelberg 1988

alization lavas show elevated K, Ca, and low Na concentration as typical for low-grade seafloor weathering to zeolite facies alteration.

Secondary mineral stabilities, stable isotope, fluid inclusion and sulfide "thermometry" data have shown that alteration temperatures within the Sheeted Dikes ranged between 300–400°C. A steep temperature gradient from the bottom (350°C) to the top (150°C) of the mixing "chamber" and high water/rock ratios ($\geq 10:1$) forced pervasive alteration and abundant sulfide precipitation.

Modes of alteration and sulfide formation in the ancient hydrothermal system Agrokipia "B" are similar to processes observed in sub-Recent (DSDP/ODP Hole 504B) and modern (Galapagos Rift) oceanic crust.

Introduction

The Agrokipia "B" stockwork-type sulfide deposit is located in the northern extrusive series of the Troodos ophiolite on Cyprus (Fig. 1) and represents a typical example of sulfide mineralization formed by sub-seafloor hydrothermal activity in ancient oceanic crust. It is supposed that the Agrokipia hydrothermal system operated above a late Cretaceous subduction zone (Rautenschlein et al. 1985).

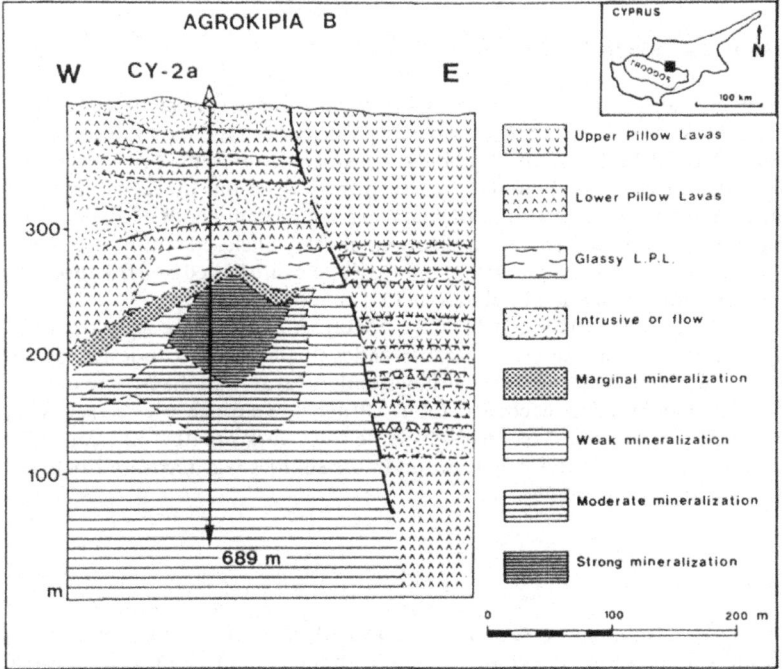

Fig. 1. Location and cross-section of the Agrokipia "B" deposit showing the position of research borehole CY-2a. (After Adamides 1984)

Present-day analogs of this type of deposit are believed to occur below the seafloor at discharge sites of low-temperature fluids which previously cooled during sub-surface contact with seawater, as discovered at the Galapagos Spreading Center (Corliss et al. 1979) and the Mid-Atlantic Ridge (Thompson et al. 1985), and are supposed be found in a more comparable tectonic setting at back-arc spreading centers of the western Pacific margin.

Agrokipia "B" was selected as the drillsite for Hole CY-2a of the Cyprus Crustal Study Project (Robinson et al. 1987), which focused on drilling a complete profile through all major stratigraphic units of the Troodos ophiolite complex. The 689 m "ore deposit hole" first intersected slightly altered uppermost lavas, followed by the stockwork mineralization of Agrokipia "B" and then penetrated about 400 m into the extensive alteration pipe beneath the deposit (Fig. 1). This profile provided the unique opportunity for a systematic study of metasomatic processes and metal mobilization and deposition within the lower oceanic layer 2B and the upper part of the Sheeted Dike Complex (layer 2C). A similar section in young oceanic crust was recently obtained by the DSDP/ODP Hole 504B south of the Costa Rica Rift in the eastern equatorial Pacific (Anderson et al. 1985).

This paper presents data on alteration silicates, sulfide minerals, and bulk rock composition from Hole CY-2a in order to define the mineralogical, chemical, and thermal characteristics of sulfide formation in ancient oceanic crust.

Geological and Lithological Setting

The Agrokipia deposit, near the village of Agrokipia, Cyprus, consists of the "A" and "B" ore bodies, which represent massive and stockwork-type mineralization, re-spectively (Herzig 1986). Adamides (1984) considered Agrokipia "A" an exhala-tive deposit, located at the boundary between the Upper and Lower Pillow Lava unit, in contrast to Agrokipia "B", which occurs at a present depth of about 150 m and represents a replacement mineralization, entirely enclosed within the Lower Pillow Lavas (Fig. 1). It is supposed that both ore bodies are linked by a fracture zone and were therefore generated by the same hydrothermal fluids.

Hole CY-2a is located at 35°02'40"N, 33°08'55"E above the Agrokipia "B" deposit and penetrated a series of basaltic-andesitic (0–90 m) to andesitic-dacitic (90–689 m; Bednarz et al. 1987) sheet flows, pillow lavas, and dikes. The oceanic layer 2B/2C boundary was reached at about 530 m below surface, where a sequence of parallel dikes appeared. Low- to high-grade stockwork ore (0.3–12% Zn + Cu), with abundant fracture, cavity, and groundmass sulfides was recovered between about 150 m and 300 m depth.

Methods

Silicate alteration minerals were identified and analyzed using petrographic, X-ray diffraction, and microprobe techniques. Sulfide minerals were studied in polished thin sections and by electron microprobe. Bulk rock major element compositions were determined by X-ray fluorescence carried out on glass beads of ignited samples fused with lithium tetra- and metaborate (rock/flux ratio 1:10). Trace element analyses were performed on pressed powder pellets. Further details of analytical methods are given elsewhere (Herzig and Friedrich 1987).

Alteration Mineralogy

Based on the alteration mineral distribution occurring in the core, four major alteration zones were distinguished in Hole CY-2a (Fig. 2):

1. an uppermost low-grade alteration zone (0–154 m);
2. a silicified and argillized stockwork zone (154–300 m);
3. a propylitic transition zone (300–400 m) and;
4. a "lower greenschist facies" alteration zone (400–689 m).

Zone 1 (0–154 m)

Smectite-vermiculite, sepiolite, celadonite, zeolites and calcite dominate as alteration products in zone 1. They occur chiefly in veins and void spaces and less frequently replacing the groundmass. Glassy and chilled margins are mostly altered

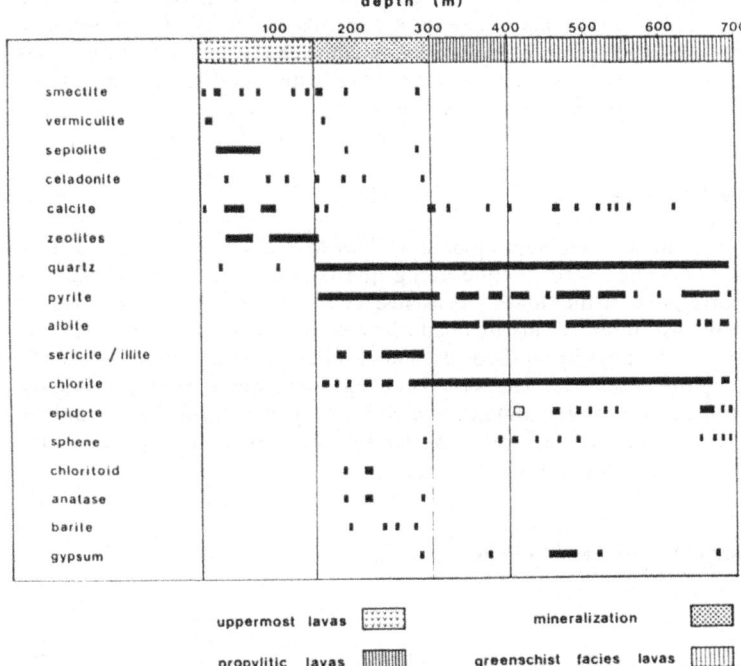

Fig. 2. Alteration mineral distribution and alteration zones in Hole CY-2a

to smectite and celadonite, but fresh glass is locally preserved. Zeolites were not always distinguishable by X-ray diffraction, but it is suggested that heulandite and laumontite are the principal zeolites within the upper 154 m of CY-2a (cf. Sunkel et al. 1987).

Zone 2 (154–300 m)

Zone 2 represents an intensively altered and silicified horizon with abundant sulfide mineralization. Pyrite-quartz veins, locally associated with sphalerite and chal-copyrite, occur in a matrix of chlorite, sericite, and illite. Massive pyrite ore is restricted to brecciated and highly altered glassy intervals, whereas disseminated pyrite dominates within originally massive lavas. In some places, red jasper and hematite-bearing quartz replace sulfides, but pyrite disseminated within red jasper is also observed. Additional but rare alteration minerals are chloritoid, barite, gypsum, and prehnite. Smectite-vermiculite, sepiolite, celadonite, and calcite are still found in zone 2.

Zone 3 (300–400 m)

A propylitic style of alteration is developed in zone 3. Chlorite, quartz, and disseminated pyrite are the main alteration minerals, but the first occurrence of albite (oligoclase-albite composition) at 300 m is the characteristic feature of this transition zone. Vesicles are lined or partly filled with chlorite, whereas quartz, pyrite, minor calcite, and gypsum occur in fissures and veins.

Zone 4 (400–689 m)

In zone 4, the mineral assemblage epidote + chlorite + albite + sphene + calcite is observed, associated with fracture-filling quartz and minor sulfides. This as-semblage corresponds to the "lower greenschist facies" of Elthon (1981). Veins in the deeper part of this alteration zone are lined with chlorite, epidote, quartz, and pyrite, followed by late-stage calcite and gypsum. Vesicle fillings commonly consist of early chlorite lining the vesicles and later epidote overgrowths. Some vesicles are almost completely filled with chlorite and minor epidote, or quartz + chlorite + epidote. Fe-Ti oxides are generally altered to skeletal crystals containing sphene, rutile needles, leucoxene, and sulfides.

Composition of Chlorite and Epidote

Chlorite

Chlorite mainly occurs as groundmass replacement, but also lines or partly fills vesicles and veins. Optically, it is light green and shows brownish to bluish bire-fringence. Selected microprobe analyses are presented in Table 1. Table 2 contains

Table 1. Representative microprobe data of chlorite from drillcore CY-2a

	1	2	3	4	5	6	7	8	9	10	11	12	13	14	15	16	17	18
								Analyses as oxides, wt.%										
SiO_2	27.26	27.48	27.42	27.39	27.86	27.29	27.92	28.73	27.96	27.26	27.52	27.51	28.49	27.96	27.28	27.94	27.66	26.72
Al_2O_3	19.31	19.51	19.27	20.40	18.60	19.90	18.48	17.55	17.90	19.12	19.11	18.60	21.24	20.52	20.84	19.90	18.81	19.77
FeO	28.17	27.75	28.63	21.98	22.24	22.41	27.09	25.94	26.35	21.30	21.02	21.02	25.04	24.35	24.48	24.94	24.14	23.81
MnO	0.82	0.80	0.76	0.67	0.73	0.59	1.10	0.88	0.94	0.73	0.70	0.73	0.67	0.68	0.65	0.79	0.65	0.75
MgO	11.81	11.42	10.73	17.59	18.34	17.46	15.20	16.43	16.02	20.67	20.34	20.08	16.70	17.10	16.97	15.98	17.18	16.13
CaO	0.28	0.20	0.20	0.08	0.04	0.09	0.08	0.09	0.08	0.06	0.05	0.05	0.06	0.04	0.04	0.06	0.03	0.08
Total	87.65	87.16	87.00	88.12	87.81	87.73	89.87	89.62	89.25	89.13	88.75	87.99	92.21	90.64	90.25	89.62	88.46	87.26
								Structural formula based on 28 oxygens										
Si	5.835	5.893	5.919	5.634	5.770	5.655	5.797	5.936	5.826	5.546	5.609	5.660	5.658	5.645	5.544	5.732	5.739	5.622
Al(IV)	2.165	2.107	2.081	2.366	2.230	2.345	2.203	2.064	2.174	2.454	2.391	2.340	2.342	2.355	2.456	2.268	2.261	2.378
Al(VI)	2.707	2.824	2.822	2.537	2.310	2.515	2.320	2.210	2.223	2.131	2.200	2.171	2.630	2.529	2.536	2.544	2.199	2.525
Fe^{2+}	5.043	4.977	5.169	3.872	3.851	3.884	4.704	4.482	4.592	3.624	3.583	3.617	4.159	4.112	4.161	4.279	4.189	4.190
Mn	0.149	0.145	0.139	0.117	0.128	0.104	0.193	0.154	0.166	0.126	0.121	0.127	0.113	0.116	0.112	0.137	0.114	0.134
Mg	3.768	3.650	3.453	5.394	5.661	5.393	4.705	5.060	4.976	6.268	6.180	6.159	4.944	5.147	5.141	4.887	5.313	5.059
Ca	0.064	0.046	0.046	0.018	0.008	0.020	0.018	0.020	0.018	0.013	0.011	0.011	0.013	0.008	0.008	0.013	0.006	0.018
Total	19.729	19.641	19.629	19.891	19.959	19.915	19.940	19.926	19.975	20.161	20.095	20.084	19.856	19.912	19.958	19.861	19.960	19.926

Total Fe expressed as FeO.

Samples 1–3 are from 161.00 m depth; samples 4–6 are from 297.90 m depth; samples 7–9 are from 386.45 m depth; samples 10–12 are from 487.88 m depth; samples 13–15 are from 541.15 m depth; samples 16–18 are from 670.99 m depth.

188   P.M. Herzig

Table 2. Average and range of composition of chlorite from drillcore CY-2a; (wt.%)

Depth (m)	n	SiO₂ mean	min	max	Al₂O₃ mean	min	max	FeO mean	min	max	MnO mean	min	max	MgO mean	min	max	CaO mean	min	max	Total mean	min	max
161.00	7	27.59	25.97	30.10	19.21	18.89	19.51	27.35	25.94	28.63	0.77	0.72	0.82	11.12	10.65	11.81	0.26	0.20	0.31	86.29	84.27	87.65
297.90	6	27.38	26.81	27.86	19.78	18.60	20.40	21.67	20.64	22.41	0.67	0.59	0.73	18.12	17.46	18.88	0.08	0.04	0.10	87.68	87.36	88.12
386.45	6	27.83	26.66	28.73	17.92	17.19	18.48	26.42	25.71	27.18	0.97	0.85	1.10	15.89	15.20	16.43	0.13	0.08	0.32	89.16	88.52	89.87
487.88	6	27.36	26.87	27.82	18.51	17.79	19.12	20.90	20.60	21.30	0.72	0.69	0.74	20.09	19.23	20.67	0.07	0.01	0.22	87.64	85.87	89.13
541.15	7	27.52	26.69	28.49	20.93	20.09	21.36	24.45	23.98	25.04	0.66	0.65	0.69	16.60	15.89	17.10	0.05	0.04	0.07	90.22	88.62	92.12
670.99	9	26.83	24.98	27.94	19.02	18.54	19.90	23.22	22.15	24.94	0.70	0.65	0.79	16.68	15.18	17.21	0.07	0.03	0.10	86.50	84.94	89.62

Total Fe expressed as FeO. Number of analyses = n.

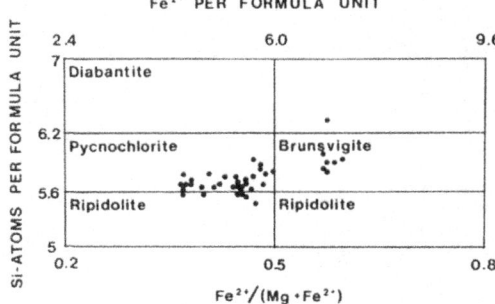

Fig. 3. Classification diagram (after Hey 1954) for chlorite from Hole CY-2a

mean, maximum, and minimum values. Chlorite displays a strong compositional homogeneity and shows only very limited variation in the Mg/(Mg + Fe) ratio with depth. According to the classification used by Riverin (1977) and Exley (1982), all chlorites can be classified as Mg-rich varieties. In the classification diagram after Hey (1954) the chlorites exhibit $Fe^{2+}/(Mg + Fe^{2+})$ ratios of 0.37–0.58 and plot mostly within the compositional field of pycnochlorite with a subordinate trend toward brunsvigite (Fig. 3). The chlorites have a high average MnO content of 0.75 wt.%, reaching a maximum of 1.10 wt.% at 386 m depth.

Epidote

Epidote occurs mainly as fan-shaped and hexagonal, prismatic crystals up to 100 μm across in vesicles, where it is often associated with chlorite. Greenish-yellow to bluish-pink interference colors are characteristic. Representative analyses are listed in Table 3. Table 4 contains mean, maximum, and minimum values. Epidote varies in Fe_2O_3 and Al_2O_3 contents between 12.04 and 19.14 wt.% (mean 15.19 wt.%) and 19.25–26.44 wt.% (mean 22.82 wt.%), respectively. High MnO content (av. 0.36 wt.%, max. 0.77 wt.%) is characteristic of epidote as for chlorite. The Fe/Al ratio, expressed as $[Fe^{3+}/(Fe^{3+} + Al)]$. 100 after Holdaway (1972), within the Sheeted Dike Complex (i.e., below 530 m) considerably decreases from 36 at 555 m to 23 at 681 m (Fig. 4). Data between 488 m and 516 m indicate an inconsistent variation in the Fe/Al ratio, which may be attributed to the change in lithology from a more permeable sequence of flows, pillows, and dikes to a less permeable Sheeted Dike series. However, it is possible that some of this variation in epidote composition may be due to zonation within individual crystals, as described by Viereck et al. (1982).

Ore Mineral Paragenesis

The ore mineral paragenesis of Hole CY-2a (Fig. 5) consists predominantly of pyrite, accompanied by varying amounts of sphalerite, less chalcopyrite, and minor but widespread pyrrhotite. Marcasite and galena are accessory minerals. No abrupt changes in the ore mineral paragenesis with depth were observed.

Table 3. Representative microprobe data of epidote from drillcore CY-2a

	1	2	3	4	5	6	7	8	9	10	11	12	13	14	15	16	17	18	19	20	21	22	23	24	25	26	27
Analyses as oxydes, wt.%																											
SiO_2	38.77	38.39	37.68	37.00	36.76	38.39	38.16	37.31	38.61	37.96	37.28	38.07	36.92	36.09	38.15	37.45	36.68	39.06	38.16	38.59	38.88	38.81	38.96	38.55	38.28	38.26	36.94
Al_2O_3	22.55	22.24	21.62	22.40	21.28	21.52	22.34	22.81	24.14	23.72	23.79	20.38	20.55	19.26	22.22	19.86	21.70	25.60	25.32	25.71	25.72	23.81	27.03	25.76	26.64	25.30	24.83
Fe_2O_3	15.76	15.94	17.03	15.74	17.41	17.04	16.10	15.58	13.82	14.50	15.14	17.47	16.87	18.55	15.20	17.58	15.59	13.19	13.38	12.34	13.12	14.32	10.08	12.27	11.27	12.69	13.54
MnO	0.34	0.35	0.24	0.40	0.18	0.43	0.72	0.33	0.31	0.40	0.16	0.33	0.28	0.14	0.34	0.26	0.36	0.37	0.77	0.46	0.51	0.54	0.49	0.61	0.63	0.49	0.11
MgO	1.00	0.41	0.21	0.32	0.25	0.05	0.11	0.06	0.09	0.10	0.08	0.19	0.22	0.14	0.10	0.11	0.14	0.09	0.10	0.14	0.13	0.14	0.02	0.22	0.02	0.14	0.02
CaO	21.39	21.89	22.55	22.33	22.46	22.46	22.09	22.71	22.90	22.67	23.11	22.71	22.81	22.98	22.32	22.16	22.09	22.62	23.17	23.34	22.86	22.92	22.61	22.23	22.26	22.18	22.79
Total	99.81	99.23	99.32	98.20	98.33	99.89	99.52	98.80	99.87	99.35	99.56	97.40	95.97	95.31	98.33	97.42	96.56	100.94	100.90	100.59	101.22	100.53	99.19	99.64	99.10	99.05	98.22
Structural formula based on 25 Oxygens																											
Si	6.062	6.057	5.981	5.925	5.914	6.053	6.040	5.932	6.033	5.968	5.872	6.070	5.985	5.934	6.071	6.080	5.974	5.992	5.895	5.949	5.957	6.022	6.021	5.980	5.952	5.979	5.859
Al	4.156	4.136	4.045	4.228	4.035	3.999	4.143	4.274	4.445	4.396	4.416	3.830	3.926	3.732	4.168	3.800	4.165	4.629	4.611	4.672	4.645	4.355	4.924	4.710	4.882	4.661	4.642
Fe^{2+}	1.854	1.893	2.035	1.897	2.108	2.022	1.906	1.864	1.625	1.716	1.794	2.096	2.058	2.295	1.820	2.148	1.911	1.523	1.556	1.432	1.513	1.672	1.172	1.432	1.319	1.493	1.616
Mn	0.045	0.047	0.032	0.054	0.025	0.057	0.096	0.044	0.041	0.053	0.021	0.045	0.038	0.020	0.046	0.036	0.050	0.048	0.101	0.060	0.066	0.071	0.064	0.080	0.083	0.065	0.015
Mg	0.233	0.096	0.050	0.076	0.060	0.012	0.026	0.014	0.021	0.023	0.019	0.045	0.054	0.025	0.024	0.027	0.034	0.021	0.023	0.032	0.030	0.032	0.004	0.051	0.004	0.033	0.004
Ca	3.583	3.701	3.836	3.832	3.872	3.794	3.724	3.869	3.767	3.819	3.900	3.880	3.961	4.048	3.806	3.855	3.855	3.718	3.836	3.855	3.753	3.811	3.744	3.695	3.708	3.714	3.873
Total	15.933	15.928	15.978	16.012	16.013	15.937	15.935	15.998	15.932	15.975	16.022	15.967	16.022	16.053	15.935	15.946	15.989	15.931	16.020	15.999	15.963	15.963	15.929	15.948	15.947	15.943	16.010

Total Fe expressed as Fe_2O_3 (recalculated from FeO).
Samples 1–5 are from 487.88 m depth; samples 6–8 are from 508.09 m depth; samples 9–11 are from 516.09 m depth; samples 12–14 are from 554.54 m depth; samples 15–17 are from 631.20 m depth; samples 18–22 are from 666.49 m depth; samples 23–27 are from 680.60 m depth.

Table 4. Average and range of composition of epidote from drillcore CY-2a; (wt.%)

Depth (m)	n	SiO$_2$ mean	min max	Al$_2$O$_3$ mean	min max	Fe$_2$O$_3$ mean	min max	MnO mean	min max	MgO mean	min max	CaO mean	min max	Total mean	min max
487.88	6	37.53	36.57 38.77	21.93	21.28 22.55	16.43	15.74 17.41	0.28	0.18 0.40	0.40	0.19 1.00	22.21	21.39 22.64	98.78	97.75 99.81
508.09	12	37.99	37.31 38.40	22.39	21.52 22.97	16.09	15.42 17.04	0.43	0.33 0.72	0.06	0.02 0.11	22.50	22.09 22.71	99.45	98.80 100.08
516.09	17	37.95	37.28 38.61	23.77	23.31 24.21	14.42	13.61 15.14	0.31	0.14 0.47	0.10	0.07 0.14	22.80	22.47 23.22	99.34	98.61 100.08
554.54	8	37.03	36.09 38.07	19.86	19.25 20.55	17.75	16.08 19.14	0.20	0.14 0.33	0.23	0.13 0.72	22.70	21.81 23.07	95.99	94.59 97.43
631.20	17	37.43	36.68 38.15	21.66	19.86 22.87	15.41	13.64 17.82	0.35	0.26 0.48	0.12	0.05 0.16	22.10	21.69 22.57	97.07	95.64 98.33
666.49	15	38.70	37.69 39.67	24.46	22.55 25.72	13.72	12.34 15.74	0.45	0.17 0.77	0.13	0.09 0.18	22.89	22.05 23.46	100.35	98.21 101.22
680.60	5	37.40	36.36 38.55	25.70	24.83 26.44	12.53	12.04 13.54	0.47	0.11 0.69	0.10	0.02 0.22	22.25	21.87 22.79	98.44	97.54 99.64

Total Fe expressed as Fe$_2$O$_3$. Number of analyses = n.

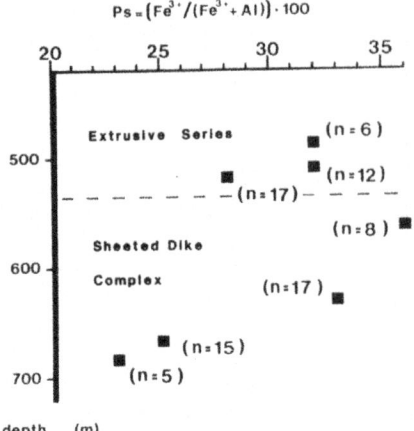

$$Ps = (Fe^{3+}/(Fe^{3+} + Al)) \cdot 100$$

Fig. 4. Downhole variation in the Fe/Al ratio of epidote in the lithological transition zone and the Sheeted Dike Complex (Hole CY-2a)

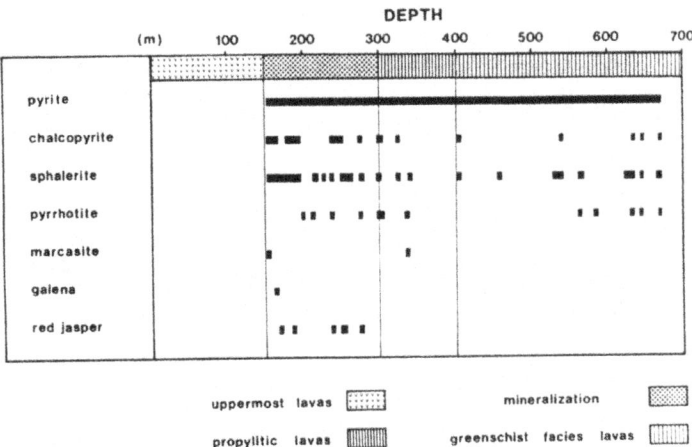

Fig. 5. Distribution of ore minerals and red jasper with depth in Hole CY-2a

Pyrite is concentrated in the stockwork zone between 154 m and 300 m, but occurs continuously down to 689 m either in veins and vesicles or disseminated in the groundmass. Colloform pyrite occurs within large grains of massive pyrite (Fig. 6a) and subconcentric shrinkage cracks are commonly filled with sphalerite and silicate minerals. Pyrite is often replaced by quartz and red jasper, but euhedral quartz crystals in pyrite and pyrite disseminated in red jasper are also found. Both chalcopyrite (Fig. 6b) and sphalerite (Fig. 6c) apparently marginally replace pyrite.

Sphalerite and *chalcopyrite* are also mainly restricted to the interval between 154 m and 300 m. In the alteration zone beneath they are chiefly found as minute

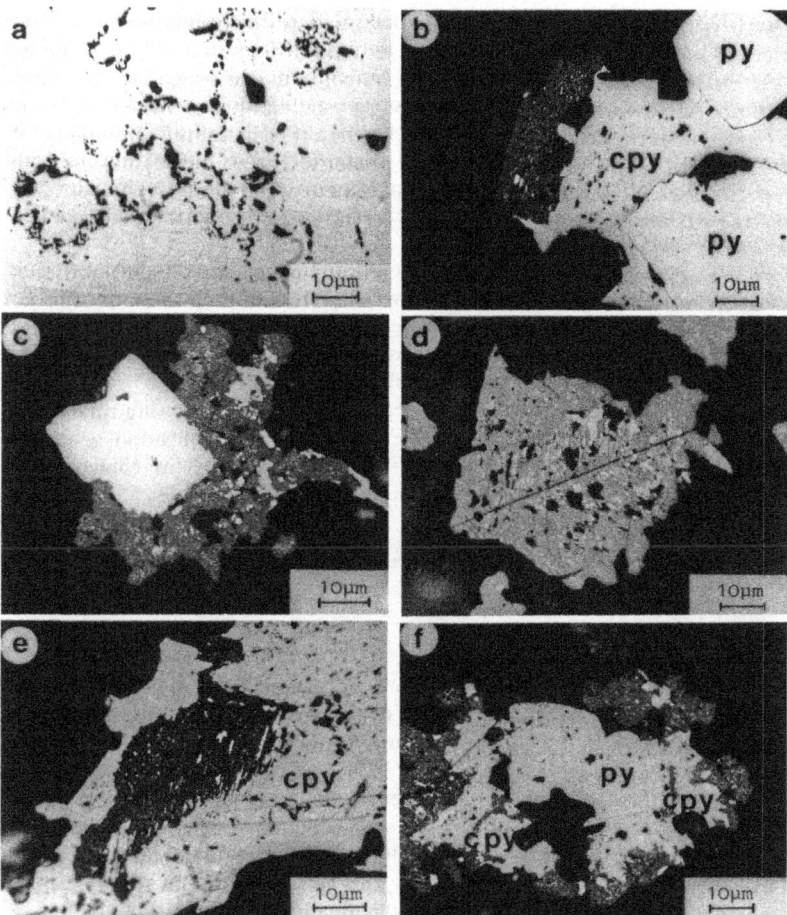

Fig. 6a-f. Photomicrographs of sulfides from Hole CY-2a (PPL, oil immersion). **a** 192.95 m: Colloform gel-pyrite showing shrinkage cracks, embedded in massive pyrite. **b** 166.42 m: Subhedral pyrite (*py*) partly replaced by chalcopyrite (*cpy*) which is intergrown with sphalerite (*dark grey*) within quartz gangue. Sphalerite carries abundant blebs and spindles of chalcopyrite. **c** 166.42 m: Sphalerite (*dark grey*) replacing euhedral pyrite. Sphalerite carries abundant chalcopyrite inclusions. **d** 166.42 m: Sphalerite twin showing a herring-bone texture of chalcopyrite laths and blebs. **e** 166.42 m: Comb-like intergrown texture of anhedral sphalerite (*dark grey*) and chalcopyrite (*cpy*). Sphalerite carries orientated chalcopyrite blebs and laths. **f** 166.42 m: Anhedral pyrite (*py*), partly replaced by chalcopyrite (*cpy*) and sphalerite (*dark grey*). Chalcopyrite in turn is replaced by sphalerite carrying chalcopyrite inclusions

inclusions in pyrite. Small (<1 μm), star-shaped sphalerite inclusions in chalcopyrite are observed between 298 m and 300 m. Between 161 m and 198 m, complex comb-like intergrowths of sphalerite and chalcopyrite are suggested to reflect replacement of chalcopyrite by sphalerite along crystallographic planes (Fig. 6e), as well as along fractures and grain margins. On the other hand, the abundance of chalcopyrite blebs, laths and dust (≪ 1 μm) in sphalerite (Fig. 6c and 6d) may indicate "chalcopyrite disease" (Barton 1978), i.e., replacement of sphalerite by chalcopyrite. Chalcopyrite is commonly replaced by quartz, as indicated by quartz-filled fractures in chalcopyrite and marginal corrosion.

Pyrrhotite occurs exclusively as discrete minute inclusion (<30 μm) in pyrite, particularly from 198 m to 334 m and below 550 m. In addition, small pyrrhotite-chalcopyrite inclusions were found in pyrite below 580 m depth.

Traces of galena appear in only two polished sections (161 m and 166 m), where galena replaces pyrite and is replaced by sphalerite.

Sulfide veins associated with red jasper and/or hematite-bearing quartz are common between 165 m and 297 m. The intensity of mineralization generally decreases with depth, although pyrite with minor inclusions of chalcopyrite, sphalerite, and pyrrhotite is present down to the bottom of the hole.

Sulfide Chemistry

Pyrite

Pyrite shows a variation of 32.1–37.3 atomic % Fe (mean 33.9%, std. dev. 0.87) and corresponds to the average formula $Fe_{1.03}S_2$ (Table 5). Cobalt concentration ranges up to 0.30 wt.%, whereas Ni is generally around the detection limit of 100 ppm. The Co/Ni ratio varies between 2.0 and 8.0 (av. 0.05 wt.% Co, 0.01 wt.% Ni), which is typical for pyrite formed under volcanogenic conditions (Bralia et al. 1979). A significant concentration of As up to 1.87 wt.% is characteristic for pyrites from CY-2a. Gold values of 0.12–0.33 wt.% were detected in individual pyrite grains from 198 m to 279 m and at 457 m depth which may be related fo fine Au-inclusions. Selenium does not reach concentrations above the detection limit of 400 ppm.

The average depth related concentrations of Ni, Co, and As in pyrite of CY-2a is shown in Fig. 7 (cf. Table 5). Arsenic varies over a narrow range of 0.64 to 1.20 wt.%, with a mean of 0.87 wt.% (std. dev. 0.20). Cobalt exhibits a slight increase in concentration down hole, which may be attributed to increasing formation temperature (Loftus-Hill and Solomon 1967). Nickel shows erratically elevated values up to 700 ppm in the most intensively mineralized interval between 150 m and 300 m.

Sphalerite

Copper is found in several sphalerite samples in concentrations up to 9.5 wt.% or 7.2 atomic %, respectively (Table 6), whereas others carry only minor Cu or are completely free of Cu. Copper and Fe in the cupriferous sphalerite show generally positive correlation, indicating that elevated Cu concentration results from minute

Table 5. Average microprobe analyses of pyrite from drillcore CY-2a

Depth (m)	n	Analyses, wt.%						Atomic %				
		Fe	S	As	Co	Ni	Total	Fe	S	As	Co	Ni
153.60	25	47.82	50.84	0.76	0.04	0.01	99.47	34.91	64.65	0.41	0.03	0.01
161.00	11	47.62	50.98	0.74	0.05	0.01	99.40	34.75	64.81	0.40	0.03	0.01
171.05	10	46.37	53.22	0.95	0.05	0.02	100.61	33.16	66.29	0.51	0.03	0.01
185.70	32	46.63	52.89	0.68	0.04	0.01	100.35	33.47	66.13	0.36	0.03	0.01
197.75	17	47.60	50.50	1.05	0.06	0.01	99.22	34.89	64.48	0.57	0.04	0.01
223.10	12	46.75	53.16	0.66	0.04	0.01	100.62	33.42	66.20	0.35	0.03	0.01
255.10	10	47.52	51.08	1.20	0.05	0.01	99.86	34.57	64.74	0.65	0.03	0.01
278.50	13	45.04	53.48	0.92	0.05	0.02	99.51	32.41	67.04	0.49	0.03	0.01
297.90	13	46.69	52.80	0.67	0.06	0.03	100.25	33.53	66.05	0.36	0.04	0.02
323.20	17	46.88	53.89	0.69	0.06	0.01	101.53	33.17	66.42	0.36	0.04	0.01
401.40	9	47.52	49.62	1.16	0.05	0.01	98.36	35.23	64.08	0.64	0.04	0.01
456.95	18	47.28	52.33	0.99	0.07	0.01	100.68	33.95	65.46	0.53	0.05	0.01
462.05	10	46.75	53.30	0.90	0.04	0.01	101.00	33.32	66.17	0.48	0.03	0.01
541.15	12	47.33	51.27	1.15	0.05	0.01	99.81	34.41	64.93	0.62	0.03	0.01
561.03	18	46.83	53.47	0.97	0.06	0.01	101.34	33.27	66.17	0.51	0.04	0.01
631.20	16	47.26	51.70	0.64	0.08	0.01	99.69	34.28	65.32	0.35	0.05	0.01
666.49	12	46.39	52.95	0.64	0.04	0.01	100.03	33.34	66.29	0.34	0.03	0.01
min		44.23	48.20	0.56	0.02	0.00	98.36	32.14	62.27	0.30	0.01	0.00
max	255	50.22	54.55	1.87	0.30	0.07	101.53	37.25	67.31	1.02	0.20	0.05
mean		46.96	52.22	0.84	0.05	0.01	100.10	33.89	65.61	0.45	0.04	0.01

Number of Analyses = n.

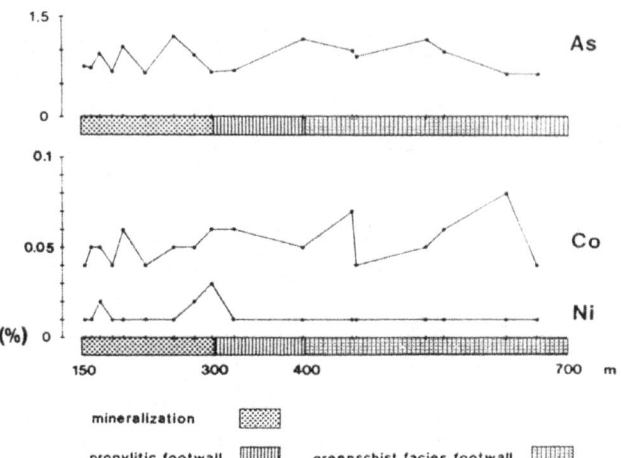

Fig. 7. Downhole distribution of As, Co, and Ni in pyrite from Hole CY-2a (average values)

chalcopyrite inclusions. A general feature of sphalerite from CY-2a is a high Fe content varying between 2.3–16.0 wt.% (av. 7.6 wt.%) or 4.0–28.9 hol % FeS (av. 14.0 mol%), respectively (Fig. 8). Throughout the core, the Fe content of sphalerite exhibits a bimodal distribution with an average of about 4.5 atomic % Fe (9.7 mol % FeS) above and 10.0 atomic % Fe (19.7 mol % FeS) below 270 m (Fig. 9). High Fe content in sphalerite is generally not associated with high Cu concentration. Fe-rich sphalerite (>10.0 wt.% Fe) typically shows extremely low Cu values of about 0.02 wt.% on average. In some cases, elevated S values probably indicate submicroscopic pyrite inclusions, but the majority of sphalerite reveals normal S, low to zero Cu and significant elevated Fe content. The average formula corresponds to $Zn_{0.85}Fe_{0.13}S$. Samples with chalcopyrite inclusions (positive Cu-Fe correlation) also show a negative correlation of Zn and Fe, which suggests that Fe also substitutes for Zn in sphalerite.

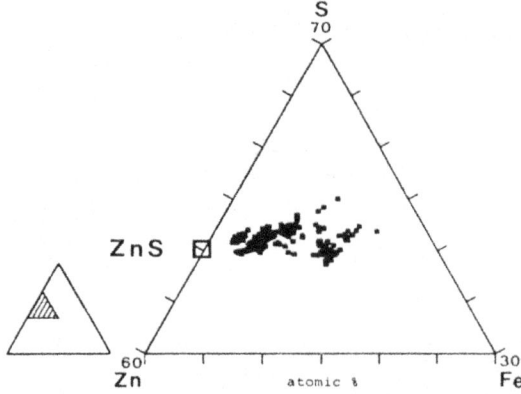

Fig. 8. Zn-Fe-S ternary diagram showing the variation in the Fe-content of sphalerite

Table 6. Average microprobe data of sphalerite from drillcore CY-2a

Depth (m)	n	Analyses, wt.%							FeS mol %	Atomic %					
		Zn	Fe	S	Cd	Mn	Cu	Total		Zn	Fe	S	Cd	Mn	Cu
153.60	13	61.98	4.36	33.93	0.11	0.08	0.01	100.47	7.60	45.43	3.74	50.71	0.05	0.07	0.01
161.00	33	57.15	6.49	33.44	0.08	0.09	2.79	100.05	11.72	42.04	5.59	50.15	0.03	0.08	2.11
166.42	13	58.32	4.30	33.25	0.13	0.04	3.55	99.60	7.94	43.22	3.73	50.25	0.06	0.04	2.71
177.30	12	58.02	6.22	34.91	0.11	0.09	0.31	99.65	11.14	42.36	5.32	51.97	0.05	0.08	0.23
192.95	12	61.51	4.13	32.62	0.17	0.00	0.17	98.60	7.28	46.20	3.63	49.96	0.07	0.00	0.13
197.75	9	58.15	5.72	33.67	0.08	0.11	0.99	98.71	10.32	43.17	4.97	50.97	0.03	0.10	0.76
249.88	13	57.25	8.75	34.04	0.10	0.08	0.02	100.24	15.16	41.77	7.47	50.64	0.04	0.07	0.02
251.60	10	62.12	3.31	34.20	0.10	0.15	0.28	100.18	5.87	45.59	2.84	51.18	0.04	0.13	0.21
272.30	10	54.19	11.96	33.65	0.07	0.55	0.02	100.43	20.52	39.41	10.18	49.89	0.03	0.48	0.01
323.20	10	52.77	12.42	32.86	0.06	0.98	0.01	99.11	21.59	38.94	10.73	49.44	0.03	0.86	0.01
405.66	6	53.33	12.61	32.32	0.08	0.36	0.00	98.70	21.66	39.66	10.98	49.01	0.03	0.32	0.00
526.45	7	53.65	11.87	32.86	0.11	0.26	0.05	98.81	20.56	39.75	10.29	49.64	0.05	0.23	0.04
530.93	4	54.31	10.63	35.65	0.05	0.10	0.02	100.75	18.63	38.90	8.91	52.07	0.02	0.09	0.01
561.03	9	59.57	6.23	33.44	0.17	0.12	0.02	99.55	10.90	44.03	5.39	50.39	0.07	0.11	0.02
631.20	6	49.15	13.69	34.47	0.11	0.46	2.22	100.12	24.57	35.52	11.58	50.80	0.05	0.40	1.65
643.89	6	51.34	13.66	34.69	0.18	0.04	0.02	99.93	23.73	37.14	11.57	51.17	0.08	0.03	0.01
666.49	7	55.69	6.21	34.24	0.50	0.05	3.59	100.27	11.54	40.70	5.31	51.03	0.21	0.04	2.70
min		46.20	2.25	31.51	0.00	0.00	0.00	98.60	4.04	33.44	1.96	48.52	0.00	0.00	0.00
max	180	63.19	15.95	36.06	0.59	2.97	9.54	100.75	28.87	47.19	13.51	52.41	0.25	2.61	7.17
mean		57.12	7.55	33.68	0.12	0.19	1.09	99.72	14.02	42.01	6.48	50.47	0.05	0.16	0.83

Number of analyses = n.

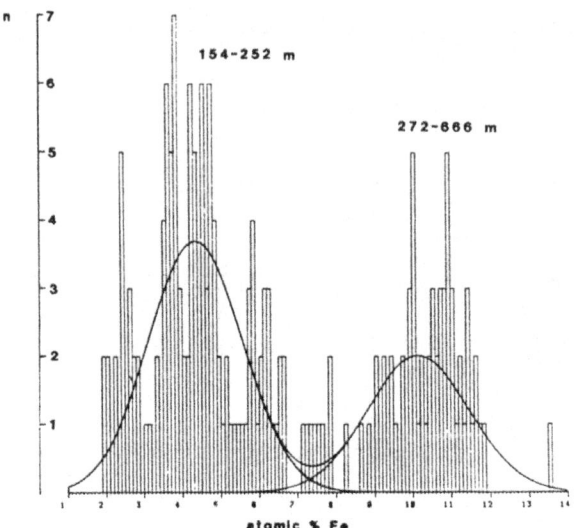

Fig. 9. Bar diagram showing the bimodal distribution of Fe in sphalerite in CY-2a (n = number of samples)

The distribution of average Cu, Mn, Cd, and Fe concentrations in sphalerite with depth in CY-2a is shown in Fi. 10. Significant amounts of Cu are restricted to the highest and lowest levels in the core. The Mn concentration in sphalerite from below 250 m is generally higher than in those from 154 m to 250 m. Up to 3.0 wt.% Mn is found in sphalerite at 323 m. The average Cd content in sphalerite increases with depth, reaching a maximum of 0.50 wt.% at 666 m.

Chalcopyrite

Chalcopyrite is generally quite homogeneous in composition (Table 7), with low Co and Ni concentrations (both averaging 0.02 wt.%) and only a slight variation in chemistry with depth. However, elevated As values occur in the uppermost zone of the stockwork (Fig. 11). Gold was not detected in concentrations above 0.10 wt.%.

Samples from 197.75 m depth, are characterized by a low average Cu value of 17.95 atomic % and a slight enrichment in Fe and S relative to all other samples. These phases may represent iss-related exsolution products with a maximum substitution of Cu by Fe of about 2.5%. The average formula calculated for all analyzed chalcopyrites except those from 197.75 m is $Cu_{0.96}Fe_{1.01}S_2$.

Pyrrhotite

Copper concentration in pyrrhotite locally reaches up to 1.5 wt.%, probably due to intimate association with chalcopyrite. In addition, Co up to 0.07 wt.%, Ni up to 0.04 wt.% and a maximum As concentration of 0.31 wt.% are detected. The Fe content of

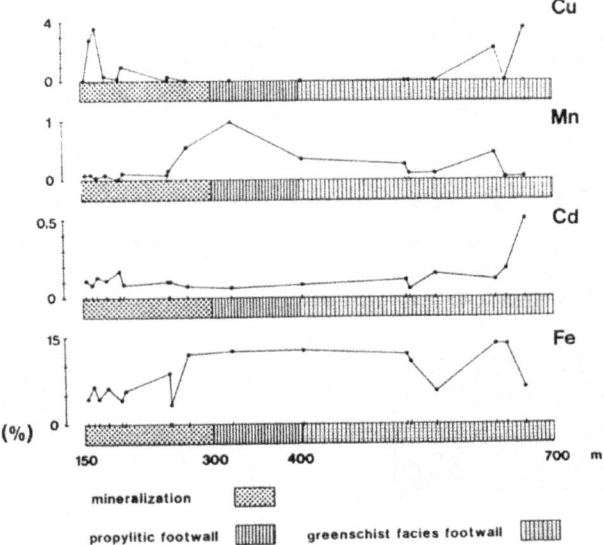

Fig. 10. Downhole distribution of Cu, Mn, Cd, and Fe in sphalerite from Hole CY-2a (average values)

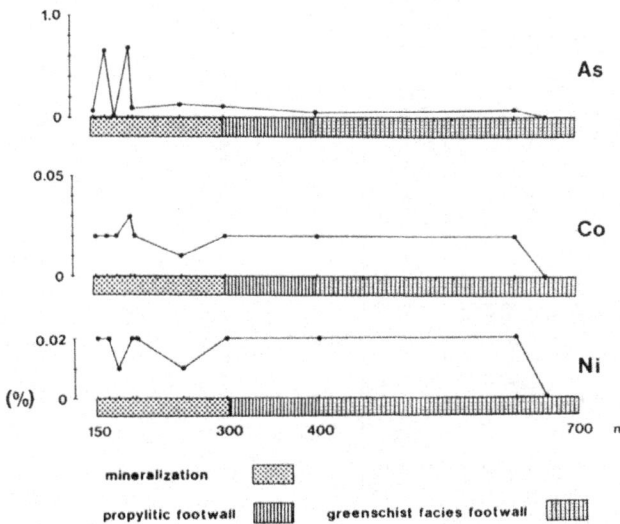

Fig. 11. Downhole distribution of As, Co, and Ni in chalcopyrite from Hole CY-2a (average values)

Table 7. Average microprobe data of chalcopyrite from drillcore CY-2a

Depth (m)	n	Fe	Cu	S	Co	Ni	As	Total	Fe	Cu	S	Co	Ni	As
		Analyses, wt.%							Atomic %					
153.60	12	30.46	33.46	34.63	0.02	0.02	0.07	98.66	25.32	24.45	50.15	0.02	0.02	0.04
166.42	13	30.22	33.40	34.47	0.02	0.02	0.66	98.83	25.15	24.43	49.98	0.02	0.02	0.41
177.30	3	30.67	32.53	35.53	0.02	0.01	0.00	98.75	25.31	23.59	51.07	0.02	0.01	0.00
192.95	3	31.38	31.76	34.12	0.03	0.02	0.69	98.00	26.31	23.40	49.82	0.02	0.02	0.43
197.75	7	35.19	25.38	38.28	0.02	0.02	0.09	98.98	28.31	17.95	53.65	0.02	0.02	0.05
251.60	6	30.72	33.95	35.31	0.01	0.01	0.13	100.11	25.14	24.42	50.34	0.01	0.01	0.08
301.18	10	31.29	33.05	34.81	0.02	0.02	0.11	99.31	25.84	23.99	50.07	0.02	0.02	0.07
405.66	6	31.65	33.71	35.18	0.02	0.02	0.05	100.61	25.81	24.16	49.97	0.02	0.02	0.03
631.20	9	30.59	33.71	35.02	0.02	0.02	0.08	99.44	25.23	24.42	50.28	0.02	0.02	0.05
666.49	3	30.88	32.13	36.00	0.00	0.00	0.00	99.27	25.35	23.18	51.47	0.00	0.00	0.00
min		29.22	19.65	33.63	0.00	0.00	0.00	98.00	24.48	13.47	49.34	0.00	0.00	0.00
max	72	37.28	34.57	42.42	0.05	0.06	0.76	100.61	30.37	25.47	57.64	0.04	0.05	0.48
mean		31.19	32.53	35.21	0.02	0.02	0.21	99.20	25.71	23.59	50.54	0.01	0.01	0.13

Number of analyses = n.

pyrrhotite from 234 m and 272 m averages 46.95 and 46.34 atomic %, respectively, whereas pyrrhotite from 584 m, 631 m, and 644 m shows average Fe concentrations of 47.27, 47.12 and 47.16 atomic %.

Whole Rock Geochemistry

Major oxides and trace elements in Hole CY-2a have been determined for 330 samples between 3.60 m and 688.88 m depth. Each analysis listed in Table 8 corresponds to the average of 5–16 samples analysed per 20 m unit. In Tables 9 and 10 the concentration of selected major and trace elements are summarized for the alteration zones as defined by the secondary mineral distribution.

Data Normalization

Assuming that Al is relatively immobile during hydrothermal alteration of oceanic basalts (Humphris and Thompson 1978; Honnorez 1981), the major elements have been normalized with respect to the Al content of almost unaltered Lower Pillow Lavas (Desmet et al. 1979) according to the method described by Krauskopf (1979). The downhole gains and losses of the normalized elements compared to the "standard values" are shown in Fig. 12. Data normalization using Ti as the stable element (Pearce and Cann 1973; Finlow-Bates and Stumpfl 1981) reveals a similar distribution curve to that obtained using Al, but individual values are generally slightly lower, i.e., plot below the Al normalized graph. Normalization based on Ti from glass analyses of the horizon 113.20 m to 141.60 m in CY-2a (Bednarz et al. 1987) gives a downhole element distribution, which, except for the uppermost 90 m, is almost identical to that shown in Fig. 12. The values above 90 m are generally higher than those in Fig. 12 because the "glass standard" reflects only the primary composition of rocks below 90 m depth.

Major Elements

Pronounced K and Ca depletions and Na enrichment are typical characteristics of alteration within the propylitic and greenschist facies footwall rocks in CY-2a. In contrast, the mineralized and highly altered zone between 154 m and 300 m is strongly depleted in Ca and Na. Some samples are almost entirely free of Ca and Na, containing only 0.01 wt.% CaO and 0.05 wt.% Na_2O (Table 9, non-normalized analytical data). An outstanding feature of the sulfide zone is a distinct K anomaly with maximum K_2O values reaching 2.73 wt.%. No distinct gains or losses are detected for Mg, either in the alteration pipe or in the sulfide zone. Silica and Fe are enriched within the feeder channel and show highest concentrations within the mineralized section. Highest SiO_2 values occur just below the sulfide stockwork and at the top of the main ore zone. The hanging wall lavas are characterized by elevated K, Ca, and low Na concentrations. The higher Mg content within the upper 90 m is caused by the predominance of olivine-bearing basaltic lavas which are altered to smectite.

Table 8. Major and trace element content of drillcore CY-2a in form of 20-m interval averages

Sample	Depth	SiO_2	TiO_2	Al_2O_3	Fe_2O_3	MnO	MgO	CaO	Na_2O	K_2O	P_2O_5	Cr_2O_3	V_2O_5	LOI	Total
1	10.00	52.53	0.55	15.54	9.21	0.11	7.86	7.56	1.94	1.25	0.05	0.02	0.04	2.85	99.52
2	30.00	50.25	0.69	13.79	9.73	0.15	8.61	8.63	1.98	1.35	0.06	0.04	0.04	4.41	99.73
3	50.00	50.48	1.06	15.96	11.03	0.12	6.59	5.49	2.37	1.92	0.10	0.01	0.06	4.77	99.95
4	70.00	47.43	0.59	14.32	8.01	0.12	7.02	10.92	2.01	1.14	0.06	0.03	0.04	5.42	97.08
5	90.00	50.52	0.74	15.28	8.82	0.13	7.05	8.74	1.93	1.40	0.06	0.02	0.04	3.87	98.61
6	110.00	50.84	1.18	16.32	10.66	0.19	6.02	4.98	2.62	1.70	0.13	0.01	0.05	3.27	97.97
7	130.00	55.16	1.20	14.84	10.78	0.27	4.33	4.65	2.68	0.75	0.08	0.01	0.03	4.25	99.04
8	150.00	58.06	0.99	11.73	10.77	0.17	4.29	4.42	2.43	0.95	0.07	0.03	0.05	3.52	97.49
9	170.00	53.22	1.10	12.53	15.47	0.23	6.29	0.56	0.80	1.22	0.10	0.04	0.03	7.62	99.20
10	190.00	33.29	0.76	8.65	35.17	0.09	2.89	0.60	0.39	1.25	0.07	0.02	0.03	18.19	101.38
11	210.00	49.12	0.89	9.45	24.38	0.07	2.96	0.16	0.44	1.38	0.06	0.01	0.04	12.51	101.47
12	230.00	38.77	0.75	7.98	32.70	0.10	3.26	0.15	0.46	0.97	0.06	0.01	0.03	16.49	101.73
13	250.00	48.03	0.68	7.84	26.13	0.12	3.28	0.19	0.48	0.89	0.07	0.03	0.03	13.41	101.17
14	270.00	49.78	0.65	7.49	25.45	0.11	2.96	0.20	0.44	0.91	0.07	0.01	0.03	13.21	101.32
15	290.00	62.89	0.74	10.73	12.00	0.30	4.97	1.86	0.79	0.71	0.07	0.01	0.03	6.28	101.38
16	310.00	58.44	0.94	13.72	13.05	0.48	5.07	1.65	1.62	0.87	0.09	0.01	0.04	5.14	101.10
17	330.00	57.24	1.10	14.68	10.25	0.28	5.25	2.27	2.55	0.73	0.08	0.01	0.05	5.02	99.51
18	350.00	55.79	1.20	14.58	12.30	0.30	5.35	1.88	3.46	0.56	0.09	0.01	0.05	5.10	100.66
19	370.00	52.34	1.21	15.14	12.43	0.22	6.50	3.83	3.29	0.33	0.07	0.01	0.07	5.22	100.67
20	390.00	58.24	1.03	14.59	11.20	0.36	5.05	1.64	4.48	0.35	0.09	0.01	0.03	3.87	100.94
21	410.00	56.31	1.22	14.53	12.74	0.54	5.82	1.16	3.65	0.22	0.10	0.01	0.05	4.77	101.10
22	430.00	54.99	1.16	13.52	12.14	0.67	5.61	2.01	3.44	0.11	0.09	0.01	0.05	4.76	98.53
23	450.00	60.32	1.15	13.00	11.13	0.25	4.02	1.91	4.61	0.10	0.10	0.01	0.03	4.43	101.05
24	470.00	53.45	1.03	12.87	14.42	0.28	4.55	3.11	3.86	0.13	0.08	0.01	0.05	5.77	99.62
25	490.00	48.81	1.10	12.81	17.66	0.28	5.56	3.24	3.23	0.12	0.08	0.01	0.05	6.95	99.89
26	510.00	58.51	1.10	12.53	13.26	0.25	4.35	2.29	3.48	0.17	0.08	0.01	0.04	4.83	100.91
27	530.00	49.00	1.15	13.13	20.26	0.26	4.84	1.58	3.06	0.16	0.08	0.01	0.04	7.88	101.43
28	550.00	49.17	1.00	12.47	19.13	0.22	4.37	2.52	2.88	0.26	0.07	0.01	0.04	8.49	100.61
29	570.00	42.69	1.03	12.59	24.14	0.35	4.49	2.14	2.57	0.21	0.08	0.01	0.05	9.88	100.20
30	590.00	42.56	1.07	11.98	25.77	0.34	4.81	1.57	2.34	0.10	0.07	0.01	0.05	10.01	100.70
31	610.00	49.29	1.14	15.37	17.53	0.50	5.88	1.40	3.37	0.10	0.07	0.01	0.07	4.78	99.52
32	630.00	52.42	1.04	13.47	16.19	0.41	5.29	1.67	2.68	0.29	0.07	0.01	0.07	5.22	98.84
33	650.00	46.50	1.15	12.42	19.48	0.30	5.60	2.55	3.04	0.09	0.07	0.01	0.08	7.97	99.27
34	670.00	45.24	1.13	12.16	20.26	0.31	5.71	2.39	2.84	0.17	0.09	0.01	0.07	8.56	98.94
35	690.00	52.74	1.24	13.21	14.93	0.32	5.82	4.30	2.83	0.14	0.11	0.01	0.08	5.70	101.44

Sample	Depth	Pb	Ba	Sb	As	Ga	Zn	Cu	Ni	Co	Zr	Y	Sr	Rb
1	10.00	10	31	20	10	15	143	280	39	39	40	13	84	13
2	30.00	10	30	20	10	14	98	47	101	38	48	18	82	20
3	50.00	10	36	20	10	18	96	58	31	36	69	28	90	17
4	70.00	10	32	20	10	15	64	64	52	31	45	18	87	17
5	90.00	10	39	20	10	16	69	32	49	33	52	21	85	17
6	110.00	10	53	20	10	20	111	35	20	31	79	34	87	19
7	130.00	10	185	20	10	18	129	40	20	26	81	30	90	15
8	150.00	62	143	24	29	17	10987	304	57	29	57	29	79	32
9	170.00	32	138	21	47	20	2960	1383	20	24	72	31	23	13
10	190.00	34	480	24	214	18	942	77	24	35	61	26	12	18
11	210.00	25	634	20	184	16	795	145	20	29	52	22	12	16
12	230.00	14	673	21	105	17	865	97	23	41	51	20	12	13
13	250.00	12	302	20	26	15	283	41	22	31	54	23	11	12
14	270.00	11	426	21	78	14	1644	77	22	36	56	24	14	12
15	290.00	10	160	20	35	14	284	90	25	24	55	22	30	10
16	310.00	10	60	20	10	17	220	235	20	21	74	31	33	10
17	330.00	10	30	20	12	17	228	38	20	27	75	33	55	10
18	350.00	12	34	20	10	18	378	218	20	27	77	34	60	10
19	370.00	10	30	20	10	18	219	64	20	37	67	29	73	10
20	390.00	11	30	20	10	18	602	230	20	24	77	34	73	10
21	410.00	12	32	20	11	19	834	219	20	25	85	35	57	10
22	430.00	339	30	20	36	15	2626	132	20	27	82	34	73	10
23	450.00	63	30	21	162	15	331	46	20	25	84	32	67	10
24	470.00	24	30	25	111	17	296	49	20	31	70	30	72	10
25	490.00	10	30	20	13	17	218	27	20	35	70	29	66	10
26	510.00	10	34	20	15	16	161	68	20	27	70	28	70	10
27	530.00	10	30	20	17	19	174	21	20	38	75	30	57	10
28	550.00	10	30	20	13	17	126	26	20	57	68	27	54	10
29	570.00	10	35	20	10	18	158	173	20	57	65	25	41	10
30	590.00	10	30	20	11	19	173	747	21	66	73	23	41	10
31	610.00	10	30	20	10	19	309	297	20	26	69	27	60	10
32	630.00	10	30	20	10	17	224	193	20	25	65	26	64	10
33	650.00	10	30	20	11	19	192	79	20	34	70	27	60	10
34	670.00	10	30	20	13	19	183	45	20	47	74	29	52	10
35	690.00	10	30	20	10	18	132	124	20	29	82	32	83	10

Total Fe expressed as Fe_2O_3.

Oxydes in wt.%, traces in ppm. Lower detection limits for $P_2O_5 = 0.05\%$, $Cr_2O_3 = 0.01\%$; Pb, As, and Rb = 10 ppm; Sb and Ni = 20 ppm; Ba = 30 ppm. For element concentrations below the lower detection limit, the value of the lower detection limit is listed.

Table 9. Average content of Fe$_2$O$_3$, CaO, K$_2$O, SiO$_2$, MgO and Na$_2$O (wt.%) in 330 rock samples from uppermost lavas (0–150 m), mineralization (150–130 m), propylitic lavas (300–400 m) and greenschist facies lavas (400–689 m) in the drillcore CY-2a

		Fe$_2$O$_3$	CaO	K$_2$O	SiO$_2$	Al$_2$O$_3$	MgO	Na$_2$O
Uppermost lavas	x̄	9.64	7.35	1.41	51.53	14.91	6.75	2.17
(0–150 m)		3.46	0.49	0.14	19.33	3.05	1.66	0.48
n = 68	r	16.52	24.23	4.71	81.61	17.44	15.50	4.32
	s	1.88	3.58	1.14	6.23	2.38	2.37	0.86
Mineralization	x̄	21.79	0.82	1.00	49.61	9.65	4.10	0.70
(150–300 m)		8.27	0.01	0.06	1.13	0.12	0.09	0.05
n = 87	r	61.33	7.01	2.73	88.41	16.59	10.22	2.98
	s	14.63	1.54	0.72	18.17	5.82	3.07	0.65
Propylitic lavas	x̄	11.86	2.25	0.56	56.42	14.54	5.44	3.10
(300–400 m)		5.58	0.46	0.09	43.39	12.08	1.16	0.53
n = 49	r	18.54	6.50	1.55	66.31	19.14	10.24	6.17
	s	2.20	1.51	0.33	4.98	1.41	1.79	1.37
Greenschist facies	x̄	16.99	2.20	0.16	51.03	13.12	5.13	3.22
lavas (400–689 m)		8.48	0.41	0.08	19.60	1.61	1.47	0.47
n = 126	r	43.74	11.76	0.93	70.67	18.35	9.40	6.10
	s	6.92	1.57	0.14	8.88	2.66	1.46	1.11

x̄ = average, wt.%; r = range (min./max.); s = standard deviation.

The element combination Ca-Na-K was found to be discriminative for different degrees of hydrothermal alteration in CY-2a (Fig. 13). Samples of the greenschist facies and propylitic alteration zone group toward the Na$_2$O apex, whereas the uppermost lavas, which are almost unaffected by hydrothermal processes tend to have high CaO values. Samples from the mineralized and highly altered horizon form an individual group with low CaO and Na$_2$O, but high K$_2$O values.

Trace Elements

The trace element distribution in Hole CY-2a is given in the form of averages for each 20 m interval in Table 8. Average, range and standard deviation for each alteration zone are listed in Table 10.

As shown in Fig. 14, the stockwork zone is characterized by higher Zn (av. 2.5 wt.%, max. 11.0 wt.%; cf. Table 10) than Cu values (av. 0.03 wt.%, max. 1.5 wt.%; cf. Table 10). The Cu/Zn average ratio decreases from 0.43 between 300–689 m to 0.13 between 154 m and 300 m. Copper maxima occur at the top of the mineralized zone, as well as within the Sheeted Dike Complex, corresponding to the Cu distribution in sphalerite (cf. Fig. 10), whereas Zn maxima are restricted to the interval between 150 m and 430 m. However, a distinct zoning is not developed. Arsenic shows a strong correlation with the distribution of Zn, which suggests that As is hosted within the sphalerite.

Below 154 m Ni shows low average concentrations of 27 ppm (154 m to 300 m) and less than 20 ppm (detection limit) between 300 m and 689 m. Elevated Ni values up to 325 ppm are restricted to the uppermost, olivine-phyric lavas. Cobalt contents

Table 10. Average content of Ba, As, Zn, Cu, Ni, Co, Sr and Rb (ppm) in 330 rock samples from uppermost lavas (0–150 m), mineralization (150–300 m), propylitic lavas (300–400 m) and greenschist facies lavas (400–689 m) in the drillcore CY-2a

		Ba	As	Zn	Cu	Ni	Co	Sr	Rb
Uppermost lavas (0–150 m) n = 68	x̄	46.43	10.20	111.71	77.23	46.29	33.14	85.12	19.88
	r	30	10	46	20	20	20	16	10
		803	24	438	1522	325	55	106	187
	s	93.64	1.69	74.84	189.92	51.45	7.19	14.38	22.81
Mineralization (150–300 m) n = 87	x̄	367.21	89.40	2561.10	326.50	26.94	31.67	20.14	13.18
	r	30	10	21	20	20	20	10	10
		2130	807	110428	15220	412	144	130	33
	s	366.58	170.47	12235.00	1664.90	42.63	16.77	22.17	5.54
Propylitic lavas (300–400 m) n = 49	x̄	36.54	10.40	348.88	181.59	20.00	26.77	58.92	10.04
	r	30	10	23	20	20	20	10	10
		303	30	2525	1140	20	44	97	11
	s	37.92	2.77	379.58	275.42	0.00	7.72	20.88	0.19
Greenschist facies lavas (400–689 m) n = 126	x̄	30.81	29.40	426.35	144.47	20.08	36.56	60.51	10.00
	r	24	10	52	20	20	20	16	10
		68	863	17508	2603	26	196	168	10
	s	4.73	98.85	1572.00	342.13	0.61	30.45	20.50	0.00

x̄ = average, ppm; r = range (min./max.); s = standard deviation.

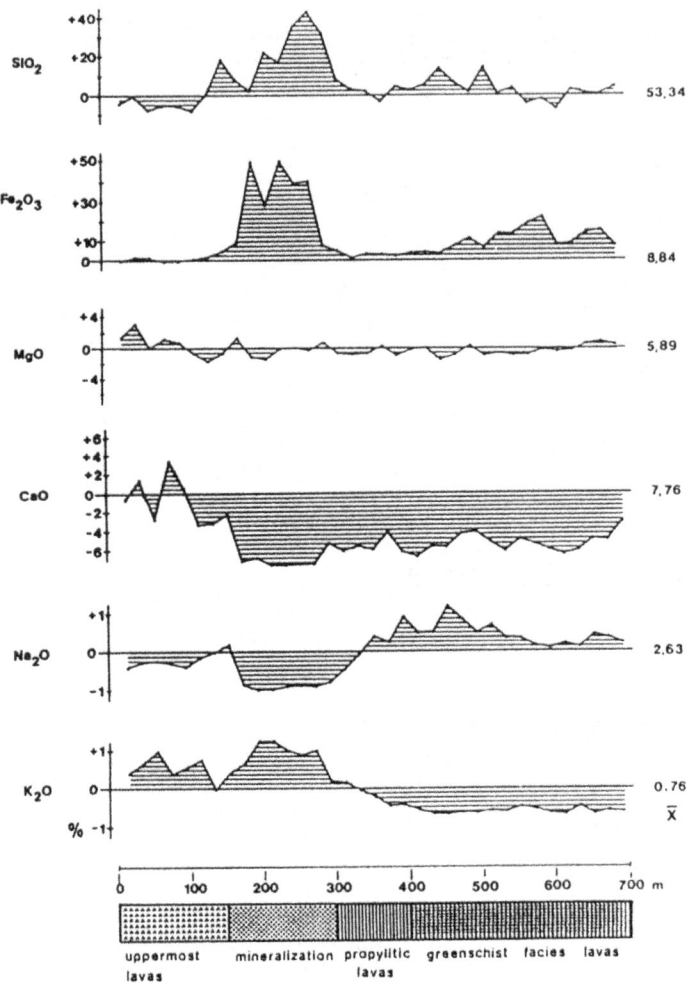

Fig. 12. Major element gains and losses downhole CY-2a in relation to fresh lava composition ($\bar{x}$) (Al-normalized values)

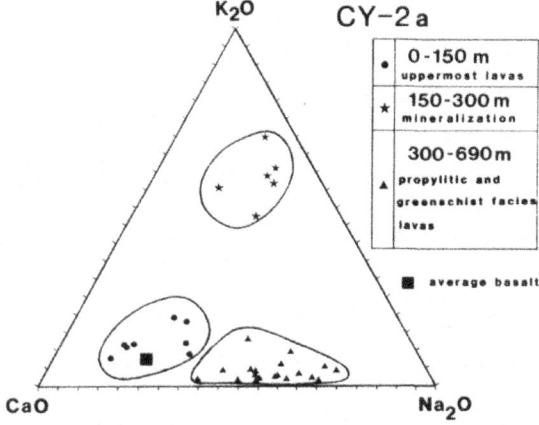

Fig. 13. K_2O-CaO-Na_2O ternary diagram showing the position of less altered uppermost lavas, mineralized lavas and propylitic to greenschist facies altered lavas in relation to average basalt

are the opposite, showing an increase in concentration below 500 m (max. 196 ppm), which reflects the elevated Co content of pyrite within this interval (cf. Fig. 7).

Strontium is slightly depleted in the greenschist facies and propylitized zones (av. 61 and 59 ppm, respectively) and strongly leached from the sulfide-bearing interval (av. 20 ppm). The Rb values exceed the detection limit of 10 ppm only within the mineralized and uppermost lavas (av. 13 ppm and 20 ppm, respectively). Therefore the sulfide zone is characterized by high Rb/Sr ratios.

A strong positive correlation was found between the sulfide mineralization and the distribution of Ba (Fig. 14). Maximum values of more than 2000 ppm (av. 367 ppm) occur in the mineralized interval between 150 m and 300 m, whereas Ba reaches only an average value of 46 ppm in the hanging wall and 30–40 ppm in the propylitic and greenschist facies zones between 300 m and 689 m.

The concentrations of Bi, Sb and Mo are mostly below the detection limits of 20 ppm for Bi and Sb, and 10 ppm for Mo. Only within the "roof" zone of mineralization (154 m to 200 m), up to 24 ppm Bi, 62 ppm Sb and 42 ppm Mo are found. Elevated Pb values occur between 155 m and 235 m (up to 270 ppm) and between 416 m and 462 m (up to 1200 ppm). Gallium varies in the whole core between 10–30 ppm, whereas Y and Zr concentrations are between 10–50 ppm and 20–115 ppm, respectively.

Discussion

Alteration Temperature

The uppermost lavas (0 m to 154 m) are characterized by the occurrence of celadonite, heulandite, laumontite, smectite, sepiolite, and calcite. The absence of very low-temperature zeolites, e.g., phillipsite, indicates a minimum alteration temperature of about 50°C (Evarts and Schiffmann 1983). Oxygen isotope data from seafloor basalts show that celadonite formation takes place at temperatures of about

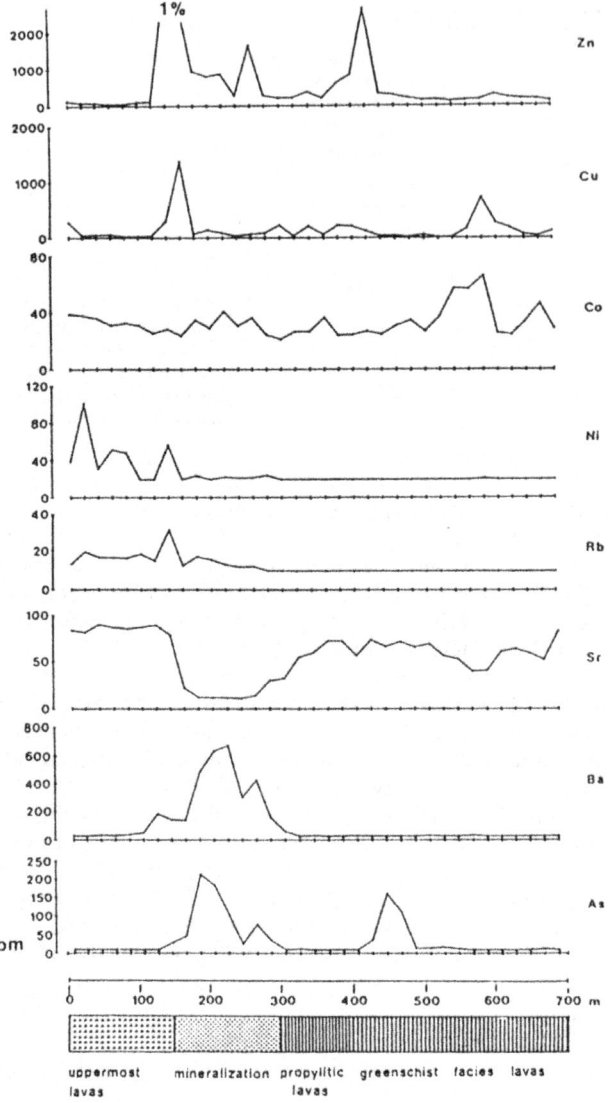

Fig. 14. Trace element distribution downhole CY-2a

40°C (Stakes and O'Neil 1982; Bohlke et al. 1984). Stable isotope data of Hole CY-2a (Rommel and Friedrichsen 1987) coincide with these data, showing alteration temperatures of about 60°C for the top zone of the uppermost lavas (Fig. 15). Since celadonite contains both Fe^{2+} and Fe^{3+}, varying oxygen fugacities are suggested.

Heulandite occurring at 126 m and laumontite at 153 m in CY-2a indicate a steep temperature gradient toward the top of the stockwork zone. Henley and Ellis (1983) found heulandite to be stable between 100°–115°C and measured lowest stability temperatures for laumontite of about 140°C in active geothermal systems in New Zealand. Palmason et al. (1979) assigned a stability range of 100°–200°C for laumontite found in geothermal areas of Iceland, whereas experimental equilibrium reactions indicate stability temperatures of 150°–300°C (Liou 1971).

However, oxygen and hydrogen isotope ratios give alteration temperatures of about 150°C for the lowermost part of the lava sequence at 154 m depth, i.e., above the stockwork mineralization in Hole CY-2a (Rommel and Friedrichsen 1987) and fit with data obtained from natural hydrothermal systems.

The stockwork zone (154 m to 300 m) is characterized by abundant sulfide-quartz mineralization, largely associated with chlorite and sericite-illite. Red jasper and/or hematite-bearing quartz are found between 165 m and 297 m depth. Based on the first occurrence of chlorite at 160 m, a minimum alteration temperature of 200°–230°C can be estimated, considering chlorite stability in thermal brines (Tomasson and Kristmannsdottir 1972). The first occurrence of illite at 177 m and prehnite at 185 m support this temperature estimation (Ellis 1979; Exley 1982), whereas stable isotope data indicate a temperature of 330°–360°C (Rommel and Friedrichsen 1987). Fluid inclusions in quartz from 275 m depth have yielded average pressure (250 bar) corrected filling temperatures of 335°C.

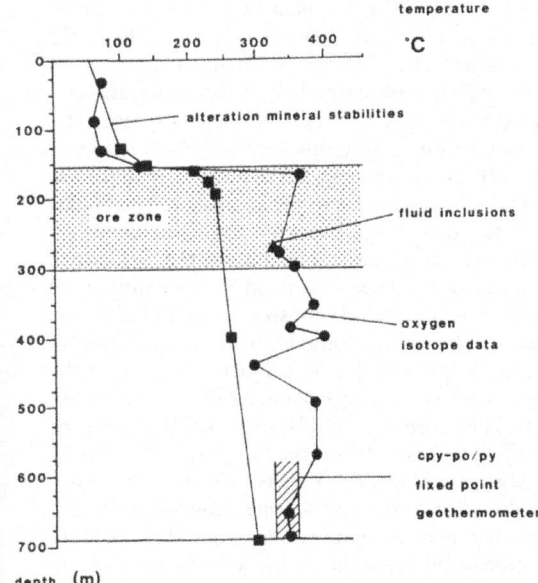

Fig. 15. Alteration temperature profile downhole CY-2a comparing secondary mineral stability, oxygen isotope, fluid inclusion and sulfide "thermometry" data

Low-temperature alteration minerals (celadonite, sepiolite, smectite) found in the stockwork zone are considered to be late stage products, superimposed on the previously formed high-temperature mineral assemblage possibly during the waning stage of hydrothermal activity.

The occurrence of epidote at 400 m depth indicates alteration temperatures of at least 260°–280°C according to experimental reactions and observations in natural systems (Liou 1973; Palmason et al. 1979). Due to the absence of actinolite, alteration temperatures at the base of the hole should not have exceeded about 320°C (Elthon 1981). Isotope data of Rommel and Friedrichsen (1987), however, indicate temperatures of about 350°C for the bottom of Hole CY-2a.

Chalcopyrite-pyrrhotite inclusions occurring in pyrite below 580 m depth can be used as a "fixed-point" geothermometer for the lower dike section in CY-2a. According to Sugaki et al. (1975) the tie line of pyrite + intermediate solid solutions in the Cu-Fe-S system switches to chalcopyrite + pyrrhotite at 334°C + /–17°C, indicating that initial fluid temperatures exceeded this temperature.

Considering the experimental data and conclusions drawn by Scott and Barnes (1971) and Scott and Kissin (1973), the increase of the FeS content in sphalerite with depth can be explained in terms of increasing formation temperature assuming that sphalerite was formed in equilibrium with coexisting pyrite and pyrrhotite which served to buffer the sulfur fugacity of the system.

Thus, stable isotope data and in part sulfide "thermometry" generally yield higher alteration temperatures than obtained from secondary mineral stabilities.

Chemical Changes and Fluid Composition

As indicated by experimental data (cf. Mottl 1983), the most prominent process of seawater-basalt interaction at temperatures between 300° and 350°C is the rapid and complete depletion of seawater in Mg and SO_4^{2-}. Dissociation of seawater results in precipitation of Mg^{2+} and OH^- as $Mg(OH)_2$ predominantly in clay minerals (e.g., chlorite) and a drastic decrease in pH (to about pH 4) due to the related excess in H^+. Oxidation of Fe^{2+} to Fe^{3+} of the rock-forming minerals forces the reduction of seawater SO_4^{2-} to S^{2-}. The resulting acidic and reducing high temperature fluids are capable of leaching the rock due to instability of silicate minerals, magmatic sulfides and titanomagnetite. Interaction with such fluids caused significant chemical changes within the dike section and the transition zone below the stockwork mineralization in Hole CY-2a. Potassium and Ca depletion and Na enrichment are typical characteristics of this sequence. The inverse relationship of Ca and K versus Na is explained by the introduction of Na by seawater and hydrothermal liberation of Ca and K during alteration of plagioclase feldspar to oligoclase-albite. In the experiment, leaching of Ca from feldspar and pyroxene balances the uptake of Mg and OH from seawater (Mottl 1983). Therefore, it would be expected that zones of high-temperature seawater-basalt reaction should be marked by a loss of Ca and a gain of Mg. Instead, the whole rock Mg distribution in CY-2a is quite homogeneous, showing no distinct gains within the alteration or stockwork zone. However, Mg-rich chlorite is the dominant phyllosilicate throughout. It may be speculated that a more significant increase in Mg concentration occurs near the base of CY-2a, where large amounts of downwelling seawater were rapidly heated by the steep temperature gradient within the less permeable zones of the Sheeted Dike Complex during the initial evolution of the system. However, Alt and Emmermann (1985) have shown for DSDP/ODP Hole 504B that the pillow lavas altered at about 100°–150°C are a significant sink for Mg.

High MnO contents of chlorite and epidote and a low Fe/Al ratio of epidote in the lower Sheeted Dike section of CY-2a indicate strongly evolved, reducing hydrothermal fluids. Also sphalerite below 250 m depth shows elevated Mn concentration, although whole rock data indicate low and quite uniform Mn values throughout the core. The systematic increase of the epidote Fe/Al ratio upward in the dike section may be attributed to slightly higher oxygen fugacities (Liou 1973) in the more permeable dike-pillow lava transition zone.

Leaching of the Sheeted Dikes and the deeper basement resulted in high SiO_2 and metal concentrations of the upwelling fluids. Solubility of quartz increases with temperature and pressure, except near the critical point of water. Since the pressure difference between dike section and the sulfide-quartz stockwork is negligible (in the order of about 50 bar), SiO_2 precipitation is largely controlled by the fluid temperature. A temperature decrease from, e.g., 320°C to 280°C results in precipitation of about 25% of SiO_2 in solution (Kennedy 1950). Such cooling may have caused initial SiO_2 precipitation in the more permeable zone directly above the dike sequence between 400 m and 530 m in depth. However, the silicification of the alteration pipe may be attributed to quartz dumping during late-stage cooling of the entire hydrothermal system (Bowers et al. 1985).

According to Sr isotope data, reactions within the stockwork zone were dominated by high water/rock ratios of about 10:1 to 15:1 (Rommel and Friedrichsen 1987) causing strong alteration of glass layers and massive lavas, as well as abundant sulfide and quartz precipitation due to fluid-seawater mixing and subsequent drop in temperature. Complete destruction of primary minerals, e.g., plagioclase feldspar is documented by extremely low Ca, Na, and Sr values. Extensive mixing of fluids with seawater is also obvious by the absence of a temperature zoning of sulfide mineral formation within the stockwork zone, e.g., chalcopyrite at the base and sphalerite at the top. The bottom of the mixing zone is evidently marked by the lowermost occurrence of red jasper and hematitic quartz at 297 m depth which coincides with the offset of the main ore zone. Mg-chlorite occurring in the stockwork zone does not necessarily indicate downwelling seawater which became depleted in Mg, since Mg-precipitates have also been produced experimentally by mixing of hydrothermal fluids with seawater (Seyfried and Bischoff 1977).

The strong K enrichment within the stockwork zone most probably originates from leaching of plagioclase within the Sheeted Dike Complex and the deeper basement and fixation in sericite-illite. Major K sources are probably deep-seated plagiogranites and trondjemites which were highly depleted in K at temperatures of about 650°C according to fluid inclusion data (J. Malpas pers. commun. 1986). Hydrothermally transported Ba substitutes for K in sericite-illite, which is indicated by a significant positive whole rock correlation of K_2O and Ba. Barium at least partly originates from leaching of the Sheeted Dike diabase (Herzig and Friedrich 1987). However, the formation of sericite-illite and the consequent uptake of K within the high-temperature stockwork zone is not consistent with experimental results showing that K is mobilized at temperatures $> 150°C$ (cf. Mottl 1983). This would indicate a fixation of K during the waning low-temperature stage of the hydro-thermal system, and implies that also the substitution of K in sericite-illite by Ba is a late-stage process. Therefore it has to be considered that seawater downwelling after the formation of the mineralization represents an additional, probably even a major source of K and Ba.

The hanging wall lavas are characterized by elevated K and Ca and low Na concentrations as typical for low-grade seawater-basalt reaction, i.e., zeolitization (Honnorez 1981). Sr isotope ratios indicate that alteration in the uppermost lavas took place at water-rock ratios of 1:1 to 5:1 (Rommel and Friedrichsen 1987). It has to be noted that due to post-depositional erosion the present pillow lava surface is situated about 100–200 m below the original seafloor.

Summary and Conclusions

The Agrokipia "B" deposit represents a typical example of sulfide formation in ophiolitic pillow lavas, i.e., in oceanic layer 2. It was formed by sub-surface mixing of ascending about 350°C hot hydrothermal fluids with seawater in Cretaceous ocean crust. High permeability, a prerequisite for the evolution of a sub-surface mixing zone, was provided by abundant hyaloclastite layers which are intercalated within a series of andesitic-dacitic lava flows. High water/rock ratios of $\geq 10:1$ forced the predominant replacement of glass layers by sulfides and quartz. The stockwork deposit is covered by about 150 m of pre-mineralization lavas comprising olivine-bearing basaltic andesites (0–90 m) and andesitic dacites (>90 m). Pyrite + sphalerite + /−chalcopyrite mineralization associated with abundant quartz occurs between 150 m and 300 m below surface, but minor mineralization can be traced down to a total depth of about 700 m.

The stockwork itself is characterized by pervasive argillitic alteration of the host rock which caused the formation of abundant Mg-chlorite and sericite-illite, now representing the host for the sulfide-quartz mineralization. Most pronounced characteristics of the stockwork zone are a strong enrichment of K and Ba, whereas Ca, Na, and Sr are almost totally depleted due to complete decomposition of primary minerals. Barium almost exclusively occurs as substituent of K within the sericite-illite structure. Apparently Ba as well as K dominantly originate from leaching of Sheeted Dike diabase and plagiogranites/trondjemites of the deeper basement and were fixed subsequently to hydrothermal transport within the stockwork zone due to fluid-seawater mixing. Since experimental data predict that K is mobilized at temperatures $> 150°C$ (which prevailed within the stockwork), the K enrichment probably represents a late-stage low-temperature process, as indicated by the abundance of low-temperature minerals (celadonite, smectite-vermiculite) super-imposed on the high-temperature mineral assemblage of the stockwork zone. Also gypsum and calcite are considered to be of late-stage origin. Therefore, however, it cannot be excluded that K and also Ba at least partially derive from seawater downwelling into the mixing zone after the sulfide stockwork was formed.

Sulfide mineralogy displays a complex intergrowth pattern. Various re-placement relationships indicate multiple stages of sulfide formation. Intensive fluid-seawater mixing caused local variations in temperature, oxygen fugacity, and metal concentration, which is documented by different successions of sulfide precipitation and replacement and the occurrence of red jasper and hematitic quartz.

The alteration zone directly below the sulfide stockwork shows a more propylitic type of alteration and mainly consists of Mg-chlorite + albite + quartz with minor sulfides, predominantly pyrite. The downward-following dike sequence which onsets about 200 m below the stockwork displays a greenschist facies style of alteration comprising Mg-chlorite + albite + quartz associated with epidote + sphene and minor sulfides. Due to strong albitization of plagioclase feldspar, Ca, K, and Sr are distinctly depleted, whereas Na was added by down-welling seawater.

The pre-mineralization hanging wall lavas were predominantly affected by mixed, cooled fluids upwelling from the stockwork and low-temperature seawater-basalt reactions. Therefore they show only a slight seafloor weathering to zeolite facies type of alteration and are characterized by smectite-vermiculite, sepiolite, celadonite, and zeolites, as well as by elevated K, Ca, and low Na concentrations.

The chemical composition of various sulfides has shown only very limited depth-related variation. While chalcopyrite is quite homogeneous in chemistry, erratically high Co contents in pyrite and elevated Cd concentrations in sphalerite were found below 600 m depth, probably indicating higher formation temperatures. Elevated As values are characteristic for pyrite throughout. Sphalerite, however, shows a distinct bimodal downhole distribution in the Fe content reaching an average of 9.7 mol % FeS above and 19.7 mol % FeS below 270 m depth, which apparently is related to an increase in formation temperature.

Secondary mineral stabilities and stable isotope data have shown that alteration temperatures within the Sheeted Dike Complex ranged between about 300–400°C causing alteration to greenschist facies mineralogy in the dikes and propylitic alteration in the transition zone above. Upon upwelling of high-temperature metal- and Si-rich hydrothermal fluids into the high permeable glass-rich sub-surface zone, intensive fluid-seawater mixing led to a steep temperature gradient from about 350°C at the bottom to 150°C at the top of the mixing "chamber" forcing abundant alteration and sulfide-quartz precipitation. The uppermost pre-mineralization lavas interacted with (1) circulating bottom seawater during regional seafloor weathering and (2) cooled hydrothermal fluids, ascending from the sub-surface mixing zone below and suffered alteration at temperatures between about 60°C and 150°C. Since celadonite contains both ferric and ferrous iron, varying oxygen fugacities are suggested for the alteration of this sequence.

Mineralogy, chemistry, and alteration temperature data have documented the modes of sulfide formation for the Agrokipia "B" deposit by hydrothermal activity in Cretaceous oceanic crust. In relation to the sulfide stockwork encountered by DSDP/ODP Hole 504B in sub-Recent 5.9 m.y. old crust, Agrokipia "B" represents the fossil analog of submarine sulfide formation in layer 2. The recent pendants of Agrokipia "B" are presumed to occur below the seafloor in presently active low-temperature hydrothermal mound areas at oceanic spreading centers (e.g., Galapagos Rift) probably representing economic sulfide deposits.

Acknowledgments. The author gratefully acknowledges funding of studies on core material of Hole CY-2a by grants of the European Community Research & Development Program "Raw Materials" (MSM-015-D) and the Deutsche Forschungsgemeinschaft (Fr 240/40-1) given to Professor Günther Friedrich, who initiated the project and encouraged writing this paper. The author is also grateful to the International Crustal Research Drilling Group for the opportunity to participate in the Cyprus Crustal Study Project. Special thanks are due to Drs. John Malpas (St. John's, Canada) and Jeffrey C. Alt (St. Louis, USA), who read and discussed an early draft of this paper which was written on board JOIDES Resolution during ODP Leg 111 to Hole 504B. Thanks are extended to Dipl.-Min. Dietmar Schöps for his contribution to data management and interpretation.

References

Adamides NG (1984) Cyprus volcanogenic sulfide deposits in relation to their environment of formation. PhD Thesis, University of Leicester

Alt JC, Emmermann R (1985) Geochemistry of hydrothermally altered basalts: DSDP Hole 504B, Leg 83. Init Rep DSDP 83:249-262

Anderson RN, Honnorez J, Becker K et al. (1985) Initial Reports DSDP 83. Washington, US Govt Printing Office

Barton PB Jr (1978) Some ore textures involving sphalerite from the Furutobe mine, Akita Prefecture, Japan. Miner Geol 28:293-300

Bednarz U, Sunkel G, Schmincke H-U (1987) The basaltic andesite-andesite and the andesite-dacite series from ICRDG drill-holes CY-2 and CY-2a. I. Lithology, petrology, and geochemistry. In: Robinson PT, Gibson IL, Panayiotou A (eds) Cyprus Crustal Study Project Init Rep Hole CY-2/2a, Geol Survey Canada

Bohlke JK, Alt JC, Muehlenbachs K (1984) Oxygen isotope-water relationships in altered deep sea basalts: Low temperature mineralogical controls. Can J Earth Sci 21:67-77

Bowers TS, Damm KL von, Edmond JM (1985) Chemical evolution of mid-ocean ridge hot springs. Geochim Cosmochim Acta 49:2239-2252

Bralia A, Sabatini G, Troja F (1979) A revaluation of the Co/Ni ratio in pyrite as geochemical tool in ore genesis problems. Miner Dep 14:353-374

Corliss JB, Dymond J, Gordon LI et al. (1979) Submarine thermal springs on the Galapagos rift. Science 203:1073-1083

Desmet A, Gagny CL, Lapierre H, Rocci G (1979) Organisation spatio-temporelle du complexe filonien du Troodos: Son enracinement dans la chambre magmatique. In: Panayioutou A (ed) Ophiolites. Proceedings Int Ophiolite Symp Cyprus, Cyprus Geol Surv Dep: 66-72

Ellis AJ (1979) Explored geothermal systems. In: Barnes HL (ed) Geochemistry of hydrothermal ore deposits, 2nd edn. Wiley, New York, pp 632-683

Elthon D (1981) Metamorphism in oceanic spreading centers. In: Emiliani C (ed) The Sea. The Oceanic Lithosphere 7:285-303

Evarts RC, Schiffmann P (1983) Submarine hydrothermal metamorphism on the Del Puerto ophiolite, California. Am J Sci 283:289-340

Exley RA (1982) Electron microprobe studies of the I.R.D.P. high-temperature hydrothermal mineral geochemistry. J Geophys Res 87:6547-6557

Finlow-Bates T, Stumpfl EF (1981) The behaviour of so-called immobile elements in hydrothermally altered rocks associated with volcanogenic submarine-exhalative ore deposits. Miner Dep 16:319-328

Henley RW, Ellis AJ (1983) Geothermal systems ancient and modern: a geochemical review. Earth Sci Rev 19:1-50

Herzig PM (1986) Hydrothermale Alteration und Sulfidmineralisation der Lagerstätte Agrokipia, Troodos Ophiolith Komplex, Zypern. PhD Thesis, Rheinisch-Westfälische Technische Hochschule, Aachen

Herzig PM, Friedrich GH (1987) Sulphide mineralization, hydrothermal alteration and chemistry in the drill hole CY-2a, Agrokiopia, Cyprus. In: Robinson PT, Gibson IL, Panayiotou A (eds) Cyprus Crustal Study Project Init Rep Hole CY-2/2a, Geol Survey Canada

Hey MH (1954) A review of the chlorites. Miner Mag 30:277

Holdaway MJ (1972) Thermal stability of Al-Fe-epidote as a function of f_{o_2} and Fe content. Contrib
    Mineral Pet 37:307-340
Honnorez J (1981) The aging of the oceanic crust at low temperature. In: Emiliani C (ed) The Sea. The
    Oceanic Lithosphere 7:525-587
Humphris SE, Thompson G (1978) Hydrothermal alteration of oceanic basalts by seawater. Geochim
    Cosmochim Acta 42:107-125
Kennedy GC (1950) A portion of the system silica-water. Econ Geol 45:629-653
Krauskopf KB (1979) Introduction to geochemistry. Mc Graw-Hill, Tokyo
Liou JG (1971) Synthesis and stability relations of prehnite, $Ca_3Al_2Si_3O_{10}(OH)_2$. Am Miner 56:507-531
Liou JG (1973) Synthesis and stability relations of epidote, $Ca_2Al_2Fe Si_3O_{12}$ (OH). J Pet 14:381-413
Loftus-Hills G, Solomon M (1967) Cobalt, nickel and selenium in sulphides as indicators of ore genesis.
    Miner Dep 2:228-242
Mottl MJ (1983) Metabasalts, axial hot springs, and the structure of hydrothermal systems at mid-ocean
    ridges. Geol Soc Am Bull 94:161-180
Palmason GS, Arnorsson IB, Fridleifsson H et al. (1979) The Iceland crust: Evidence from drillhole data
    on structure and processes. In: Talwani M, Harrison CG, Hayes DE (eds) Deep drilling-results in
    the Atlantic Ocean: ocean crust. Maurice Erwing Series 2. Am Geophys Union: 43-65
Pearce JA, Cann JR (1973) Tectonic setting of basic volcanic rocks determined using trace element
    analyses. Earth Planet Sci Lett 19:290-300
Rautenschlein M, Jenner GA, Hertogen J, Hofmann AW, Kerrich R, Schmincke H-U, White WM (1985)
    Isotopic and trace element composition of volcanic glasses from the Akaki Canyon, Cyprus:
    implications for the origin of the Troodos ophiolite. Earth Planet Sci Lett 75:369-383
Riverin G (1977) Wall-rock alteration at the Millenbach mine, Noranda area, Quebec. PhD Thesis,
    Queen's University, Kingston
Robinson PT, Gibson IL, Panayiotou A (1987) Cyprus Crustal Study Project Initial Reports Hole
    Cy-2/2a, Geol Survey Canada
Rommel U, Friedrichsen H (1987) The alteration chemistry of CY-2a: A stable and radiogenic isotope
    study. In: Robinson PT, Gibson IL, Panayiotou A (eds) Cyprus Crustal Study Project Init Rep Hole
    CY-2/2a, Geol Survey Canada
Scott SD, Barnes HL (1971) Sphalerite geothermometry and geobarometry. Econ Geol 66:653-669
Scott SD, Kissin SA (1973) Sphalerite composition in the Zn-Fe-S system below 300°C. Econ Geol
    68:475-479
Seyfried WE, Bischoff JL (1977) Hydrothermal transport of heavy metals by seawater: The role of
    seawater/basalt ratio. Earth Planet Sci Lett 34:71-77
Stakes DS, O'Neil JR (1982) Mineralogy and stable isotope geochemistry of hydrothermally altered
    oceanic rocks. Earth Planet Sci Lett 57:285-304
Sugaki A, Shima H, Kitakaze A, Harada H (1975) Isothermal phase relations in the system Cu-Fe-S under
    hydrothermal conditions at 350°C and 300°C. Econ Geol 70:806-823
Sunkel G, Bednarz U, Schmincke H-U (1987) The basaltic andesite-andesite and the andesite-dacite
    series from the ICRDG drillholes CY-2 and CY-2a. II. Alteration. In: Robinson PT, Gibson IL,
    Panayiotou A (eds) Cyprus Crustal Study Project Init Rep Hole CY-2/2a, Geol Survey Canada
Thompson J, Mottl MJ, Rona PA (1985) Morphology, mineralogy, and chemistry of hydrothermal
    deposits from the TAG area, 26°N, Mid-Atlantic Ridge. Chem Geol 49:243-257
Tomasson J, Kristmannsdottir H (1972) High temperature alteration minerals and thermal brines,
    Reykjanes, Iceland. Contrib Miner Pet 87:123-134
Viereck LG, Griffin BJ, Schmincke H-U, Pritchard RG (1982) Alteration of pyroclastic rocks in the
    Reydarfjordur drill core. J Geophys Res 87:6459-6476

Kuroko-Type Ore Deposits on the Aegean Islands, Greece

M.B. HAUCK[1]

Abstract

Several sulfide-sulfate ore accumulations which show typical features of Kuroko-type deposits occur within Pleistocene pyroclastics on the Greek islands of Milos, Kimolos, Poliegos, and Antimilos, which belong to the currently active Aegean island arc. A large number of ore and country rock samples were analyzed to obtain their sulfur isotope ratios and geochemical compositions including trace elements and REE distribution. Based on this data, a metallogenic model of continuous interaction between andesitic tuffs and circulating seawater is proposed. Ore formation took place by oxidation and disproportionation at temperatures between 220° and 260°. The sulfide-sulfur is interpreted to be derived from seawater sulfate, whereas the heavy metals originate exclusively from the Plio-Pleistocene volcanics. It is concluded that these sulfide-sulfate deposits are the youngest Kuroko-type mineralization so far described, being even younger than the Miocene Japanese deposits.

Introduction

Kuroko deposits (named from the mines on Honshu Island, Japan) are generally defined as strata-bound polymetallic sulfide-sulfate deposits genetically related to felsic volcanism in island arc systems (Sato 1977; Sawkins 1984). Similar occurrences are found all over the world, ranging in age from Miocene to Devonian.

The sulfide-sulfate deposits on the islands of Milos, Kimolos, Poliegos, and Antimilos in the Southern Aegean Sea, Greece, are especially suitable for a study of Kuroko-type deposits in an active volcanic arc in Europe, because they are very young (Pliocene-Pleistocene) and have suffered only negligible tectonic deformation.

Field observations reveal a typical zonation of the ore bodies as known from Japanese examples: siliceous (Keiko) ore, gypsum ore, black (Kuroko) ore, and barite ore. Based on stable isotope and both trace element and REE data, the proposed genetic model appears to be comparable to that of the Miocene Japanese deposits.

[1]Institut für Geochemie u. Petrographie, Kaiserstrasse 12, D-7500 Karlsruhe, FRG
Present address: Shell U.K. Expro Ltd., The Strand, London WC 2N ODX, Great Britain

Base Metal Sulfide Deposits
G.H. Friedrich, P.M. Herzig (Eds.)
© Springer-Verlag Berlin Heidelberg 1988

Geotectonic Framework

The plate tectonic situation of the eastern Mediterranean Sea is defined by the Aegean and Anatolian microplates occurring between the major Eurasian and African plates (Mc Kenzie 1978; Dewey and Şengör 1979). The Aegean microplate overrides the NE-moving African plate along the Hellenic Trench (Fig. 1); a well-defined seismic plane of earthquake centers (Benioff-Wadati Zone) dips at an angle of 35° from the trench to the NE. Several geophysical, geological, and geochemical features of a typical but relatively small island arc-marginal sea system can be observed. Behind the trench, the Hellenic metamorphic arc extends from the Peleponnes to the islands of Crete and Rhodes. The back arc basin is characterized by high heat flow and a shallow Moho. The recent calc-alkaline volcanoes of the Aegean volcanic arc are situated about 120–150 km above the Benioff-Wadati Zone and some 220 km behind the Hellenic Trench (Fig. 2).

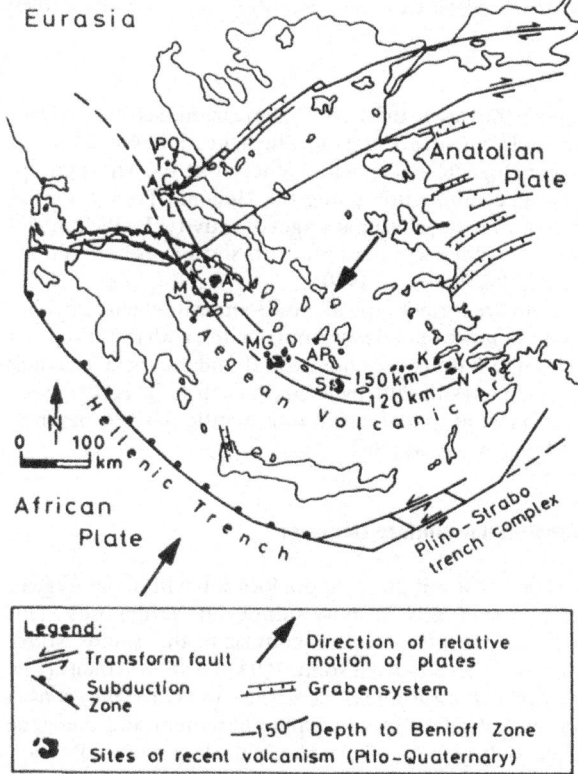

Fig. 1. Summary of present Aegean tectonics showing plate setting and the occurrences of Plio-Quaternary volcanics within the Aegean volcanic arc (Mc Kenzie 1978; Keller 1982). *PO* Porphyrion; *T* Theben; *AC* Achilleion; *L* Likhades; *C* Crommyonia; *M* Methana; *A* Aegina; *P* Poros; *AP* Antiparos; *MG* Milos group; *S* Santorini; *K* Kos; *Y* Yali; *N* Nisyros

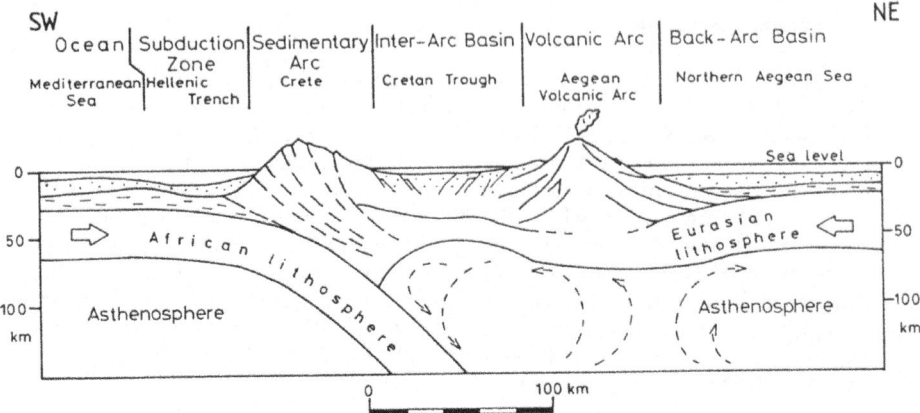

Fig. 2. Schematic SW-NE section across the Aegean region and proposed geodynamic model (no vertical exaggeration)

From the Greek mainland in the west to the Turkish shore in the east, the Aegean volcanic arc comprises several volcanic sites (Fig. 1), of which the islands of Santorini and Nisyros are at present volcanogenic active, while others, like the Milos group, show only solfataric activity. The subduction along the Hellenic Trench started about 13.5 m.y. ago, but seems now to be in its final stages of activity (Le Pichon and Angellier 1979; Keller 1982). The oldest volcanic products show ages between 2.7 and 3.4 m.y (Fytikas et al. 1976; Ferrara et al. 1980).

The volcanics of the Aegean arc form a typical basalt-andesite-dacite-rhyolite calc-alkaline suite with a predominance of andesites and dacites, with the exception of Santorini, where high alumina basalts are the most abundant rocks (Puchelt 1978). The derivation of the magma is still controversial (Puchelt 1978; Keller 1982; Barton et al. 1983); however, most authors believe in a mantle-derived parental magma with contamination by continental crust.

Main Features of Investigated Sulfide-Sulfate Deposits

The Milos group of islands is the most suitable onshore location within the Aegean volcanic arc for the investigation of Kuroko-type sulfide-sulfate deposits. This group, forming the largest site of recent volcanism, consists of the islands Milos, Kimolos, Poliegos, and Antimilos and is located some 150 km SSE of Athens (Fig. 1). The geological constitution of all these islands is very much the same; they consist of mainly Pliocene-Pleistocene volcanics. Metamorphic basement and Neogene sediments outcrop sporadically only in the southern part of the largest island, Milos. The geological map of Milos showing the investigated mines is given in Fig. 3. The volcanic sequence which is more than 700 m thick can be ascribed to three or four phases of activity (Fytikas 1976) beginning with both subaerial and subaquatic pyroclastics of Upper Pliocene to Lower Pleistocene age on all islands of the Milos

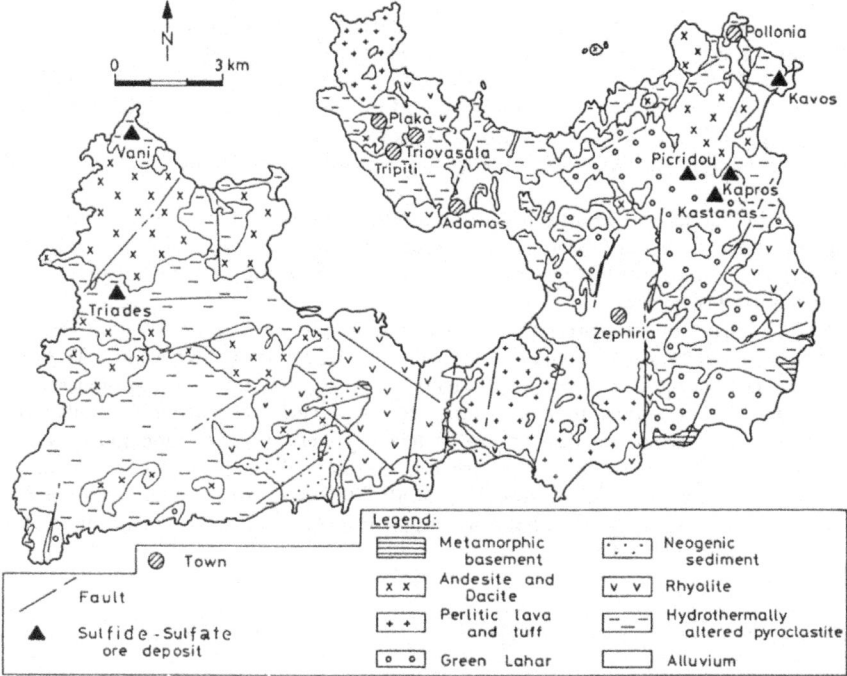

Fig. 3. Simplified geologic map of the island of Milos with location of ore deposits (Fytikas 1976; Hauck 1984)

group. Some ignimbrites also belong to this early phase, which was followed by the formation of huge lava domes and flows of dacitic to andesitic composition. Subsequently, large areas of the main island Milos were covered by lahar deposits up to 60 m thick. In the last phase of activity, perlitic pyroclastics and lavas of rhyolitic composition were deposited. Their age is considered to be 0.48 m.y. (Fytikas et al. 1976). The final volcanic phase was followed by phreatomagmatic eruptions and strong solfataric activities.

Some elements of the tectonic regime in the back arc basin, which is characterized by spreading and extension, can be observed on the island of Milos, such as open and mineralized fractures in the mines, normal faulting in NE-SW direction and continuous hydrothermal activity caused by high heat flow. The hydrothermal activity led to strong alteration of some of the early tuffs, creating small but economically valuable mineral deposits. At present, mining on the islands of Milos and Kimolos mainly concentrates on bentonite, kaoline, and barite.

With the exception of the ancient manganese-barite mine of Vani (NW part of Milos), which is a volcano-sedimentary deposit, all other ore bodies show indications of vertically zoned ore facies, similar to those described from Japanese Kuroko deposits (Lambert and Sato 1974; Shimazaki 1974; Sato 1977): siliceous ore, gypsum ore, black ore, and barite ore. Galena and barite are by far the most abundant phases but several other minerals are present (Table 1; see also Fig. 4).

The siliceous ore consists mainly of cryptocrystalline quartz with minute veinlets of barite and forms part of the footwall and in some cases the immediate roof of barite ore bodies. The lesser-developed gypsum zone consists of lenses and nodules of gypsum and anhydrite which are well exposed in the mines of Kavos and Kastanas.

In the Kavos and Triades mines, a very fine-grained assemblage of abundant sphalerite, galena, and barite, and minor amounts of silver minerals such as pyrargyrite and proustite were observed, which are most similar to the Kuroko black ore reported from Japan.

The barite zone is represented by large bodies of barite ore mainly found in the eastern part of Milos. It occurs in two forms as impregnations (Kavos and Triades mines), consisting of massive hard barite within quartz, and as metasomatic ore (Picridou and Kastanas mines), forming chimneys, pipes and irregular bodies of loosely packed, white to reddish barite with layers of crypto-crystalline quartz at the footwalls. Both type of ore bodies are 150–200 m in diameter and contain small amounts of disseminated sulfides (Fig. 5). Besides these two genetic types, barite veins and veinlets of younger age with minor sulfide mineralization are abundant on all islands of the Milos group.

Table 1. Observed ore minerals on the Milos group of islands (Mack 1977; Hauck 1984).

Elements		
	native sulfur	S
Sulfides		
	galena	PbS
	sphalerite	ZnS
	chalcopyrite	$CuFeS_2$
	bornite	Cu_5FeS_4
	pyrite	FeS_2
	covellite	CuS
	pyrargyrite	Ag_3SbS_3
	proustite	Ag_3AsS_3
Carbonates		
	cerussite	$PbCO_3$
	malachite	$Cu_2[(OH)_2CO_3]$
	rhodochrosite	$MnCO_3$
Oxides		
	psilomelane	$Ba_3(OH)_6Mn_8O_{16}$
	hematite	Fe_2O_3
	pyrolusite	MnO_2
	cryptomelane	$K Mn_8O_{16}$
	hollandite	$Ba Mn_8O_{16}$
Sulfates		
	barite	$BaSO_4$
	gypsum	$CaSO_4 x 2H_2O$
	celestite	$SrSO_4$
	anglesite	$PbSO_4$

Fig. 4. Scanning electron microscope photo of secondary formed anglesite within fracture in galena. Ancient underground mine. island of Poliegos

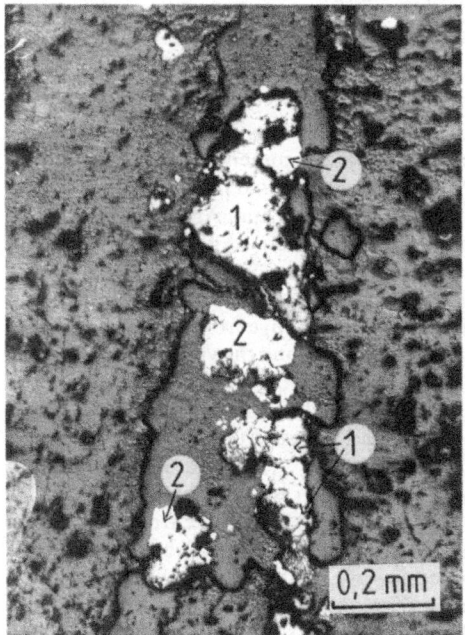

Fig. 5. Photomicrograph of disseminated galena (*1*) and chalcopyrite (*2*) in baryte ore, Triades mine, Milos. Plane polarized reflected light

Overall reserves are estimated in the order of 2 million tons of barite (Mack 1977). The top zones of the ore bodies are controlled by impermeable overlaying silicified tuffs, while kaolinitic zones form the footwall (Hauck 1984).

Analytical Methods

Trace and rare earth element contents were determined in various ore types using instrumental neutron activation analyses (INAA) and atomic absorption spectrometry. The parameters of the INAA are described in Puchelt and Kramar (1976).

Some 240 sulfide and sulfate ore samples from about 20 localities all over the Milos and Poliegos islands have been selected for sulfur isotope studies. The preparation procedure and the measurements with a VG-Micromass 1202 S-mass spectrometer are described in detail by Hauck (1984). The results are reported in the common per mil (‰) deviation from the Cañon Diablo Troilit standard with a maximum overall error of ± 0.2‰.

A radionuclide-excited energy-dispersive X-ray fluorescence (EDXRF) device has been used to determine the $SrSO_4$ content in some 160 barite samples.

In order to examine the source of the metals occurring in the Kuroko-type deposits, 25 trace and rare earth elements have been analyzed from both hydrothermally altered and fresh volcanics, using X-ray fluorescence (XRF) and instrumental neutron activation analyses (INAA).

Results and Data Presentation

Selected examples of trace and rare earth element contents in various ore types are listed in Table 2. Besides pure barite ore with low metal content, some barite samples contain up to several thousand ppm Pb and several hundred ppm Zn, and may also contain a considerable amount of Ag and As (up to 250 ppm Ag and 125 ppm As). Both the black ore and the manganese ore are characterized by high Pb- (max. 10 wt.%), Zn- (max. 3 wt.%) and Cu contents (max. 0.5 wt.%).

Results of the sulfur isotope studies on sulfide and sulfate ore samples are presented in Fig. 6. $\Delta^{34}S$ values for all barite samples range from + 7‰ to + 27‰, with an average of + 19‰. Both the impregnated ore and the vein ore have isotope ratios between + 18‰ and + 27‰ with an average of + 21‰, while the metasomatic barites gave values between + 7‰ and + 20‰, averaging + 14‰.

$\Delta^{34}S$ values close to + 19‰ were detected in celestite and bassanite, while gypsum, alunite, and alunogene show values between + 3‰ and + 8‰, similar to the isotope ratios in native sulfur from other Aegean islands (Puchelt et al. 1971; Hubberten et al. 1977). While sulfur isotope ratios in sphalerite from Milos were found to be in the range of + 2‰ to + 6‰, the $\delta^{34}S$ values for galena from Poliegos show a strong enrichment of the light ^{32}S-isotope, with ratios between −2‰ and −15‰. Isotope temperatures calculated from this data range between 220° and 260°C, and are slightly higher than the temperatures derived from oxygen isotope geothermometry of three analyzed barite samples (Kalogeropoulos and Mitropoulos 1983).

Table 2. Examples of trace element and REE contents in ores from Milos and Poliegos. Values are in ppm where not indicated.

Mine	type of ore	Pb	Zn	Cu	Ag	As	Sb	Br	La	Ce
Kavos	BO	10 %	20	1000	–	b	–	–	–	–
Triades	BO	1.5%	2 %	0.5%	–	600	–	–	–	–
Kavos	SO	4 %	50	300	–	–	–	–	–	–
Triades	SO	550	0.5%	0.3%	–	70	–	–	–	–
Vani	MG	10%	0.5%	1000	–	500	–	–	–	–
Kavos	MG	500	3 %	400	–	300	–	–	–	–
Picridou	BMO	520	7	–	2	1	10	1.7	3.3	8.5
Agrilies	BMO	600	50	300	250	125	83	72	5.8	7.7
Kastanas	BMO	1020	7	–	11	b	4	2	1.6	3.3
Kavos	BIO	1500	170	–	12	54	75	6	5.6	2.4
Triades	BIO	3200	1200	–	28	55	54	4	10.4	2.6
Plaka	BIO	3500	250	135	–	b	–	–	–	–
Kavos	BVO	1830	10	–	2	3	6	2	2.4	1.9
Vani	BVO	7000	150	–	3	72	73	b	2.8	b
Poliegos	BVO	160	20	–	1	1	b	5	4.0	3.7

BO: Black ore; SO: Siliceous ore; MG: Manganese ore;
BMO: Metasomatic barite ore; BIO: Impregnative barite ore;
BVO: Barite vein;
b: below limit of detection; –: not analysed.

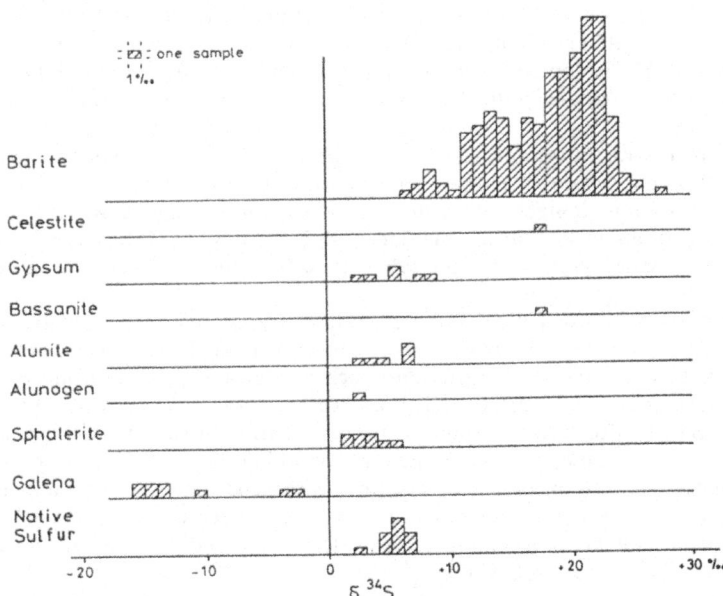

Fig. 6. Sulfur isotope ratios of various minerals from the islands of Milos, Kimolos, and Poliegos

Table 3. SrSO₄ content in various baryte ores from Milos and Poliegos islands.

Type of ore	Number of analyses	Range %	Mean Value %
Milos:			
Impregnation	35	1.5–4.5	2.7
Metasomatism	86	0.1–2.5	1.1
Veins	30	1.4–3.2	2.1
Poliegos:			
Veins	10	2.1–6.4	4.1

SrSO₄ content in barite was determined in some 160 samples. A summary of the results is listed in Table 3. The different types of barite ore are clearly reflected in the mean SrSO₄ content. Metasomatic ores show a relatively low value (1.1 wt.% SrSO₄) whereas vein samples from Poliegos are characterized by a higher value (4.1 wt.%).

Data Interpretation

Interpretation of the determined sulfur isotope ratios in sulfide and sulfate ores leads to the conclusion that seawater with a $\delta^{34}S$ value around $+20‰$ has to be seen as the major source for the entire sulfur. The observed isotopic trend can be explained by the inorganic reduction of seawater sulfate, resulting in low values for the sulfides and in high values for some corresponding barites (Shanks et al. 1981). Besides this, mixing of seawater sulfate with magmatic sulfide-sulfur has to be seen as source for the native sulfur. The isotope ratios are related to depth and time of ore formation, which may reflect variations in the degree of mixing between hydrothermal fluids and cold seawater.

In agreement with the interpretation obtained from sulfur isotope ratios are the observed SrSO₄ contents in the investigated barites. Consistent horizontal and vertical concentration changes within the orebodies indicate that seawater participated in the origin of the barites. Significant differences in the SrSO₄ contents may also reflect variation in the mixing ratios of pore solution and cold seawater during precipitation.

The comparative geochemistry of hydrothermally altered rocks and fresh volcanics may be a hint to the source of metals occurring in the Kuroko ores (Table 4). Silicified tuffs, kaoline, and bentonite which are the residues of strong hydrothermal leaching near the surface and groundwater level are depleted in the trace elements Sr, Ba, Mn, Cu, Pb, Zn, La, and Ce compared with their parental rock, fresh andesitic volcanics. On the other hand, the more immobile elements Ti and Zr are enriched in these alteration products. Together with geochemical mass balance calculations (Hauck 1984), the observed differences in trace element contents can be interpreted by hydrothermal leaching of fresh volcanics and redistribution of some of the mobile elements (Ba, Mn, Cu, Pb).

Chondrite normalized REE data sets for volcanic rocks and their alteration products (Fig. 7) show a strong enrichment in light REE and a negative Eu anomaly, caused by plagioclase-controlled fractionation of the parent magma (Innocenti et al.

Table 4. Selected analyses of fresh volcanics and alteration products from Milos, Kimolos and Poliegos islands.

	Andesitic pumice tuff	Rhyolite	Kaoline	Bentonite	Silicified tuff
Na$_2$O (%)	3.62	3.98	0.07	0.08	0.13
K$_2$O	1.82	3.09	0.08	1.85	0.14
CaO	5.21	1.29	0.12	0.60	0.01
MgO	1.33	0.25	0.05	1.22	0.12
Fe$_2$O$_3$	2.39	0.60	0.09	0.68	0.25
FeO	2.15	0.54	0.08	0.61	0.22
Al$_2$O$_3$	18.40	12.80	21.10	13.30	0.90
TiO$_2$	0.87	0.16	0.79	0.43	1.14
P$_2$O$_5$	0.15	0.04	0.19	0.06	0.01
MnO (ppm)	700	600	b	100	b
Sr	620	78	158	129	24
Ba	900	530	40	250	80
Cr	35	4	10	87	19
Co	12	b	b	b	b
Ni	21	4	1	18	7
Cu	35	b	7	1	2
Pb	35	12	5	13	26
Zn	188	28	7	41	12
Rb	79	91	2	93	9
Zr	162	95	230	115	420
La	108	–	34	32	0.5
Ce	174	–	52	71	2.0

b: below limit of detection; –: not analysed.

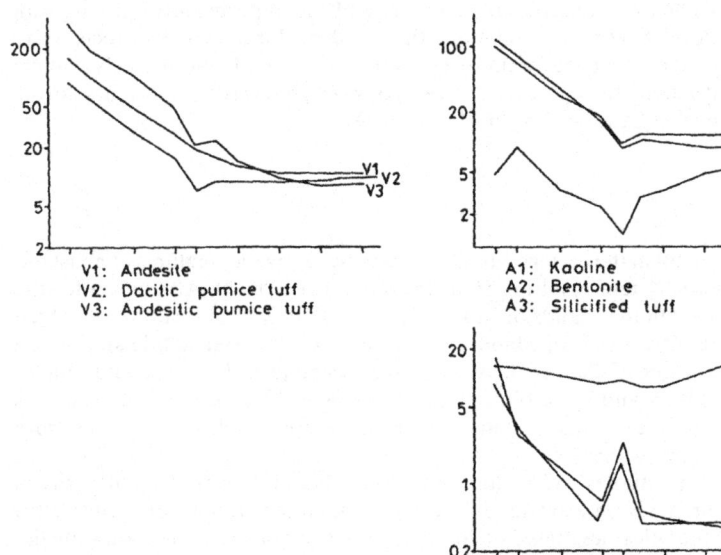

Fig. 7. Chondrite normalized REE pattern for fresh volcanics (V1-V3), alteration products (A1-A3) and barite ores (B1-B3)

V1: Andesite
V2: Dacitic pumice tuff
V3: Andesitic pumice tuff

A1: Kaoline
A2: Bentonite
A3: Silicified tuff

B1: Impregnation ore
B2: Metasomatic ore
B3: Vein ore

1981; Barton et al. 1983). The pattern of the silicified tuffs is characterized by slight enrichment of the heavy REE with a positive Ce anomaly resulting from the insolubility of Ce^{4+} and the heavy REE during hydrothermal leaching of the tuffs. The normalized pattern for barite therefore shows a distinctive positive Eu anomaly caused by an enrichment of the soluble Eu^{2+} in the ore solution (Guichard et al. 1979).

Origin of Ore Solution and Metallogenesis

Based on geochemical and geological studies, the following metallogenetic model of the Aegean islands Kuroko-type deposits is proposed. After subaqueous and subaerial deposition of the first pyroclastic series in the Upper Pliocene/Lower Pleistocene, seawater, circulating within the porous course-grained pumice flows, was gradually heated to a temperature of about 260°C at a relatively shallow depth, and convective flows were created in the vicinity of dacitic intrusives.

Continuous interaction between pore solutions (seawater) and volcanic tuffs led to hydrothermal leaching of the rocks which was controlled by the temperature, the redox conditions, and the pH-value. Subsequently, a typical alteration pattern was formed with the strongest leached rocks (impermeable, silicified tuffs) in the hanging wall layers.

The metal ions, which are interpreted as having been derived exclusively from the volcanic rocks, were probably transported in a solution of intermediate oxidized sulfur species (Oberste-Padtberg 1982), until they precipitated by oxidization during mixing with cold seawater in the upper part of the tuff series. The precipitation being spatially controlled by the groundwater level and the impermeable hanging wall tuff. The sealevel fluctuations of more than 140 m during the Pleistocene are therefore of great importance for the origin of these deposits. It is most likely that the hydrothermal activity took place over a period of only a few thousand years during the Calabrian transgression (Lower Pleistocene).

Summary

The Aegean region in the eastern Mediterranean Sea shows all features of an island arc-marginal sea system resulting from the subduction of the African under the European plate. On the islands of Milos, Kimolos, Poliegos, and Antimilos, which are part of the active Aegean island arc, several hydrothermal sulfide-sulfate ore deposits were formed. They occur within Pleistocene pyroclastics of andesitic to dacitic composition and show typical features of Kuroko-type deposits: siliceous (Keiko) ore, gypsum ore, black (Kuroko) ore, barite ore, which are associated with a strong alteration pattern.

Sulfur isotope studies and Sr determination in barite show that the majority of the sulfur originates from seawater-sulfate with a small contribution of pyrite-sulfur leached from the volcanics. Based on this data, the temperature of ore formation lies within 220° and 260°C. A comparison of geochemical data of ore, fresh volcanic rocks and alteration products show that Ba, Pb, Zn, and Cu were leached from the volcanic tuffs by continuous interactions with circulating seawater. Precipitation of the ores took place probably by oxidation of intermediate oxidized sulfur species caused by disequilibrium mixing of cold seawater with hot solutions close to the seafloor, and is controlled by a variable groundwater level during the Calabrian transgression.

Sulfide-sulfate deposits on the Milos group of islands are therefore considered as the youngest Kuroko-type mineralization, being even younger than the Miocene deposits in Japan.

Acknowledgments. This paper forms part of a Ph.D. thesis, carried out at the University of Karlsruhe (F.R.G.), Institut für Geochemie und Petrographie. I therefore wish to thank Prof. H. Puchelt, head of the Department, for his support. Many thanks are due to Prof. R. Stellrecht, Karlsruhe for fruitful discussions. Field work was carried out with kind permission of Silver and Baryte Ores Mining Co., Athens. I am also grateful to A. Hillier and B. Howe, London, who improved the English of the manuscript. The studies were financially supported by the Studienstiftung des Deutschen Volkes.

References

Barton M, Salters VJM, Huijsmans JPP (1983) Sr isotope and trace element evidence for the role of continental crust in calc-alkaline volcanism on Santorini and Milos, Aegean Sea, Greece. Earth Plan Sci Lett 63(2):273-291

Dewey JF, Şengör AMC (1979) Aegean and surrounding regions: Complex multiplate and continuum tectonics in a convergent zone. Geol Soc Am Bull 90:84-92

Ferrara G, Fytikas M, Giuliani O, Marinelli G (1980) Age of the formation of the Aegean active volcanic arc. In: Doumas C (ed) Thera and the Aegean world II. London, pp 37-41

Fytikas M (1976) Geology and geothermal research in the island of Milos. Thesis, University of Thessaloniki, Thessaloniki

Fytikas M, Giuliani O, Innocenti F, Marinelli G, Mazzuoli R (1976) Geochronological data on recent magmatism of the Aegean Sea. Tectonophysics 31:T29-T34

Guichard F, Church TM, Treuil M, Jaffrezic H (1979) Rare earths in barites: distribution and effects on aqueous partitioning. Geochim Cosmochim Acta 43:983-997

Hauck M (1984) Die Barytlagerstätten der Inselgruppe Milos/Ägäis (Griechenland). Ph D Thesis, University of Karlsruhe, Karlsruhe

Hubberten HW, Nielsen H, Puchelt H (1977) Sulfur isotope investigations on rocks and ores of the Santorini Archipelago, Greece. Ann Geol Pays Hell 28:334-348

Innocenti F, Manetti P, Peccerillo A, Poli G (1981) South Aegean volcanic arc: geochemical variations and geotectonic implications. Bull Volcanol 44:378-391

Kalogeropoulos SI, Mitropoulos P (1983) Geochemistry of barites from Milos island (Aegean Sea). Neues Jahrb Miner Mh 1983:13-21

Keller J (1982) Mediterranean island arcs. In: Thorpe RS (ed) Andesites. Wiley, New York, pp 307-325

Lambert IB, Sato T (1974) The Kuroko and associated ore deposits of Japan: A review of their features and metallogenesis. Econ Geol 69:1215-1236

Le Pichon X, Angellier J (1979) The Hellenic arc and trench system: a key to the neotectonic evolution of the Eastern Mediterranean area. Tectonophysics 60:1-42

Mack E (1977) Die Erkundung und Bewertung der Mineralvorkommen auf der Insel Milos (Griechenland). Berg Hüttenm Mh 122:48-57

Mc Kenzie D (1978) Active tectonics of the Alpine-Himalayan belt: The Aegean Sea and surrounding regions. Geophys J R Ast Soc 55:217-254

Oberste-Padtberg R (1982) Die Bedeutung des intermediär oxidierten Schwefels für die Genese hydrothermaler Sulfat-Sulfid-Lagerstätten. Ber Bunsen Ges Phys Chem 86:1038-1041
Puchelt H (1978) Evolution of the volcanic rocks of Santorini. In: Doumas C (ed) Thera and the Aegean world Vol 1. Athens, pp 131-146
Puchelt H, Hoefs J, Nielsen H (1971) Sulfur isotope investigations of the Aegean volcanoes. Acta 1st Int Scient Congress Volc Thera, Athens, pp 303-317
Puchelt H, Kramar U (1976) Standardization in neutron activation analysis of geochemical material. In: IAEA (ed) Nuclear techniques in geochemistry and geophysics. Wien, pp 245-252
Sato T (1977) Kuroko deposits: Their geology, geochemistry and origin. Spec Publ Geol Soc (Lond) 7:153-161
Sawkins FJ (1984) Metal deposits in relation to plate tectonics. Springer, Berlin Heidelberg New York
Shanks WC, Bischoff JL, Rosenbauer RJ (1981) Seawater sulfate reduction and sulfur isotope fractionation in basaltic systems: Interaction of seawater with fayalite and magnetite at 200-350°C. Geochim Cosmochim Acta 45:1977-1995
Shimazaki Y (1974) Ore minerals of the Kuroko-type deposits. Mining Geol Spec Issue 6:311-322

New Metallogenetic Aspects Concerning the Copper Deposit of Murgul, NE Turkey

N. ÖZGÜR and H.-J. SCHNEIDER[1]

Abstract

According to the plate tectonic concept, the East Pontic metallotect represents a volcanic island arc system of Jurassic through Tertiary age which hosts a great number of base metal deposits. The Murgul copper deposit is linked to a 250-m-thick felsic (dacitic) pyroclastic member of Upper Cretaceous age. The great majority of all base metal deposits in the East Pontides occupy the same stratigraphic position indicating a time- and strata-bound formation of the mineralization.

The mineralization of the Murgul deposit is spatially and temporally combined with an intense two-stage alteration of the host rock. A first phase of the ore mineralization is represented by predominating framboidal pyrites containing microscopic inclusions of chalcopyrite and sphalerite. The younger base metal paragenesis of economic range with considerable amounts of selenium and gold in ore lodes (stockwork-like veins) is the product of subsequent phases of mobilization and recrystallization.

The occurrence of framboidal pyrites in vast quantities in a true volcanogenic environment establishes a genetic problem unless they are considered to be formed (close to the surface) during a last stage of hydrothermal activity.

This time interval is proved by a subsequent short but intense stage of erosion and weathering forming a kaolinized "marker bed" which occurs locally throughout the entire East Pontides. The formation of the mineralization must have occurred just before this time interval close to the (subaerial) surface during a stage of hydrothermal discharge linked to few centers. The mineralization at Murgul does not traverse the "marker bed" up to the overlain nonaltered and barren felsic volcanites. Therefore the mineralization and alteration of the host rock must have been completed before the erosion took place.

Introduction

The Murgul copper deposit, located near the northeastern border of Anatolia, belongs to the East Pontic metallotect which hosts a great number of base metal occurrences (Dieterle 1986). According to the plate tectonic concept, it represents a (paleo) island arc system developed south of the Pontic microplate during the Lias

[1]Institut für Angewandte Geologie, Freie Universität Berlin, Wichernstr. 16, D-1000 Berlin 33, FRG

Base Metal Sulfide Deposits
G.H. Friedrich, P.M. Herzig (Eds.)
© Springer-Verlag Berlin Heidelberg 1988

through Miocene time period (Akın 1978). The East Pontides consist of a 2000 to 3000-m-thick sequence of volcanites with minor intercalations and lenses of marine sediments. The host rocks of the Murgul deposit are dacitic pyroclastics of Senonian age which have undergone an intense multistage alteration associated with the (subvolcanic) mineralization.

There appears to be a general correlation between $Cu/(Pb+Zn)$ ratios of the ore deposits and the intensity of volcanic activity, both of which increase from west to east. The feature of the ore deposits also tends to vary systematically. In the western part they are predominantly stratiform (Maucher 1960; Maucher et al. 1962), whereas in the eastern part they are generally strata-bound (stockworks, veins, and disseminations). Both types are confined to dacitic and andesitic volcanites (Özgür 1985).

Regional studies (Akın 1978; Dieterle 1986) have shown that the stratigraphical position of all base metal deposits (e.g., Keltaş, Lahanos, Madenköy, Murgul, and Akarşen) is of pre-Maastrichtian age except the Başköy deposit NW of Murgul. This minor vein-type mineralization seems to represent a secondary mobilization of ore minerals into Eocene pyroclastics during volcanic activity of a younger age.

The ore mineral paragenesis of the Murgul deposit is relatively uniform. Pyrite is quantitatively predominating, chalcopyrite is widespread, whereas sphalerite and galena occur only locally and sporadically. Fahlores and native gold are proved microscopically only.

The abandoned mine of Akarşen, some kilometers SW of Murgul, represents the same ore paragenesis and stratigraphic position. Therefore it has been studied briefly only in order to obtain a geochemical comparison.

Geological Setting

The 2000 to 3000-m-thick East Pontic volcanic sequence has been divided into three volcanic cycles (Maucher 1960; Maucher et al. 1962) (Fig. 1):

1. The first cycle comprises a volcanic pile deposited between Jurassic and Upper Cretaceous. It is characterized by a basal sequence of basaltic volcanics which changes to felsic lava flows and thick pyroclastics in the middle and top part of the cycle.

2. The second cycle starts with volcanic breccias, tuffs and minor intercalations of marine sediments overlain by andesitic and rhyolitic flows. The volcanic sequence is in turn overlain by limestones of uppermost Cretaceous age (Maastrichtian).

3. The late cycle consists of a basal sequence of marine sediments of Paleocene age which are overlain by andesitic and basaltic lava flows representing Tertiary volcanic activity.

The copper ore deposit of Murgul occurs within the upper part of the first volcanic cycle and is associated with a 250-m-thick felsic pyroclastic sequence whose upper contact is marked by a thin layer of marine sediments (Özgür 1985). The pyroclastics are overlain by 200–500 m barren felsic volcanites. The age of the mineralization in the pyroclastic sequence is indicated to be pre-Maastrichtian by paleontological observations (Buser and Cvetic 1973).

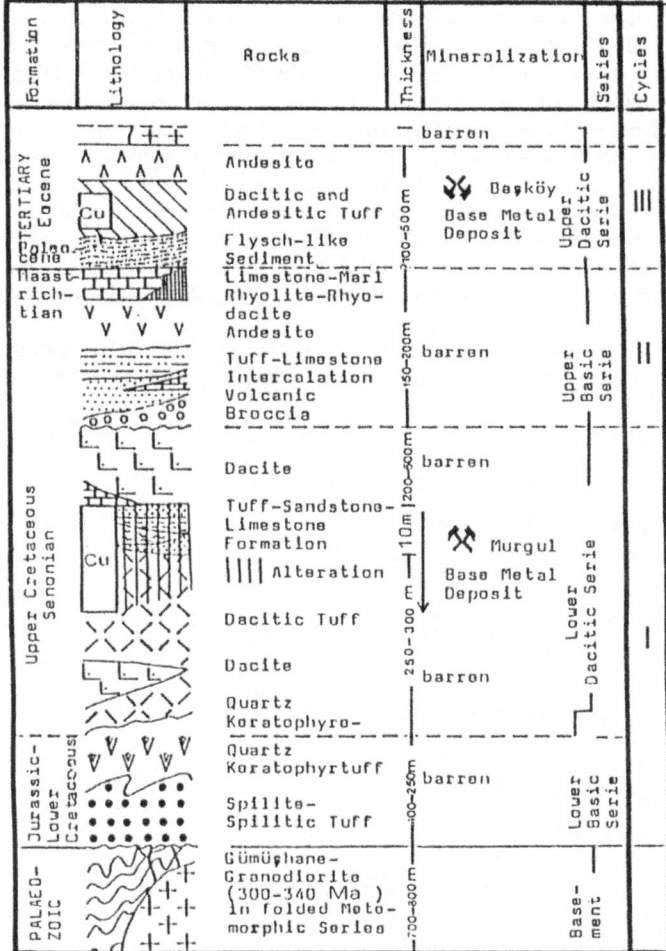

Fig. 1. Schematic column of the stratigraphic sequence in the Murgul area

The pyroclastic host rock of the mineralization shows intensive alteration. Thus, the primary mineral components of the volcanics can be observed only in few remaining parts. The fluidal groundmass contains fragments of plagioclase (An_{28-35}) and quartz phenocrysts, plagioclase microlites (An_{12-30}), relics of hornblende, biotite, quartz, and less apatite, sphene, and hematite (Özgür 1985).

Because of the general strong hydrothermal alteration, the origin and nature of the volcanic sequence can be defined only in few cases. At some places the clastic host rock is similar to an "ore-related breccia" according to the description of Sillitoe (1985).

Recent mining of the ore deposit in two open pits (Anayatak and Çakmakkaya, Figs. 2, 3) has provided excellent exposures of the lithostratigraphic record. The lithological boundary between the (mineralized) altered host rock and the overlying nonmineralized hanging wall has important significance for the genetical interpretation. The time period between the formation of the deposit and its associated alteration processes and the later transgression of the nonaltered hanging volcanites is marked by a short but intensive period of erosion. This short time interval is documented by a relatively thin sequence (max. 10 m) of strongly kaolinized tuffaceous matter with sandy and limey lenses which occurs at the top of the primary orebodies. Further evidence for this hiatus is the Bognari orebody, which can be interpreted as an erosional product of the upper parts of the Anayatak orebody (Mado 1972).

The intense host rock alteration may be divided into two stages, each of which is characterized by distinct mineral assemblages and their spatial distributions (Özgür 1985):

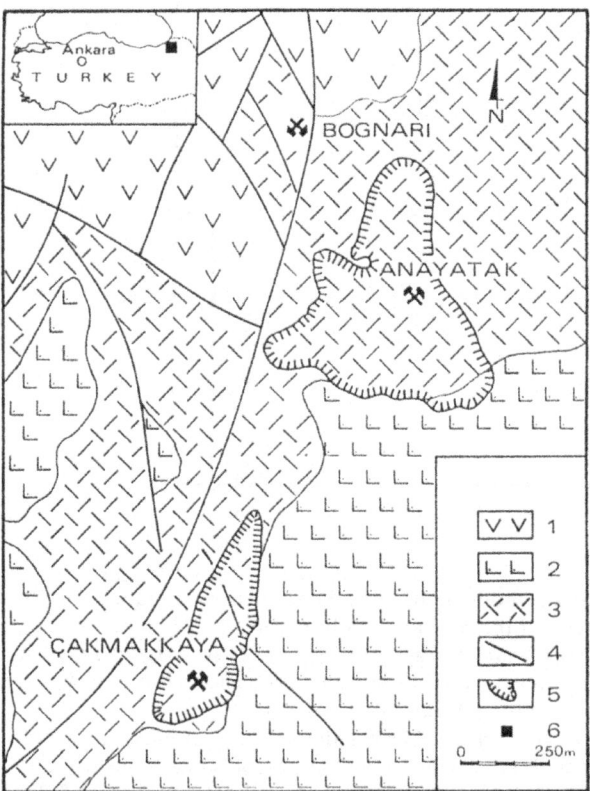

Fig. 2. Location and geological sketch map of the Murgul deposit *1* Andesitic lava flows; *2* hanging felsic volcanites; *3* dacitic tuffs: host rock of the orebodies; *4* main faults; *5* boundary of the open pits (state of mining: 1980); *6* studied area of Murgul

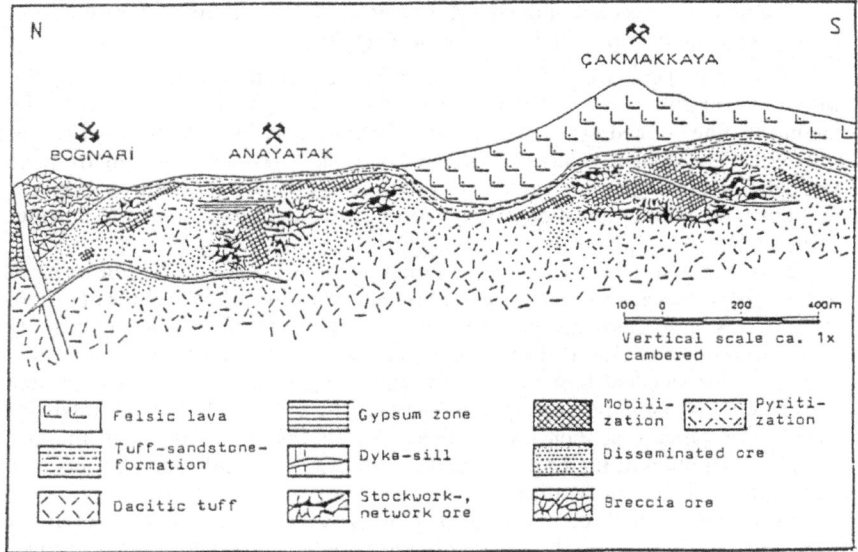

Fig. 3. Schematic cross section of the Murgul deposit generalized after Sawa and Sawamura (1970) and personal observations

1. The first stage consists of a phyllic alteration associated with both orebodies, which in turn is commonly surrounded by a halo of argillic alteration. The indicative alteration mineral assemblages consist of quartz-sericite-calzite-pyrite-chalcopyrite and quartz-montmorillonite-illite-pyrite, respectively. The disseminated generation of pyrite and chalcopyrite of this early alteration stage is of special importance for the genetical interpretation of the deposit as protore.

2. The second stage of alteration is represented by a pervasive silicification which is intimately related to the economic mineralization of the two orebodies. Its paragenesis consists of quartz-jasper-kaolinite. Pyrite, chalcopyrite, sphalerite, galena, and fahlore seem to represent the recrystallized constituents of the first phase of mineralization and form localized ore lodes.

The downward-narrowing funnel-shaped geometry of the alteration features implies a depth restriction of the mineralization in the two orebodies, as established by drilling.

Types and Features of Ore Mineralization

The Murgul deposit consists of at least two primary orebodies (Anayatak and Çakmakkaya, Figs. 2, 3) hosted in the same volcanic member, spanning a horizontal distance of about 500 m. According to former descriptions (e.g., Sawa and Sawamura 1970; Mado 1972) and our observations and interpretations of drilling profiles, the mineralization of both orebodies shows exactly the same feature.

The following account of the development and the distribution of the different ore types therefore applies to both deposits (Fig. 3):

1. An early generation of disseminated pyrite is spatially associated with the widespread argillic halo of the first stage of alteration. As observed microscopically, the mostly fine grained pyrite occasionally contains inclusions of chalcopyrite and sphalerite. Consequently this ore type shows varying Cu contents ranging from 0.2 through 0.7%. It may be interpreted as a protore stage. Analyses of drill cores reveal that this type of mineralization extends downwards for about 150 m and narrows with depth.

2. In addition to the disseminated mineralization, a second type consisting of mineralized lenses and veinlets occurs. This mineralization forms the bulk of the ore at Anayatak and Çakmakkaya with Cu contents of 1.0–2.5%. The high grade mineralization is stockwork-like (Fig. 3), indicating a phase of tension and disruption of the silicified host rock which might be generated by various volcanic activities. This ore type corresponds spatially to the distribution of intensive silicification (alteration stage two) which must be, however, older than the mechanical disruption of the host rock and its low grade mineralization because the stockwork veins are crosscutting it.

Furthermore, the breccia ores of Bognari with average Cu contents of about 1.0% may be added to this type according to its economic grade. As an erosional product of ore type two from Anayatak, it represents a younger formation of local importance only.

3. A third type is made up by local ore lodes – mostly relatively short and open veins – containing 5–10% Cu. It seems to constitute a last stage of remobilization after repeated tectonic movements. The late mineralization is characterized by a local growth of chalcopyrite and pyrite crystals in a range of about 10 mm.

The temporal succession of these three phases of ore formation is documented by clear crosscutting relationships. The early generation of disseminated pyrite together with its altered host rock is crosscut by the second-phase stockwork (2) mineralization; the third phase (3) finely crosscuts all the older ones.

There is no evidence to determine whether the ore constituents of types two and three represent remobilization of type one or represent a subsequent volcanic supply, unless the different Se contents of the ores are considered to be diagnostic of different hydrothermal events. The level of Se contents of chalcopyrite concentrates is mainly low in the first type (about 8 ppm), but significantly higher in the second and third type (Anayatak 60 ppm mean, Çakmakkaya 110 ppm mean) with maximum contents of about 400 ppm in ore lodes of type three (Özgür 1985).

The enrichment of Se in the second and third ore generation could alternatively be interpreted as an indication for the intensive hydrothermal alteration processes generating a multistage remobilization and recrystallization of the ore constituents (Fig. 4). This interpretation is supported by the existence of microscopic inclusions of the Se-fahlore hakite in the chalcopyrites of the second and third generation mainly from Çakmakkaya (Özgür 1985). The enrichment of gold in the two younger ore types might be regarded in the same way (Table 1, Fig. 5).

Finally, the veins of type three generally crosscut the stockwork-like mineralization in both orebodies. Because of their frequent open space crystallization,

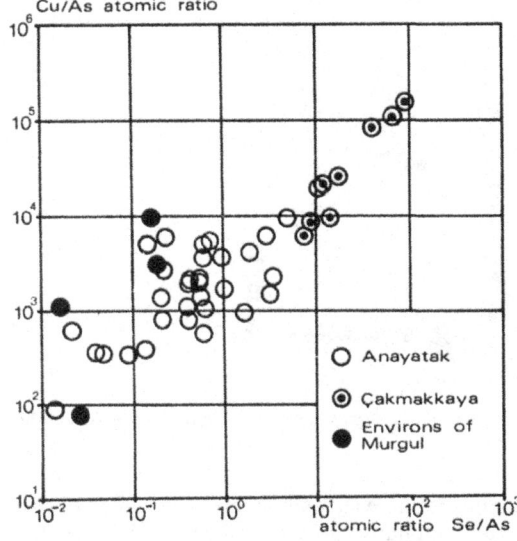

Fig. 4. Plot of Se/As vs. Cu/As in chalcopyrite from Anayatak, Çakmakkaya, and environs of the Murgul deposit (n = 44 samples)

Table 1. Au contents of chalcopyrite and pyrite in the copper deposit Murgul (Anayatak, Çakmakkaya), and the abandoned mine of Akarşen near Murgul. All samples from ore types (2) and (3) (see text)

Sample	Locality	Mineral	Au(ppb)
1	Anayatak	Chalcopyrite	280
2	Anayatak	Chalcopyrite	160
3	Anayatak	Chalcopyrite	200
4	Anayatak	Chalcopyrite	130
5	Çakmakkaya	Chalcopyrite	10
6	Çakmakkaya	Chalcopyrite	140
7	Çakmakkaya	Chalcopyrite	130
8	Çakmakkaya	Chalcopyrite	90
9	Anayatak	Pyrite	900
10	Akarşen	Pyrite	110
11	Akarşen	Pyrite	340

textures might have been formed at a relatively high level beneath the late-volcanic surface. In the subsequent erosion phase the uppermost part of the orebodies has been cut away and was locally reconcentrated, for instance in the Bognari orebody.

Ore Fabrics

Microscope investigations indicate that the ores consistently comprise a base metal mineralization. Beside the prevalent content of pyrite, the economically important chalcopyrite, in various stages of recrystallization, is ubiquitous. Sphalerite and

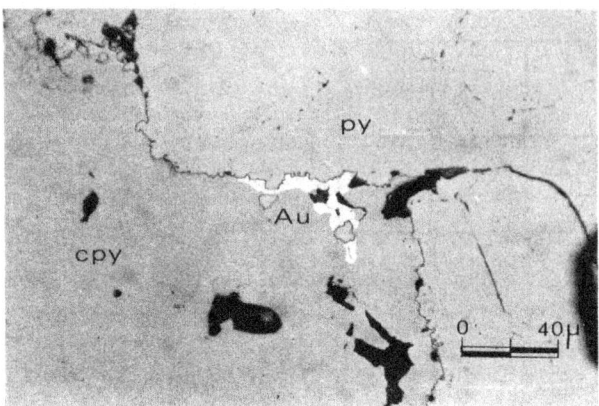

Fig. 5. Native gold (*Au*) at the boundary of pyrite (*py*) and chalcopyrite (*cpy*), Akarşen deposit. Polished section, plane polarized light, oil immersion

galena occur only sporadically, mainly in the Anayatak orebody. In some polished sections, fahlores (tennantite, hakite) and native gold were detected. In vein chalcopyrite of ore type three very small inclusions of a light fahlore are observed which probably represent the Se-fahlore hakite. Its occurrence corresponds with the maximum Se contents of the type three mineralization mainly at Çakmakkaya (Fig. 4).

A peculiar problem is established by the early pyrite generation. The genetical contradiction concerns the appearance of colloform pyrite grains in vast quantities which are interpreted generally as products of a very low thermal environment (e.g. 100°C). Nevertheless in this case they must have been precipitated in the pyroclastic sequence during the late stage of the volcanic activity in a hydrothermal range of about 280°C (Özgür 1985).

Pyrite framboids with an average size of 15 μm are very common (Fig. 6) in the silicified groundmass of the orebodies, quantitatively decreasing towards the transition into the unaltered country rock. Occasionally they occur (like "xenoliths"?) together with the stockwork type two mineralization and, also rarely, in type three lodes. Frequently, the framboids are recrystallized forming euhedral pyrite crystals. These euhedral crystals have a "clean" polished surface, whereas the framboids very often contain small inclusions of chalcopyrite and less commonly sphalerite. Obviously, during recrystallization the iron sulfide of the first generation segregates its base metal content, generating the corresponding sulfides of the second generation.

In addition to the small framboids, a second type of primary colloform iron sulfide occurs occasionally as larger grains ranging from 150 μm to 1 mm in diameter and possessing different shapes. These larger grains consist of roundish pyrite spherulites which are composed of a nodular core coated by one or two colloform layers (Fig. 7). Mostly, this type shows all stages of recrystallization into "clean" pyrite crystals. The transformation from the spherulitic to crystal habit is documented by pyrite grains which exhibit concentric arcs of very small inclusions of gangue minerals that outline the original colloidal texture.

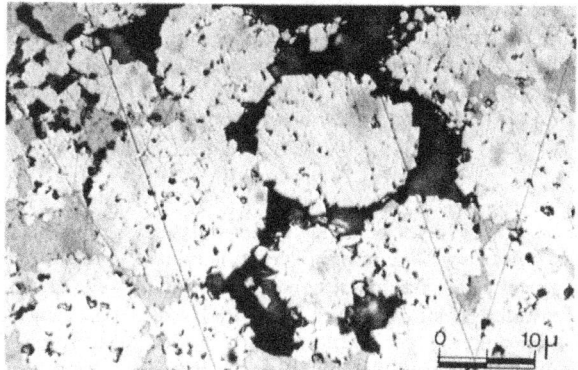

Fig. 6. Pyrite framboids of small size (ϕ 15 μm) containing small chalcopyrite inclusions in altered pyroclastics. Anayatak open pit. Polished section, plane poralized light, oil immersion

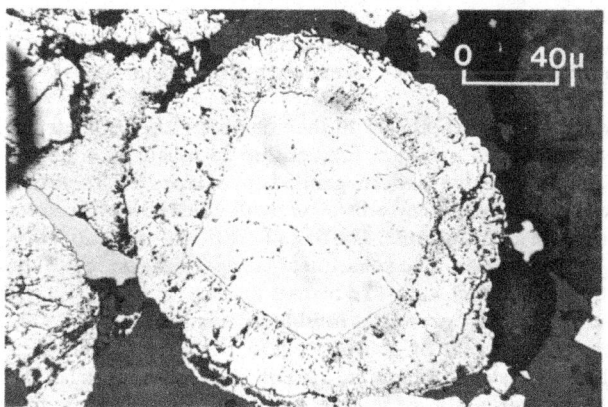

Fig. 7. Pyrite spherulite (ϕ 150 μm) with a roundish core and two younger "concentric layers" indicating colloidal formation. Akarşen deposit. Polished section, plane polarized light, oil immersion

The genesis of the colloform pyrite in the volcanic sequence presents a problem that has not yet been satisfactorily resolved. The formation of pyrite framboids has been interpreted as the result of bacterial activity in an euxenic sedimentary environment (Schneiderhöhn 1923; Love 1957). Later Love (1964) explained them as precipitates in the presence of organic matter during an early diagenetic process. In the same way Kalliokoski and Cathles (1969) considered their origin to be "probably formed by a process of coacervation and particulation due to the reaction of ferruginous humic acid and biogenic H_2S".

Love and Amstutz (1969) describe two occurrences of pyrite framboids within Permian andesites of Plötz near Halle (Germany) and in Tertiary volcanites near Antachajra (Peru), providing evidence for the formation of colloidal Fe-sulfides in volcanic rocks. The authors propose the term "concentric clusters" for these textures but did not offer any genetic interpretation.

In an ore-microscopic study of the deposits of the East Pontic metallotect, Vujanovic (1974) described pyrite framboids from many places including Murgul and Akarşen. He suggested that the pyrite framboids were proof of an exhalative-sedimentary origin.

Genetic Considerations

The Murgul deposit can be assigned to a subvolcanic formation associated with an Upper Cretaceous island arc volcanism. The mineralization is related to a distinct member of a thick dacitic sequence and may be designated, like all deposits of this type throughout the East Pontides, as strata-bound in a wider sense. The mineralization at Murgul must have taken place predominantly under subterrestrial conditions on isolated islands. The sparsely and only locally intercalated marine sediments into the thick volcanic sequence and the shallow water facies of the *Globotruncana* containing limestone lenses as well as the chemistry of the volcanites (Akın 1978) indicate the development of an island arc. The origin of the deposit can be inferred from the following observations:

The host rock, a thick pyroclastic member, has been altered and mineralized during a late stage of the dacitic volcanism by ascending hydrothermal fluids close beneath the terrestrial surface.

The shape of the two primary orebodies with their characteristic geometry of concentric alteration zones implies a spatial relationship to localized zones of ascendant hydrothermal fluids. There are, in general, two types of alteration according to microscope observations, an early distal one of phyllic and argillic zones and a later silicific alteration in a central zone. The first phase of ore mineralization occurs with the first stage of alteration. The subsequent recrystallization led to the enrichment of ore constituents in stockwork-like bodies.

The formation of the two primary orebodies must have been completed before a short interval of intense erosion cut off the uppermost part of the orebodies. Erosional products of this process are concentrated locally in the clastic part of the minor Bognari orebody. The erosional time interval is marked by a relatively thin sequence of strongly kaolinized tuffaceous matter with sandy and limey lenses at the top of the orebodies which are indicative of surficial influences. These characteristic intercalations seem to be distributed like a "marker bed" throughout the entire East Pontic metallotect but have been documented only at scattered localities.

The mineralized and intensely altered volcanics are overlain by a series of relatively low-grade altered and barren felsic volcanites. It should be noted that the mineralization nowhere traverses the "marker bed". This fact emphasizes that the deposits were formed prior to the erosional interval and the later eruption of the hanging felsic volcanites. Locally, the mineralized host rock shows features similar to the "ore-related breccias" described by Sillitoe (1985), suggesting a subsurface brecciation generated by volcanoplutonic activity contemporaneous with tuff sedimentation.

It is suggested that shortly after the period of eruption a subaerial stage of hydrothermal discharge took place causing the strong alteration of the host rock for some tens of meters below the surface. During a late stage of decreasing temperature the supply of sulfur, possibly originating from upwelling S-bearing fluids, caused the precipitation of the colloform pyrites unless the existence of organic matter is conceivable for this environment.

According to some common geological and geochemical features, the deposit is comparable with some deposits of the Fiji islands (Colley and Rice 1975; Özgür 1985). On the other hand, Maucher and Höll (1968) mention that Murgul may be interpreted as a transitional type tending to copper porphyries. In this relation the predominance of gold over molybdenum in chalcopyrite of Murgul could be regarded as a further indication to island arc position of the mineralization (Horton 1978).

In contrast to the western part of the metallotect where submarine hydrothermal activity in a volcano-sedimentary sequence caused the stratiform deposits (Maucher 1960), in the eastern part subvolcanic mineralization originated on isolated volcanic islands under mainly subterranean conditions close to the terrestrial surface. The chain of deposits between Lahanos in the west and Murgul in the east reveals all steps of transition with respect to ore paragenesis and geological environment.

References

Akın H (1978) Geologie, Magmatismus und Lagerstättenbildung im ostpontischen Gebirge/Türkei aus de Sicht der Plattentektonik. Geol Rundsch 68:253–283

Buser S, Cvetic S (1973) Geology of the environs from the Murgul copper deposit, Turkey. MTA Bull 81:22–45 (in Turkish)

Colley H, Rice CM (1975) A Kuroko-type ore deposit in Fiji. Econ Geol 70:1373–1386

Dieterle M (1986) Zur Geochemie und Genese der schichtgebundenen Buntmetall-Vorkommen in der Ostpontischen Metallprovinz/NE-Türkei. PhD Thesis, Freie Universität Berlin

Horton DJ (1978) Porphyry-type copper-molybdenum mineralization belts in Eastern Queensland, Australia. Econ Geol 73:904–921

Kalliokoski J, Cathles L (1969) Morphology, mode of formation, and diagenetic changes in framboides. Bull Geol Soc Finl 41:126–133

Love LG (1957) Micro-organisms and the presence of syngenetic pyrite. QJ Geol Soc Lond 113(4):429–440

Love LG (1964) Early diagenetic pyrite in fine-grained sediments and the genesis of sulfide ores. In: Amstutz GC (Ed) Sedimentology and ore genesis. Elsevier, Amsterdam, pp 11–17

Love LG, Amstutz GC (1969) Framboidal pyrite in two andesites. Neues Jahrb Miner Mh 3:97–108

Mado H (1972) Geology and mineralization of the copper ore deposits in the Murgul mine, north eastern Turkey. Int Rep MTA Ankara 1103:1–27

Maucher A (1960) Die Kieserze von Keltas. Ein Beispiel submariner Gleitfaden in exhalativ-sedimentären Erzlagerstätten. Neues Jahrb Miner 94:495–505

Maucher A, Höll R (1968) Die Bedeutung geochemisch-stratigraphischer Bezugshorizonte für die Altersstellung der Antimonitlagerstätte von Schlaining im Burgenland, Österreich. Miner Dep 3:272–285

Maucher A, Schultze-Westrum H, Zankl H (1962) Geologisch-lagerstättenkundliche Untersuchungen im ostpontischen Gebirge. Bayer Akad Wiss Math Naturwiss Kl Abh 109:1–97

Özgür N (1985) Zur Geochemie und Genese der Kupferlagerstätte Murgul, E-Pontiden/Türkei. Ph D Thesis, Freie Universität Berlin

Sawa T, Sawamura K (1970) Report about the Murgul ore deposits and its environs. Int Rep Etibank (Ankara) 24/300:1–24 (in Turkish)

Schneiderhöhn H (1923) Chalkographische Untersuchung des Manfelder Kupferschiefers. Neues Jahrb Miner Geol Paläontol 47:1–38

Sillitoe RH (1985) Ore-related breccias in volcanoplutonic arcs. Econ Geol 80:1467–1514

Vujanovic V (1974) The basic mineralogic, paragenetic and genetic characteristics of the sulphie deposits exposed in the eastern Black-Sea coastal region (Turkey). MTA Bull 82:21–36

Geochemical Features of Magmatic Evolution and Ore Deposition in the Pyrite Belt of Southern Spain

W. SCHÜTZ[1], P. DULSKI[2] and K. GERMANN[1]

Abstract

In the SW-Iberian Pyrite Belt, hydrothermal phases and related mineralizations are lithostratigraphically controlled. The sulfide ores occupy a lower stratigraphic position and are linked to quartz-keratophyric volcanites, whereas the overlying manganese-jasper mineralization occurs on top of a more intermediate to basic volcanic pile. The mineralizations can be located by hydrothermal alteration (e.g., chloritization) and dispersion (e.g., As-, Sb-, Tl-halos) patterns which may be very useful as exploration tools. The observed lithosetting of the Pyrite Belt mineralizations is suggestive of a subduction-related island-arc environment. A highly evolved, specialized magmatism has produced similarly advanced polymetallic mineralizations.

Introduction

The Iberian Pyrite Belt of Southern Spain and Portugal represents a metallogenetic province with one of the largest accumulations of stratiform sulfides in the world. The Belt extends over more than 200 km from near Seville in the east to the western coast of Portugal and hosts a variety of very important pyrite, base metal, and gold deposits (Fig. 1). It also contains countless occurrences of strata-bound manganese ores, some of which were being mined until the late 1970's. After more than one century of geological investigations and some 4000 years of mining, the Pyrite Belt is still the focus of geological discussions and exploration efforts.

Initially stemming from·investigations on the scientifically neglected manganese deposits, our own work later concentrated on determining the geological position of both the manganese and sulfide deposits within their volcano-sedimentary host sequence in order to derive "proximity indicators" (Möller et al. 1983a,b; Germann et al. 1986). "Position" here can have different meanings: it may describe the lithostratigraphical, geochemical, or petrological, as well as the structural control of the mineralization.

We soon became involved in some of the violent disagreements on Pyrite Belt geology [demonstrated, e.g., by the vehement discussion between Routhier et al. (1980), and Bernard and Soler (1980) on lithostratigraphy, petrology, tectonic

[1]Freie Universität Berlin, Institut für Angewandte Geologie, Wichernstraße 16, D-1000 Berlin 33, FRG
[2]Hahn-Meitner-Institut, Glienicker Str-100, D-1000 Berlin 39, FRG

Base Metal Sulfide Deposits
G.H. Friedrich, P.M. Herzig (Eds.)
© Springer-Verlag Berlin Heidelberg 1988

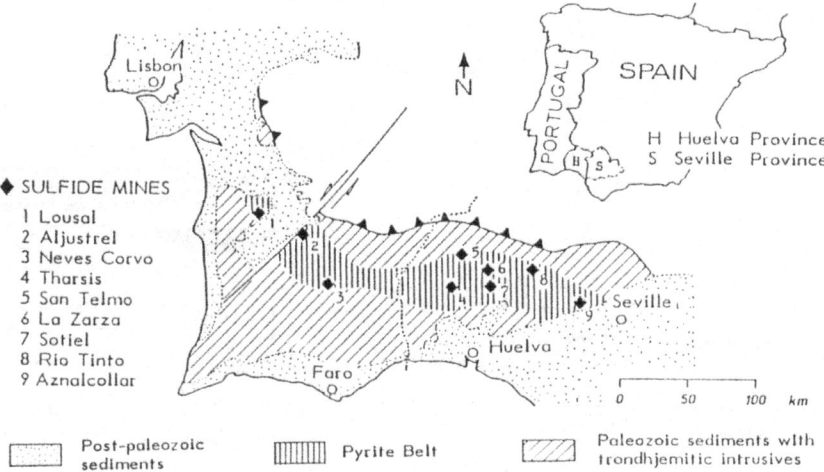

◆ SULFIDE MINES

  1 Lousal
  2 Aljustrel
  3 Neves Corvo
  4 Tharsis
  5 San Telmo
  6 La Zarza
  7 Sotiel
  8 Rio Tinto
  9 Aznalcollar

Fig. 1. Sketch map of the Pyrite Belt

setting, and geotectonic situation]. It is the intention of this chapter, therefore, to summarize some ideas on these controversial topics (see Schütz et al. 1985), based on our own geochemical and petrological findings, and to conclude with some speculations on magmatic evolution, geotectonic position, and metallogeny of the Pyrite Belt. Admittedly, each of these topics is worth a paper of its own; for a more complete range of arguments the reader is referred to Schütz (1985).

Field investigations were concentrated on the western central part of the Belt on the northern flank of the Rio Tinto Syncline, north of the Rio Tinto mine. The results may be valid for this particular area only, but in their essential aspects may be applicable to the whole metallogenetic province.

Stratigraphy and Regional Tectonic Setting

In the Pyrite Belt, stratiform massive pyrite ore bodies associated with disseminated sulfides of Cu, Zn, and Pb are hosted by a Lower Carboniferous volcano-sedimentary sequence of marine origin. There is no controversy concerning the general lithostratigraphic situation, which is illustrated in Fig. 2. The ore-bearing Volcano-Sedimentary Complex is underlain by clastic metasediments of the Devonian Phyllite Quartzite (PQ) Group and overlain by Carboniferous slates and graywackes of Culm and flysch facies (Strauss 1965; Schermerhorn 1971). Due to lateral facies changes and strong deformation, the detailed stratigraphic subdivision and correlation of the ore-bearing sequence is very difficult. The number of lithostratigraphic concepts tends to be as great as the number of geologists working on the problem (Rambaud Perez 1969; Strauss and Madel 1974; Routhier et al. 1980; Soler 1980; Grimmeisen 1983; Sierra 1984; Bernard and Soler 1985; Germann and Schütz 1985; Schütz 1985).

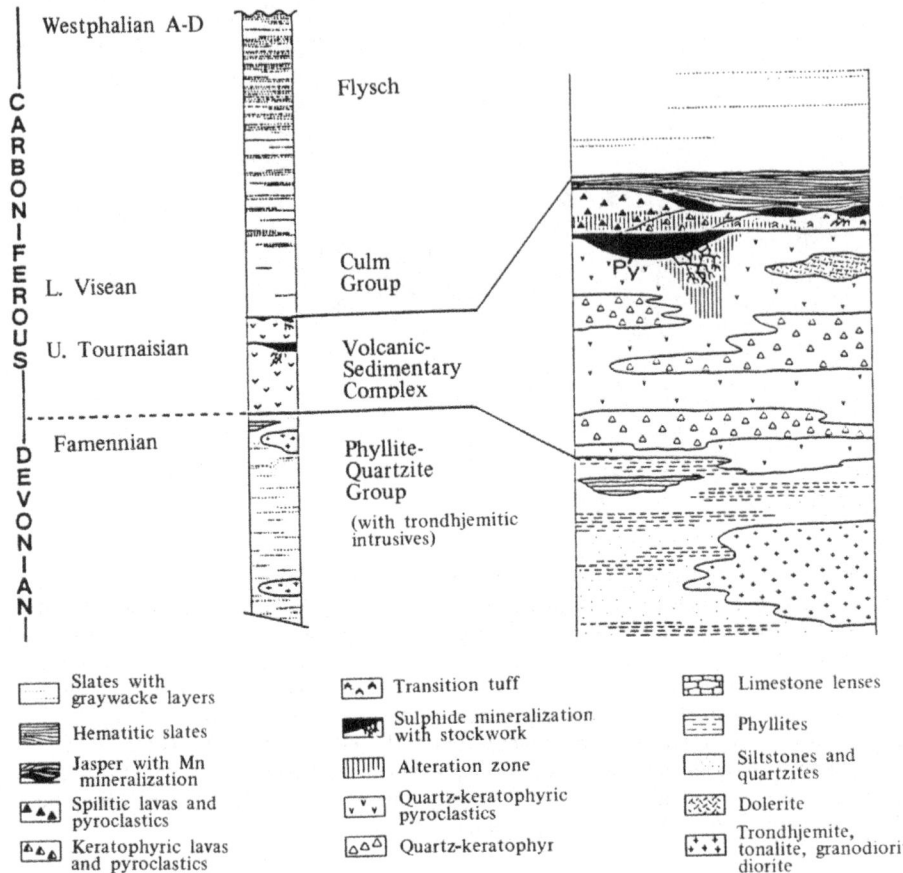

Fig. 2. Stratigraphic column for sulfide and manganese mineralization in the NW central part of the Pyrite Belt

In our study area a comparatively simple two-stage volcanic cycle exists which develops from hyperacid to acid, intermediate, and mafic products with keratophyres and spilites (Fig. 2). Sulfide ores were deposited during the waning stage of a quartz-keratophyric, mainly explosive, phase, and the jaspilitic Mn-Fe formation originated from the terminal hydrothermal event of the whole volcanic sequence. Multiple repetition of ore horizons, obvious as four east-west-trending bands of mines or mineral occurrences, is a structural repetition by intense isoclinal folding and imbrication (Fig. 3), and not a stratigraphic recurrence (Soler 1980; Schütz 1985). Regional metamorphism increases from the very low grade in the south to the low grade in the north (Schermerhorn 1975a; Munhá 1983a; Schütz 1985).

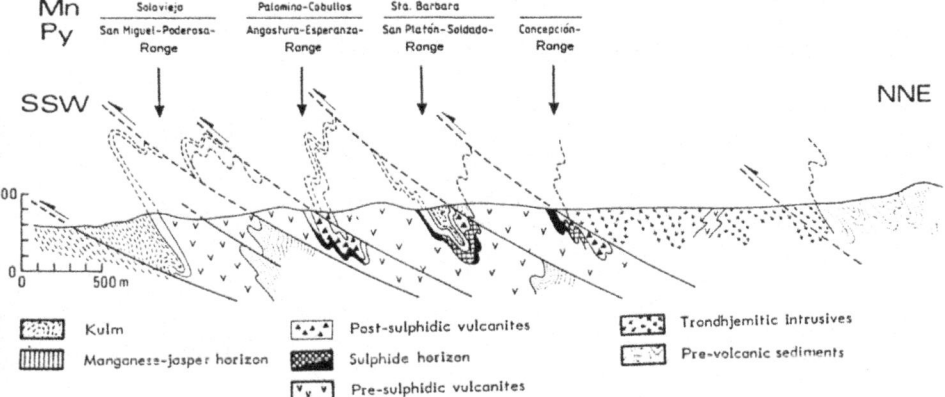

Fig. 3. Cross-section of the northern flank of the Rio Tinto-Syncline between Minas de Soloviejo (S) and Mina Concepción (N)

Magmatic Evolution and Associated Hydrothermal Formations

Magmatic evolution in the Pyrite Belt is incompletely understood if one regards the volcanites alone. Intrusive rocks such as trondhjemites, tonalites, and diorites exclusively bound to the PQ-Group are their pretectonic comagmatic precursors, the former being found locally as ejecta in quartz-keratophyric pyroclastics (Schütz et al. 1987).

The volcanic series — adequately classified as a typical spilite-keratophyre association — in the northern part of the Huelva province is completed or partly replaced by a porphyritic calc-alkaline sequence with rhyolites, dacites, andesites, and basalts, with less well-developed mineralizations.

If one analyzes the petrochemical evolution of the ore-bearing volcanic pile in more detail, in spite of an overall continuity, a chemical hiatus between the hyperacid quartz-keratophyres and the less acid volcanites becomes obvious (Figs. 4, 5). This gap at about 71% SiO_2 corresponds with a change in the lithostratigraphic succession, and marks the position of the first hydrothermal event responsible for the sulfide deposition. The second hydrothermal event, causal for the manganese enrichment, at least in some parts of the Pyrite Belt, reveals a clear spatial coincidence with the termination of the mafic volcanic activity, and, in general, with the onset of normal epiclastic sedimentary conditions. Accordingly, hydrothermal events and their products occupy particular lithostratigraphic horizons (Fig. 6).

Products of these hydrothermal events are

— stratiform sulfide and manganese ores,
— hydrothermal sediments,
— hydrothermal alterations,
— hydrothermal dispersion patterns.

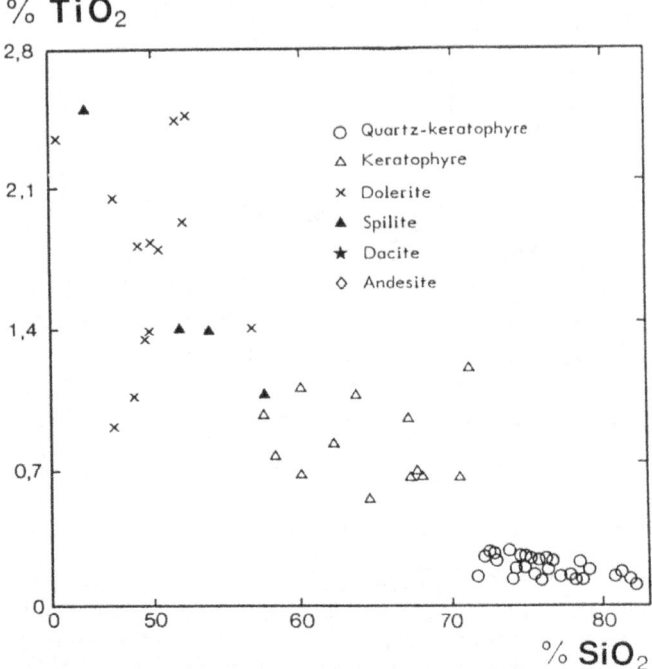

Fig. 4. Composition of Pyrite Belt volcanites in the TiO_2/SiO_2 diagram. Pre-sulfide quartz-keratophyres are separated from the post-sulfide intermediate to basic volcanites. Dacite and andesite appear in Fig. 5 only

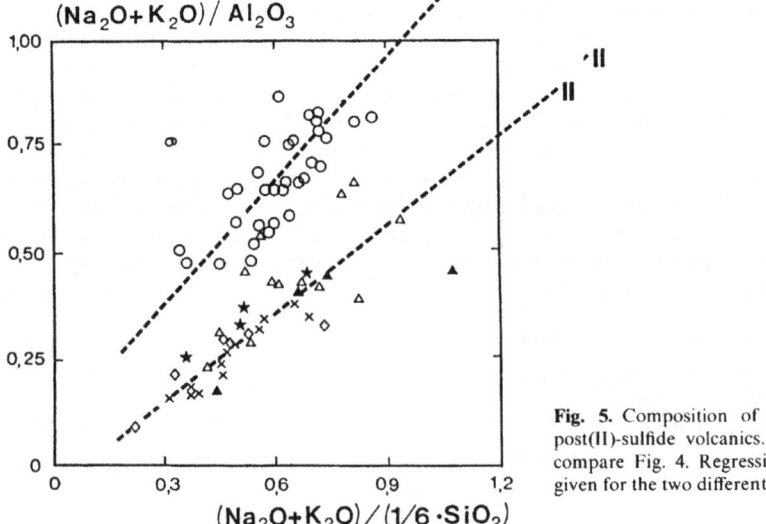

Fig. 5. Composition of pre(I)- and post(II)-sulfide volcanics. For legend compare Fig. 4. Regression lines are given for the two different populations

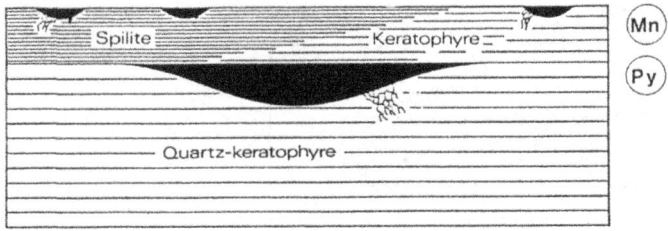

Fig. 6. Schematic representation of lithostratigraphical control of the mineralization in the Pyrite Belt

The Ores

Classification of the different sulfide ore types is best achieved on the basis of their chemical composition. In a $Cu + Zn + Pb + Ba/Fe$ diagram (Fig. 7, Schütz 1985), massive sulfur ores, the banded base-metal-rich complex ores and the nonmassive ores such as stockwork ores, siliceous ores and the low grade azufrones (Pinedo Vara 1963) are clearly distinguishable.

Stratiform barite concentrations not only in the sulfide ores but also in the Mn-Fe formation are by far more important than previously recognized in the Pyrite Belt. Locally, stratiform barite and even gypsum bodies are developed and provide close similarities with the Kuroko-type ores.

The occurrence of complex sulfide ores viz. of Zn-Pb- and Ba-rich ores with their elevated concentrations of the trace metals As, Sb, Tl, Cd, Se and Sn (Tables 1,2) could be interpreted in the sense of a more pronounced metallogenetic evolution or specification of the Rio Tinto and Kuroko ore types compared with the more "primitive" pyrite/chalcopyrite ores of the Cyprus type. Advanced metallogenetic differentiation in this case may be attributed to a higher degree of magmatic evolution.

The jaspilitic manganese deposits of the second hydrothermal event are characterized by rhodochrosite, rhodonite, braunite, and spessartite. In the jasper, up to 30% of Fe_2O_3 is concentrated in the form of hematite and more rarely as magnetite.

Hydrothermal Alteration, Element Dispersion and Their Use as Proximity Indicators

Of immediate practical interest for exploration purposes are the hydrothermal sediments, alterations, and dispersions accompanying the orebodies (Fig. 8).

Footwall, hanging wall and lateral stratigraphic equivalents of the sulfide orebodies are the sites of hydrothermal alterations resulting mainly in chloritization, sericitization, and carbonatization. Schütz (1985) combined alteration-related gains in Mg, Fe, Al and losses in Na and Ca to characterize the hydrothermal alteration path (Fig. 9). K decreases in the course of chloritization and carbonatization and increases with sericitization. Depending on the type of alteration, CO_2 and H_2O may also be enriched. Concentration of all these elements is actively changed during the alteration processes, whereas Al is passively enriched in relation to K and Na.

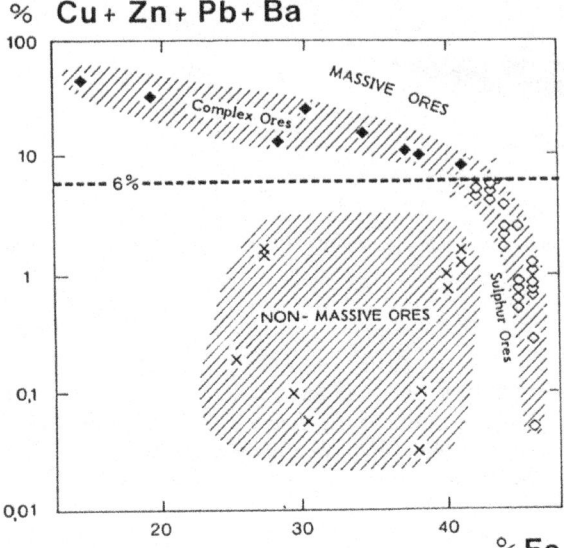

Fig. 7. Classification of Pyrite Belt sulfide ores

Table 1. Sulfur ores of Huelva and Cyprus

		Huelva		Cyprus
		Median (22 samples)	Range	Contents (1 sample)
%	SiO$_2$	0.64	0.03 – 3.42	0.04
	Al$_2$O$_3$	0.15	0.01 – 0.60	0.91
	MgO	0.03	0.002 – 1.99	0.02
	CaO	0.009	0.0006– 0.67	0.09
	P$_2$O$_5$	0.002	0.0002– 0.64	0.008
	Ba	0.004	0.0001– 4.70	0.0001
	Fe	45.00	42.00 – 46.00	46.00
	Cu	0.40	0.01 – 1.20	0.002
	Zn	0.32	0.02 – 4.40	0.005
	Pb	0.10	0.02 – 1.10	0.004
	As	0.24	0.03 – 1.20	0.01
ppm	Sb	71	4 –270	1
	Tl	10	1 – 36	2
	Cd	60	20 –160	13
	Se	17	2 – 70	1
	Sn	43	3 –350	4
	Mo	4	1 – 36	5
	In	2	1 – 51	–
	Ge	1.5	0.3 – 7.3	1
	Ga	1.5	0.3 – 9.5	–

Table 2. Complex ores of Huelva

		Median (8 samples)	Range
%	SiO_2	1.56	0.11 – 23.53
	Al_2O_3	0.07	0.02 – 0.81
	MgO	0.01	0.002 – 2.65
	CaO	0.004	0.0007– 0.05
	P_2O_5	0.0007	0.0005– 0.002
	MnO	0.025	0.007 – 0.12
	Ba	4.26	0.0001– 32.0
	Fe	32.0	14.0 – 41.0
	Cu	0.22	0.07 – 0.69
	Zn	5.45	0.03 – 12.0
	Pb	4.75	0.37 – 11.0
	As	0.18	0.09 – 1.30
ppm	Sb	235	27 –410
	Tl	33	13 –100
	Cd	130	40 –300
	Se	6	3 – 12
	Sn	290	4 –810
	Mo	6	2 – 20
	In	6	5 – 9
	Ge	1.5	0.5 – 18
	Ga	3.5	0.4 – 18

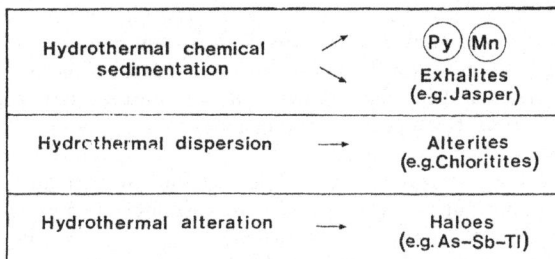

Fig. 8. Hydrothermal processes and products observed in Pyrite Belt mineralization

The much more intensive hydrothermal trend of alteration contrasts with the normal magmatic petrochemical trend. Outside the hydrothermal alteration zones, only the original, very low to low grade metamorphic rocks are encountered. Accordingly, quartz-keratophyres, keratophyres, and spilites of the Pyrite Belt cannot be conceived without distinction as allochemical products of metasomatism and alteration as claimed by Munhá and Kerrich (1980) or Munhá (1983b).

Hydrothermal supply in addition results in elevated concentrations of mobile elements such as As, Sb, Tl, Ba, and F, and positive anomalies of these elements can be detected, for example in chloritites of both hanging and footwall rocks of the sulfide mineralization. Chloritites forming lateral equivalents to the sulfide deposits, seem to be a particularly promising target for lithogeochemical exploration (see also Bernard et al. 1983).

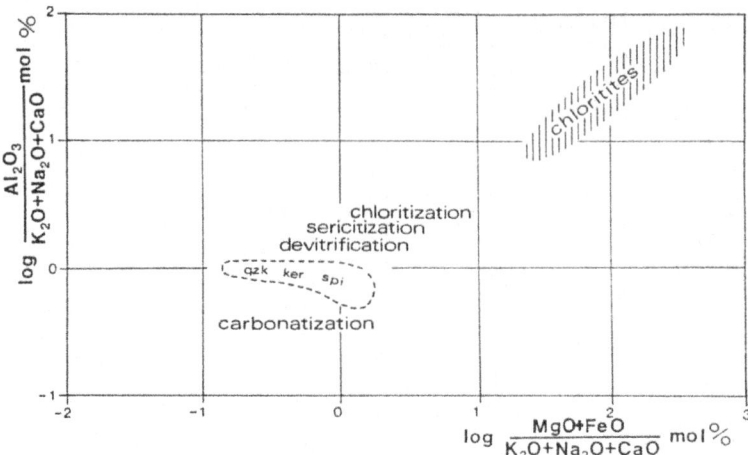

Fig. 9. Alteration diagram for Pyrite Belt volcanics

As previously shown (Möller et al. 1983a,b) a dispersion-related geochemical pattern occurs in the hematitic slate horizon accompanying and overlying the jaspilitic manganese ores. In a 35-km longitudinal section of this slate on the top of the volcanic pile, there are a number of locations with — compared to the regional background — anomalously high concentrations of As, Sb, Tl, and Ba, with correspondingly decreased concentrations of Na and Ce. Most of those anomalies more or less spatially coincide with the positions of manganese and barite orebodies. Using elemental ratios of depleted and enriched element pairs, an anomaly pattern is produced which very clearly depicts the position of the manganese-jasper bodies (Fig. 10.)

As, Sb, Hg, Tl, Ba, and F as comparatively mobile elements are generally enriched in hydrothermal solutions or hot springs of different geological environments (Corliss et al. 1979; Edmond et al. 1982; Weissberg et al. 1979). During cooling, these elements will be precipitated or absorbed by minerals near the site of hydrothermal discharge, producing aureoles of primary dispersion which can also be observed in the Pyrite Belt. Negative dispersions or (relative) losses of elements can be attributed either to original scarcity or to secondary destruction of the host mineral phases. In the case of the negative Ce anomalies, deficient incorporation of this element from Ce-deficient solutions could have been the reason for the low concentration. Decreased Ce contents in the vicinity of the Pyrite Belt's manganese/jasper deposits closely match the REE pattern produced by recent submarine hydrothermal effluents (Klinkhammer et al. 1983). Together, the enrichment of As, Sb, Tl, Ba, and decreased Na and Ce contents characterize areas or points of hydrothermal discharge.

Other elements commonly enriched in siliceous hydrothermal precipitates are Au, Ag, and W (Ridler 1973). Typically, from 20 Pyrite Belt samples in which W has been detected in a concentration range from 2 to 60 ppm, 15 belong to the siliceous

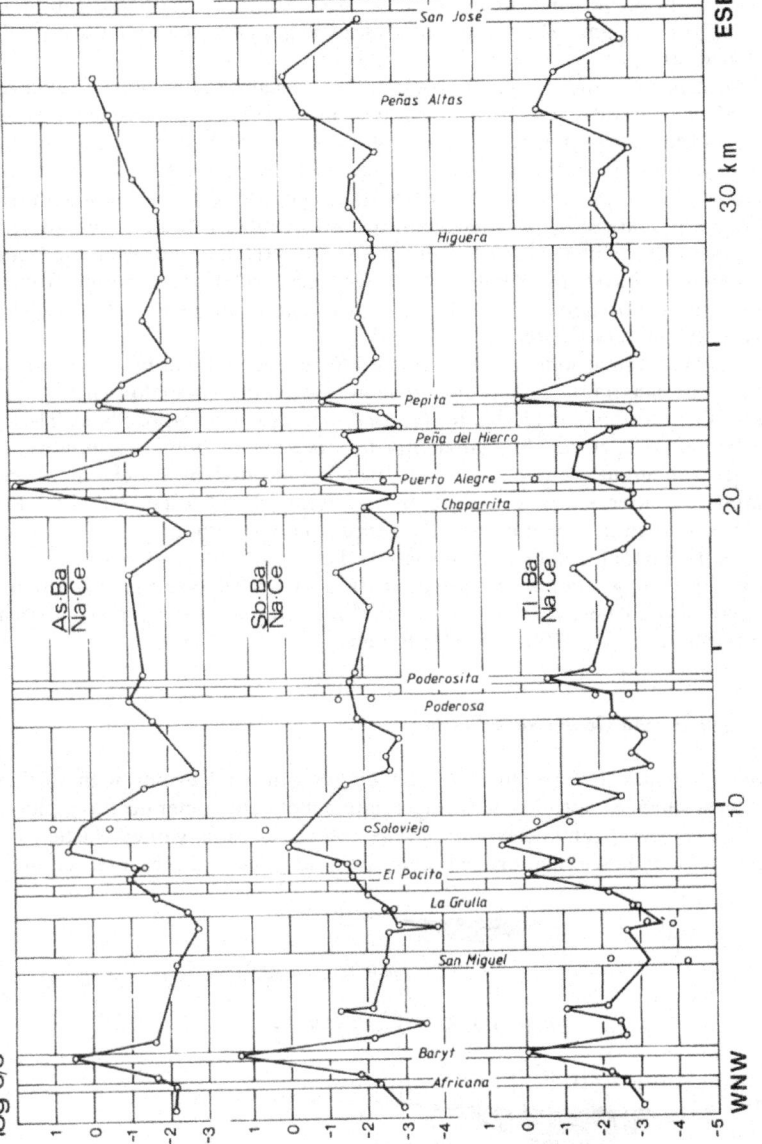

Fig. 10. Profile of ratios calculated from enriched (As, Sb, Tl, Ba) and depleted (Na, Ce) elements in hematitic slates along of the Rio Tinto-Syncline. Vertical bars mark the position of manganese and sulfide deposits

sediments of the second hydrothermal phase. W concentration is at a maximum in
the jaspers (5.8–27 ppm). The maximum value of 62 ppm was detected in a strongly
altered acid tuff.

Concerning Au and Ag in the rocks of the volcano-sedimentary sequence,
analytical data are still scarce. Both elements seem to be enriched in immediate host
rocks of sulfide ores, Au mainly in slates, chloritites, and SiO_2-rich precipitates (up
to 430 ppb), Ag in hematitic and black slates (up to 26 ppm). Au exhibits strong
correlation with Fe, As, Sb, and Pb, demonstrating its affiliation with low-temperature hydrothermal products. A geochemical correspondence, thus, is obvious with
recent mineral deposition at seafloor-spreading centers (Hannington et al. 1987) and
active geothermal systems (Henley and Hedenquist 1986), even though from a
petrochemical or geotectonic point of view these recent situations are not a model for
the Pyrite Belt mineralization.

The spatial dimensions of the anomalies have been studied by geostatistical
methods. The maximum longitudinal dimensions of the anomalous halos vary
between 700 m for Sb and 1500 m for Na, and increase in the order Sb-As-Tl-Na.

To summarize, it can be stated that the location of both the sulfide and
manganese mineralization in the Pyrite Belt is indicated by geochemical and
mineralogical patterns due to hydrothermal alteration and dispersion (Fig. 11).
Together with their lithostratigraphical position, these patterns may be very useful
as an exploration tool, as long as their spatial relations with the ore horizons are not
disturbed by strong tectonic deformation. It is this restriction that has mainly
prevented lithogeochemical exploration from receiving priority over geophysical
methods (Strauss et al. 1977) in the Pyrite Belt.

Metallogenetic and Geotectonic Setting

The observed lithostratigraphical liaison of hydrothermal events with distinct
petrological changes suggests a strong relation between mineralization phases
(metallogeny) and magmatic evolution. The absence or paucity of epiclastic components in the hydrothermal ore sediments point to rapid hydrothermal sedimen-

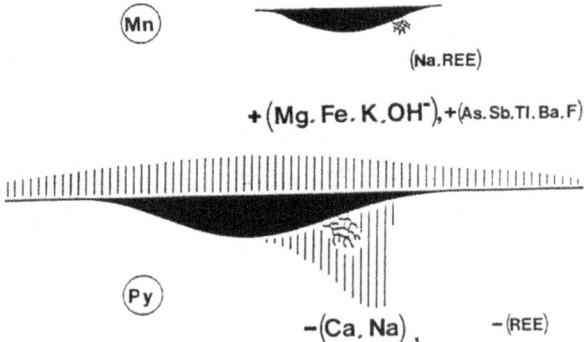

Fig. 11. Schematic representation of elemental gains and losses due to hydrothermal alteration and
dispersion in host rocks of the Pyrite Belt

tation during a comparatively short time span. Leaching of metals in long-lived hydrothermal convection cells (Munhá and Kerrich 1980; Barriga and Carvalho 1983) thus seems to be less probable than deep-seated hydrothermal supply from magmatic processes.

Regarding the geotectonic setting in which the volcanic-hosted deposits of SW Iberia have been formed, two contradictory solutions have been proposed:

1. intracontinental rifting environments: a) in applying a geosynclinal concept (Macgillavry 1961; Schermerhorn 1971, 1975b, 1981; Munhá 1979; Routhier et al. 1980; Sawkins and Burke 1980) or b) in the sense of plate tectonis: an ensialic back arc basin (Munhá 1983b; Grimmeisen 1983);

2. subduction related environments with varying conceptions concerning nature and direction of the subducting plate (Bernard and Soler 1971; Carvalho 1971; Vegas and Muñoz 1976; Bard et al. 1980; Soler 1980; Mitchell 1985; Schütz 1985).

On the basis of available information, the lithosetting of the Iberian massive sulfide deposits is more suggestive of subduction-related environments than of volcanically active intracontinental rift settings (Table 3):

— The evolution of sedimentary facies is characterized by a change from mature, presumably shallow-water to immature, deep-water sediments with flysch character. The change indicates the onset of subduction.
— Magmatic evolution in the Pyrite Belt is characterized by the well-known orogenic trends (Glikson 1972; Ringwood 1974; Donnelly and Rogers 1980):

Na-rich rocks   → K-rich rocks
low-pT-rocks   → high-pT-rocks
spumolithic facies   → effusive or intrusive facies
trondhjemitic suite   → spilite-keratophyre association
                         → calc-alkaline suite

— Mineral assemblages and sequence of mineralization starting from massive sulfides and leading to jaspilitic Mn-Fe formations point to a metallogenetic trend observed in a variety of geotectonic settings, e.g., the Archean belts or the island arc Kuroko deposits. At least for the latter, subduction is a conditio sine qua non.

Table 3. Arguments for a subduction-related geotectonic setting of the Pyrite Belt

	Intracontinental rifting	Pyrite Belt situation	Subduction
Substratum	Continental crust	Marine clastic sediments	Ocean-arc transition
Orogenic Sediments	Absent	Flysch	Flysch
Petrologic Affiliation	SiO_2-bimodal	SiO_2-unimodal AFM-unimodal	AFM-unimodal
Magmatic Evolution	Alkaline	Spilite-kerat. to calc-alkaline	Tholeiitic to calc-alkaline
Mineral Deposits	Various	Cherty Mn-Fe-Fm. massive sulphides	Porphyry coppers cherty Mn-Fe-Fm. massive sulphides

In addition, continental extension-related mineralizations like the Kupfers-chiefer or the Central African Copperbelt, in comparison to subduction-related ones, are characterized by a clearly different suite of sedimentary and magmatic host rocks.

For the remaining arguments, there is neither an established continental basement for the Southern Portuguese Zone, nor is there any typical rift volcanism with alkaline, peralkaline, or carbonatitic character. The occurrence of blueschists, melanges, and ophiolites, often required as direct proof for subduction zones (for the Belt: Schermerhorn 1975a), is substantial for modern orogens only with their steeply dipping subduction zones.

Admittedly, on the basis of the mineralization alone, the Pyrite Belt deposits exhibit some convergencies even with products from oceanic type situations.

Conclusions

In the Pyrite Belt (at least in its northwestern part) hydrothermal phases and related mineralizations are lithogeochemically and lithostratigraphically controlled, the sulfide ores in a lower stratigraphic position being linked to quartz-keratophyric volcanites, and overlying manganese-jasper mineralization occurs on top of an intermediate to basic volcanic pile. Both the hydrothermal events and their min-eralizations can be located by geochemical and mineralogical patterns due to hydrothermal alteration (e.g., chloritization) and element dispersion (e.g., halos of As, Sb, Tl). Together with their characteristic lithostratigraphic position, these patterns may be very useful as exploration tools as long as their spatial relations with the ore horizons are not disturbed by strong tectonic deformation.

The observed lithostratigraphic liaison of the hydrothermal events with distinct petrological changes suggests a strong relation between mineralization, magmatic evolution, and geotectonic setting. All arguments taken together, the lithosetting of the Iberian massive sulfide deposits and their accompanying manganese miner-alization is more suggestive of a subduction-related island-arc environment than of volcanically active intracontinental rift settings. Following the high degree of magmatic evolution, in this case a similarly advanced metallogenetic differentiation has produced metal-rich complex mineralizations. For this reason, the search for rare metals in massive sulfides and for precious metals in exhalites and alteration zones deserves increased effort.

Acknowledgments. The authors gratefully acknowledge the financial support of the Commission of the European Communities (Contract No. MSM-008-D (B)). They are greatly indebted to F. Kubanek and P. Möller of the Hahn-Meitner-Institute, Berlin, for contributing analytical data and discussing the results. L.J.G. Schermerhorn, Berlin, was very helpful in providing XRF analytical data and his expertise in Pyrite Belt geology. Not least, particular thanks are extended to the following students for their contribution to field and laboratory work: S. Cramer, J. Ebneth, M. Gerlach, M. Huch, H.-J. Kluge, R. Korthals, K.-D. Meyer, H.-P. Nehmann, B. Schürmann and T. Tomovsky. A. Paetz supplied additional geochemical and geological information on the Pyrite Belt sequence and was very helpful during preparation of the manuscript. One anonymous reviewer considerably contributed to the improvement of our paper.

References

Bard J-P, Burg JP, Matte P, Ribeiro A (1980) La chaîne hercynienne d'Europe occidentale en termes de tectonique des plaques. Proc 26 Congr Geol Int C 6:233-244

Barriga F Jas, Carvalho D de (1983) Carboniferous volcanogenic sulphide mineralizations in South Portugal (Iberian Pyrite Belt). Mem Serv Geol Port 29:99-113

Bernard AJ, Soler E (1971) Sur la localisation géo-tectonique des amas pyriteux massifs du type Rio Tinto. CR Acad Sci Paris 273D:1087-1090

Bernard AJ, Soler E (1980) Problèmes géologiques et metallogéniques en province de Huelva (Espagne). Publ 26 Congr Geol Int D 5:1-54

Bernard AJ, Soler E (1985) Discussion on the paper of P. Möller et al.: Geochemical proximity indicators of massive sulfide mineralization in the Iberian Pyrite Belt and the East Pontic Metallotect. Min Dep 20:66

Bernard AJ, Degallier G, Soler E (1983) The exhalative sediments linked to the exhalative massive sulphide deposits: a case study of European occurrences. In: Amstutz GC, El Goresy A, Frenzel G, Kluth C, Moh G, Wauschkuhn A, Zimmermann R (eds) Ore Genesis – The State of the Art. Springer, Berlin Heidelberg New York, pp 553-564

Carvalho D de (1971) The metallogenetic consequences of plate tectonics and the Upper Paleozoic evolution of southern Portugal. Estud Not Trab Serv Fom Min 20:297-320

Corliss JB, Dymond J, Gordon LI et al. (1979) Submarine thermal springs on the Galapagos Rift. Science 203:1073-1083

Donelly TW, Rogers JJW (1980) Igneous series in island arcs: The northeastern Caribbean compared with the worldwide island-arc assemblages. Bull Volcanol 42:347-382

Edmond JM, Damm KL von, McDuff RE, Measures CI (1982) Chemistry of hot springs on the East Pacific Rise and their effluent dispersal. Nature 297:187-191

Germann K, Schütz W (1985) Reply to the discussion of AJ Bernard and E Soler on the publication by Möller et al. Miner Dep 20:676-68

Germann K, Paetz A, Schütz W, Dulski P (1986) Lithogeochemistry of the ore-bearing volcanic-sedimentary complex in the SW-Iberian Pyrite Belt. In: Möller P, Brigo K, Germann K, Schneider HJ, Lithogeochemical proximity indicators for strata-bound base metal deposits. Rep Eur 10826 EN, pp 1-35

Glikson AY (1972) Early Precambrian evidence of a primitive ocean crust and island nuclei of sodic granite. Geol Soc Am Bull 83:3323-3344

Grimmeisen W (1983) Zur Geodynamik unterkarbonischer Vulkanite der südportugiesischen Zone. PhD Thesis, Eberhard-Karls-Universität, Tübingen

Hannington MD, Peter JM, Scott SD (1987) Gold in sea-floor polymetallic sulfide deposits. Econ Geol 81:1867-1883

Henley RW, Hedenquist JW (1986) Introduction to the geochemistry of active and fossil geothermal systems. In: Henley RW, Hedenquist JW, Roberts PJ (eds) Guide to the Active Epithermal (Geothermal) Systems and Precious Metal Deposits of New Zealand. Monogr Ser Miner Dep 26:1-22

Klinkhammer G, Elderfield H, Hudson A (1983) Rare-earth elements in seawater near hydrothermal vents. Nature 305:185-188

MacGillavry HJ (1961) Deep or not deep, fore-deep or "after-deep"? Geol Mijnbouw 40:133-148

Mitchell AHG (1985) Mineral deposits related to tectonic events accompanying arc-continent collision. Trans Inst Miner Metall B 94:115-125

Möller P, Dieterle MA, Dulski P, Germann K, Schneider H-J, Schütz W (1983a) Geochemical proximity indicators of massive sulphide mineralization in the Iberian Pyrite Belt and the East Pontic metallotect. Miner Dep 18:387-398

Möller P, Germann K, Schneider H-J (1983b) Geochemical proximity indicators for stratabound non-ferrous massive sulphide ore deposits. Final Report 1980-82, vol 392. Hahn-Meitner-Institut, Berlin, pp 1-75

Munhá J (1979) Blue amphiboles, metamorphic regime and plate tectonic modelling in the Iberian Pyrite Belt. Contrib Miner Pet 69:279-289

Munhá J (1983a) Low-grade regional metamorphism in the Iberian Pyrite Belt. Comunic Serv Geol Portugal 69:3-35

Munhá J (1983b) Hercynian magmatism in the Iberian Pyrite Belt. Mem Serv Geol Port 29:39-81

Munhá J, Kerrich R (1980) Sea water basalt interaction in spilites from the Iberian Pyrite Belt. Contrib Miner Pet 73:191–200

Pinedo Vara I (1963) Piritas de Huelva. Su historia, minería y aprovechamiento. Editorial Summa S.L. Madrid, p 1003

Rambaud Perez F (1969) El sinclinal carbonífero de Rio Tinto (Huelva) y sus mineralizaciones asociadas. Mem Inst Geol Min España 71:229

Ridler RH (1973) Exhalite concept a new tool for exploration. The Northern Miner Nov 29, 1973, pp 59–61

Ringwood Ae (1974) The petrological evolution of island arc systems. J Geol Soc (Lond) 130:183–204

Routhier P, Aye F, Boyer C, Lécolle M, Molière P, Picot P, Roger G (1980) La cienture sud-ibérique à amas sulfurés dans sa partie espagnole médiane. Tableau géologique et metallogénique. Synthèse sur le type amas sulfurés volcano-sedimentaires. Mem BRGM 94:265

Sawkins FJ, Burke K (1980) Extensional tectonics and mid-Paleozoic massive sulfide occurrences in Europe. Geol Rundsch 69:349–360

Schermerhorn LJG (1971) An outline stratigraphy of the Iberian Pyrite Belt. Bol Geol Miner 82:239–268

Schermerhorn LJG (1975a) Spilites, regional metamorphism and subduction in the Iberian Pyrite Belt: Some comments. Geol Mijnbouw 54:23–35

Schermerhorn LJG (1975b) Pumpellyite-facies metamorphism in the Spanish Pyrite Belt. Pétrologie 1:71–86

Schermerhorn LJG (1981) Framework and evolution of hercynian mineralization in the Iberian Meseta. Leidse Geol Medelingen 52:23–56

Schütz W (1985) Magmatismus und Metallogenese im zentralen Teil des SW-iberischen Pyritgürtels, Prov. Huelva, Spanien. Express Edition, Berlin, p 201

Schütz W, Dulski P, Paetz A, Germann K (1985) Geochemical aspects of magmatic evolution and ore deposition in the SW-Iberian Pyrite Belt. Fortschr Miner 63(1):214

Schütz W, Ebneth J, Meyer K-D (1987) Trondhjemites, tonalites and diorites in the South Portuguese Zone and their relations to the volcanites and mineral deposits of the Iberian Pyrite Belt. Geol Rundsch 76:201–212

Sierra J (1984) Geología, mineralogía y metalogenía del yacimiento de Aznalcóllar, 1 Parte. Litoestratigrafía y tectónica. Bol Geol Miner 95:553–568

Soler E (1980) Spilites et metallogénie. La province pyritocuprifère de Huelva (SW Espagne). Mem Sci Terre 39:461

Strauss GK (1965) Zur Geologie der SW-Iberischen Kiesprovinz und ihrer Lagerstätten, mit besonderer Berücksichtigung der Pyritgrube Lousal/Portugal. PhD Thesis, Ludwig-Maximilians-Universität, München

Strauss GK, Madel J (1974) Geology of massive sulphide deposits in the Spanish-Portuguese Pyrite Belt. Geol Rundsch 63:191–211

Strauss GK, Madel J, Fernandez Alonso F (1977) Exploration practice for strata-bound volcanogenic sulphide deposits in the Spanish-Portuguese Pyrite Belt: Geology, geophysics and geochemistry. In: Klemm DD, Schneider H-J (eds) Time and Strata-Bound Ore Deposits, Springer, Berlin Heidelberg 55–93

Vegas R, Muñoz M (1976) El contacto entre las Zonas Surportuguesa 4 Ossa-Morena en el SW de España. Una nueva interpretación. Comun Serv Geol Port 60:31–51

Weissberg BG, Browne PRL, Seward TM (1979) Ore metals in active geothermal systems. In: Barnes HL (ed) Geochemistry of Hydrothermal Ore Deposits, 2nd edn. Wiley, New York, pp 738–780

Tourmalinites Associated with Australian Proterozoic Submarine Exhalative Ores

I.R. Plimer[1]

Abstract

Tourmalinites and tourmaline-rich sediments are associated with exhalative mineralization in greenschist, amphibolite, and granulite metamorphic terrains. They have undergone mild to intense deformation, have stratigraphic continuity around fold structures and display sedimentary structures such as bedding, graded bedding, cross-bedding, intraformational slumps, and pull-aparts. In higher metamorphic grade areas they display metamorphic textures and are commonly cut by remobilized quartz-tourmaline veins. Tourmaline-rich rocks are present beneath, within, laterally equivalent to, and above submarine exhalative mineralization of Pb-Zn-Ag, Cu-Co, Cu-Bi, W, Sn, Au, and rare earth elements. In some places tourmalinites have no associated mineralization and in no environment are they associated with granitic rocks which could be interpreted as their source. The tourmaline composition is schorl-dravite and reflects the bulk rock composition and cannot be used as a guide to mineralization. Although tourmaline is a common mineral associated with granites, metasomatized rocks and breccia pipes, the tourmaline-rich rocks described herein are interpreted as boron-rich siliceous iron formations derived from the metamorphism of a silica-tourmaline precursor. They are a common exhalite type and most commonly are present in thick sequences of Lower-Middle Proterozoic pelitic metasediments and evaporites. It is suggested that the boron has derived from leaching of clays which contained adsorbed and substituted boron or from leaching of borate-bearing evaporites.

Introduction

The very large stability range of tourmaline (Chorlton and Martin 1978; Manning and Pichevant 1983; Werding and Schreyer 1984; Benard et al. 1985) is commensurate with its occurrence in a great diversity of rock types. Furthermore, the tourmaline family is a very complex borosilicate group which may provide some information on melt or fluid composition (e.g., P, T, f_{O_2}, a_{H_2O}). The general formula of $WX_3Y_6(BO_3)_3Si_6O_{18}(OH,F,Cl)_4$ where $W = Na$, Ca, $X = Al$, Fe^{ii}, Fe^{iii}, Li, Mg, Mn, and $Y = Al$, Fe^{iii}, Cr, Ti, Mg, V allows extensive coupled substitution and hence the major element chemistry of the tourmaline family can provide information on

---

[1]Dept. of Geology, University of Newcastle, New South Wales 2308, Newcastle, Australia

Base Metal Sulfide Deposits
G.H. Friedrich, P.M. Herzig (Eds.)
© Springer-Verlag Berlin Heidelberg 1988

the chemistry of geological systems. The most common end members of the family are elbaite (W = Na, X = Al, Li, Y = Al), schorl (W = Na, X = Fe^{ii}, Fe^{iii}, Y = Al), dravite (W = Na, X = Mg, Y = Al) and uvite (W = Ca, X = Mg, Y = Al, Mg).

Tourmaline (schorl-elbaite) is a common accessory mineral in granitic rocks, especially those associated with tin-tungsten deposits. It is more commonly noted as a stable mineral of granite alteration and altered pelitic metasediments associated with tin-tungsten deposits (e.g., Cornwall, U.K., Exley 1959; Charoy 1979). Furthermore, tourmaline-rich matrices of mineralized hydrothermal breccias generated around plutons of acid to intermediate composition emplaced at shallow depths indicate that boron is strongly enriched in the residual phase during the final stages of crystallization of some magmas. Tourmaline-rich breccia pipes are well known from the porphyry copper deposits of the Western Cordillera of the Andes (Kents 1964; Sillitoe and Sawkins 1971), subvolcanic tin deposits of the eastern Cordillera of the Central Andes of Bolivia (Grant et al. 1980), and tin-bearing breccia pipes associated with high level specialized granites (e.g., Ardlethan, Australia, Clarke et al. 1986; Wheal Remfrey, U.K., Allman-Ward et al. 1982).

Tourmaline is well documented in many clastic sedimentary rocks as a chemically and mechanically stable resistant heavy mineral or as an authigenic mineral. Adsorption of boron onto clays and phyllosilicates in sediments can be high (Harder 1959; Stubican and Roy 1962) and, upon metamorphism, can result in tourmaline-bearing pelitic metasediments. In metamorphic rocks, tourmaline is generally regarded as an accessory mineral despite the knowledge that it is both an aluminum silicate and a ferromagnesian mineral, and hence the range of coupled substitutions could be used as a petrogenetic indicator. Studies by Henry and Guidotti (1985) show that, in metamorphic rocks, AFM plots of tourmaline define different rock types in response to bulk rock chemistry and the chemistry of the detrital cores of overgrown grains can be used to determine the source region for pelitic metasediments. Although tourmaline is an extremely stable mineral, boron may be mobile during regional high grade metamorphism (Ahmed and Wilson 1981). The presence of schorl-dravite in alteration assemblages associated with metamorphic gold deposits indicates that boron is also mobile in somewhat lower grades of metamorphism (Augustithis 1967; Phillips and Groves 1983).

Borates (e.g., borax, gaylussite) and borosilicates (e.g., reedmergnerite) are not uncommon in evaporitic settings (e.g., Searles Lake, U.S.A., Smith 1979; Green River Formation, U.S.A., Milton 1971) and concentrations of dravite-uvite can be present in metamorphosed evaporites (e.g., Damara Orogen, Namibia, Behr et al. 1983). In the Houxianyu evaporite boron deposit on the Liaodong Peninsula, Manchuria, P.R. China, tourmalinite is the host to the magnesium borates suanite, szaibelyite and ludwigite (Appel pers. commun. 1986). Many generations associated with the African Copperbelt red-bed deposits (Mendelsohn 1957) and dravite-schorl inclusions in metapelites in the 3.8 Ga Isua supracrustals of Greenland are ascribed to pre-Isua evaporitic origin by Appel (1983).

Accessory tourmaline and tourmalinites are associated with many types of submarine exhalative ores (Slack 1982). Tourmaline-rich rocks can occur below (e.g., Sullivan, Canada, Ethier and Campbell 1977), within (e.g., Vihanti, Finland, Rouhunkoski 1968), stratigraphically equivalent to (e.g., Broken Hill, Australia, Plimer 1983) or above (e.g., Rosebery, Australia, Stillwell 1934) submarine exhalative ores. In some places, tourmalinites of submarine exhalative origin have no relationship to known mineralization (e.g., Eastern Creek Volcanics of Mt. Isa, Australia). Tourmaline-rich rocks are not restricted to any geological era and are present in Archean rocks (e.g., Greenland, Appel 1985, Kidd Creek, Canada, Ethier and Campbell 1977), Paleozoic rocks (e.g., Løkken, Norway, Carstens 1927) and in the Mesozoic (e.g., Sazare Besshi-type deposit, Ito pers. comm.). However, tourmaline-rich rocks and tourmalinites of exhalative origin are most common in metasediment-metavolcanic mineralized sequences of Lower-Middle Proterozoic age.

Furthermore, tourmaline-bearing rocks and tourmalinites (rocks containing >20% tourmaline) can be associated with submarine exhalative ores of Pb-Zn (e.g., Sullivan, Canada, Ethier and Campbell 1977), Cu-Zn (e.g., Elizabeth, U.S.A., Howard 1959), Cu-Co (e.g., Dome Rock, Australia), Au (e.g., Passagem de Mariana, Brazil, Fleischer and Routhier 1973), and W (Cunningham et al. 1973). These associations indicate that boron was an active constituent of the ore fluid and the extremely commonly association of calc-silicate and amphobolite-hosted statiform scheelite deposits (e.g., Broken Hill, Australia, Plimer 1980, 1983; Barnes 1983; Bohemian Massif of Austria and Czechoslovakia, Raith pers. commun.; Archean Malene supracrustals of Greenland, Appel 1985; Bergslagen area of Sweden, Parr and Rickard 1986; Bushmanland of South Africa, Joubert and Moore 1986; Damara Orogen of Namibia) strongly suggests that the boron was more than just an unpaying passenger in the ore fluid and may have been active in selective complexing of tungsten.

This study documents the field relationships, mineralogy, petrology, and geochemistry of tourmalinites associated with Lower-Middle Proterozoic submarine exhalative mineralization in Australia which has undergone greenschist (Golden Dyke Dome, Rum Jungle, Northern Territory; Eastern Creek Volcanics, Mt. Isa Block, Queensland), amphibolite (Eastern Succession, Mt. Isa Block; Olary Block of South Australia), and granulite (Broken Hill Block, N.S.W.) facies metamorphism. In many areas tourmalinites are not associated with mineralization (e.g., Olary Block, South Australia; Jervois Range, Northern Territory). Although the bulk of documented tourmalinites in Australia occur in Lower-Middle Proterozoic sequences, minor tourmalinites occur in the Archean of Western Australia, Late Proterozoic evaporitic sequences of South Australia, and associated with the Cambrian Rosebery Zu-Pb-Cu-Au deposit of Western Tasmania.

Pine Creek Geosyncline, Northern Territory

The Early Proterozoic metasediments of the Pine Creek Geosyncline are up to 14 km thick, have undergone greenschist facies metamorphism, and overlie Archean basement. Sedimentation took place between 2470 and 1870 m.y. in an intracratonic basin under alternating continental and shallow marine environments. Subsidence resulted in regional deformation and metamorphism which took place about 1870 m.y. Intrusions of granite plutons followed regional metamorphism with some volcanism. Intrusion of dolerite at about 1690 m.y. was the final igneous event prior to further erosion and the deposition of flat-lying Middle Proterozoic sediments (Page et al. 1980).

Golden Dyke Dome

The Golden Dyke area, 150 km SSE of Darwin, is a domal structure resulting from the refolding of a sequence of Lower Proterozoic metasediments. Coeval multiphase greenschist facies metamorphism and deformation occurred at about 1800 m.y. and the area is transgressed by faults and intruded by granite, dated at 1765 m.y. (Nicholson 1980).

The tourmalinites are principally within carbonaceous metapelites and pelites, especially where the host sedimentary sequence is thicker. The greatest abundance of tourmalinites and mineralization is within the Banded Ferruginous Member (Fig. 1), wherein tourmalinites are the host for gold deposits (e.g., Golden Dyke and Good Shepherd Mines) and are facies equivalents of exhalative Pb-Zn and Cu-Bi mineralization hosted by sulfide facies iron formations. Minor cassiterite and scheelite are present within some tourmaline-bearing sulfide facies iron formations and extremely rapid changes between sulfide-silicate-oxide iron formation and tourmalinite facies are common (Nicholson 1980; Plimer 1986). Less prominent iron formations and tourmalinites occur in the sequences overlying and underlying the Banded Ferruginous Member (Fig. 1). Carbonaceous pelitic metasediments enclosing tourmalinites and iron formations generally contain more than 5% tourmaline and iron sulfide/oxide minerals.

In the field, the very fine-grained tourmalinites at Golden Dyke are extremely difficult to identify as they crop out as chert-like or black beds easily confused with chert or carbonaceous metapelite. Tourmalinite occurs as narrow (0.01–10 m wide) generally discontinuous units, although horizons up to 5 km in strike have been traced around fold hinges. Bedding on the mm scale is ubiquitous and is defined by changes in the modal content of tourmaline, quartz, sulfides, graphite, chlorite, and albite. Graded bedding and slump structures are also present. Tectonism has aggregated tourmaline into ellipsoidal clots, schlieren, and axial plane tourmaline, which is commonly deformed by the second generation of deformation. Although the tourmalinites have undergone deformation, there is no lineation defined by tourmaline. In and near faults, fold hinges, and granite contacts, tourmaline occurs in quartz veins and replacing pelitic rocks.

Rum Jungle

The Rum Jungle area comprise Lower Proterozoic metasediments that either onlap or have been updomed by Archean basement complexes (Page et al. 1980). The metasediments comprise low metamorphic grade shallow-water assemblages of conglomerates, arkoses, arenites, shales, siliceous carbonates, magnesite (with minor dolomite), tuff, banded iron formations, and tourmalinite.

Tourmalinite occurs in the four principal stratigraphic units and has commonly been misidentified as amphibolite, mafic schist, chert, and scapolite (Bone pers. commun.). Tourmalinite displays layering, slump folding, and rip-up clasts and occurs as a facies equivalent of quartz-magnetite and mafic tuffs. In thin section, the tourmalinites comprise layers and en echelon clots of fine-grained acicular zoned crystals of tourmaline, quartz, and minor phyllosilicates. Tourmaline also occurs in fault breccias, transgressive quartz veins, and fractures.

Formation/Unit and Lithology	Thickness	Exhalites	
KAPALGA FORMATION phyllite, greywacke	500m	•-•-•-•-•-•-•-•-•-•-•	Iron formation (pyrite-base metals)
		•-•-•-•-•-•-•-•-•-•-•	Iron formation (pyrite-base metals)
		T-T-T-T-T-T-T-T-T-T-T-T	Tourmalinite
		T-T-T-T-T-T-T-T-T-T-T-T	Tourmalinite
GEROWIE TUFF tuffaceous chert, tuffaceous pelite, phyllite, chert	300m		
KOOLPIN FORMATION			
UPPER CARBONACEOUS METAPELITE MEMBER	80-150m	•-•-•-•-•-•-•-•-•-•-•	Iron formation (pyrite-base metals)
carbonaceous metapelite, metapelite		T-T-T-T-T-T-T-T-T-T-T-T	Tourmalinite
		T-T-T-T-T-T-T-T-T-T-T-T	Tourmalinite
		▬•▬•▬•▬•▬•▬•	Iron formation (pyrrhotite-rich metapelite with base metals)
BANDED FERRUGINOUS MEMBER			
ferruginous and carbonaceous metapelites (sulphide, silicate and oxide facies of iron formations)	50-100m	•-•-•-•-•-•-•-•-•-•-•	Iron formation
		•-•-•-•-•-•-•-•-•-•-•	Iron formation
		•-•-•-T-T-T-T-T-T-T-T	Iron formation-tourmalinite
		•-T-T-T-T-T-T-T-•-•-•-•	Tourmalinite-iron formation
		•-•-•-•-•-•-•-•-•-•-•	Iron formation
		T-T-T-T-T-T-•-•-•-•-•-•	Tourmalinite-iron formation
		•-•-•-•-•-•-•-•-•-•-•	Iron formation
LOWER CARBONACEOUS METAPELITE MEMBER		•-•-•-•-•-•-•-•-•-•-•	Iron formation
carbonaceous metapelite, greywacke	0-50m	T-T-T-T-T-T-T-T-T-T-T-T	Tourmalinite
		X – X – X – X – X – X	Garnet-biotite-chlorite-quartz-sulfide iron formation
		T-T-T-T-T-T-T-T-T-T-T-T	Tourmalinite
PHYLLITE MEMBER			
phyllite, metapelite	>200m	T-T-T-T-T-T-T-T-T-T-T-T	Tourmalinite

Fig. 1. Stratigraphy of Golden Dyke area, N.T.

Mt. Isa Orogen, Queensland

The Mount Isa Orogen of northern Queensland occupies an area of some $40,000\,km^2$ and comprises three major geological units (Plumb et al. 1980). A central belt of volcanic and granitic rocks forms basement to younger rocks which comprise vast sequences of basalt in addition to thick sequences of shallow-water sediments. These were subsequently deformed and metamorphosed to the greenschist facies as the Leichhardt River Fault Trough (Fig. 2). The Mary Kathleen Shelf, east of the Leichhardt River Fault Trough, comprises thinner sequences of mainly shallow water sediments which underwent more complex folding, metamorphism to the amphibolite facies, and more granite intrusion than the Mary Kathleen Fold Belt (Fig. 2). Sedimentation, volcanism, and basin development have been controlled by major, long acting, north-trending lineaments or zones, some of which have been the focus for reactivation of major post-depositional faults, metamorphism, and granite intrusion in the Orogen.

Although Mt. Isa and the associated Hilton Pb-Zn deposits represent one of the largest exhalative metal accumulations in the world, tourmalinites occur only

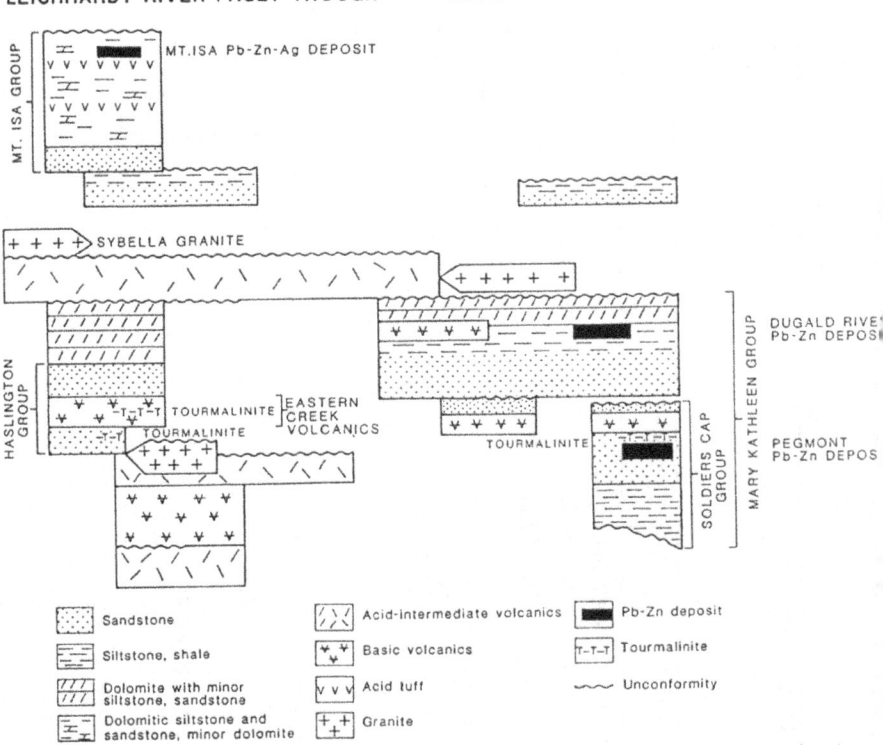

Fig. 2. Stratigraphic correlation across the Mt. Isa Orogen from the Leichhardt River Fault Trough to the Mary Kathleen Fold Belt. (Plumb et al. 1980)

accidentally in proximity to the Mt. Isa deposits. At Mt. Isa, extremely altered metabasic rocks of the Eastern Creek Volcanics are faulted against the Mt. Isa Group.

The Eastern Creek Volcanics are a suite of deformed metamorphosed altered subaerial tholeiitic flood basalts, tuff, sandstone, orthoquartzite, and altered submarine tholeiitic basalts associated with chert, jaspilite, siltstone, and sandstone (Plumb et al. 1980). Minor submarine exhalative copper mineralization occurs within the Eastern Creek Volcanics 5 km to the east of Mt. Isa. It is associated with chert, jaspilite, quartz-tourmaline rocks, tourmaline-rich siltstone, and tourmaline-rich sandstone. Chert, jaspilite, and the tourmaline-rich rocks occur 50–250 m along strike from the minor submarine exhalative copper deposits.

The quartz-tourmaline rocks are conformable with regional stratigraphy and structure, are up to 2 m wide and 60 m in length, comprise milky opaque quartz with layers and schlieren of tourmaline and could easily be mistaken for quartz veins. Bedding is coplanar with bedding in the enclosing rocks and is defined by variations in the quantity of quartz and tourmaline. Individual laminae are in the order of 1–5 mm thick with tourmaline grains generally less than 1 mm in size. Smaller quartz-tourmaline and tourmalinite can be traced around fold hinges and tourmaline within the enclosing metapelites is crenulated (Fig. 3). Minor transgressive quartz-tourmaline and tourmaline veins occur within and immediately associated with the tourmaline-rich rocks. Rare tourmaline-rich quartzose and feldspathic sandstones with cross-bedding defined by tourmaline-rich laminae are associated with the quartz-tourmaline rocks.

Minor tourmaline is also present associated with beryl, cassiterite, fluorite, Ta-Nb and U minerals in the Sybella Granite which intrudes the Eastern Creek Volcanics and is overlain by the Mt. Isa Group (Fig. 2).

A number of Pb-Zn prospects occur on the eastern margin of the Mary Kathleen Fold Belt of the Mt. Isa Orogen. The Pegmont, Jolimont, and Anitra prospects occur in the Soldiers Cap Group (Fig. 2) whereas the Cowie, Black Rock, Dingo, Mar-

Fig. 3. Folded tourmalinite with associated crenulated tourmaline-bearing pelitic metasediments, Eastern Creek Volcanics, Mt. Isa district, Qld.

ramungie, and Fairmile prospects occur in the overlying Kuridale Formation (Vaughan and Stanton 1984). Although it has been argued by Vaughan and Stanton (1984) that the Pegmont and associated deposits are similar to the Broken Hill deposit, N.S.W., such an analogy is very loose (Table 1). Nevertheless, a previously unrecognized distinct black tourmalinite horizon occurs 20 m stratigraphically above the mineralized "banded iron formation" at both Pegmont and Anitra (Fig. 4) and minor tourmaline is present in the host "banded iron formation" and enclosing metasediments. This horizon occurs as discontinuous 0.5-m-wide outcrop 50 m in strike length above the greatest concentration of sulfides delineated by drilling at Pegmont, whereas at Anitra the tourmalinite is 0.1 m wide, continuous for 250 m and discontinuous over a strike length of more than 2.5 km. At both localities, the tourmalinite is conformable, although no lineation or axial plan schistosity was evident, and has no known granite mass associated within 20 km.

The tourmalinite is bedded with the individual grains and laminae less than 1 mm in size with bedding defined by modal changes in the quartz and tourmaline content. More massive tourmalinite with more than 80 vol % tourmaline is friable and displays no sedimentary structures.

Table 1. Comparison of the Pegmont, Qld, and Broken Hill, N.S.W. deposits

	Pegmont, Qld.	Broken Hill, N.S.W.
Age	Lower-Middle Proterozoic	Lower-Middle Proterozoic
Metamorphic grade	Amphibolite	Granulite
Host lithology	Siltstone, feldspathic sandstone, arkose(?) rare pelite and subaqueous tholeiitic basalt	Turbidites, silty sandstone, rhyodacitic and tholeiitic basaltic volcanics and their alteration products
Exhalites	Banded iron formation; rare garnet quartzite, tourmalinite, quartzite	Quartz-gahnite rocks, quartzite and coticule rocks; rare banded iron formation, tourmalinite and quartz-feldspar rocks
Premetamorphic alteration	———	Addition of Si, Fe, Mn and K and depletion of Na and Ca to metavolcanics and metasediments immediately enclosing mineralization
Zoning	Pb-Zn → Cu	Cu → Zn-Pb → Pb-Zn
Depositional basin	Equidimensional	Long linear
Gangue minerals	Garnet, quartz, fayalite, pyroxene, hornblende, cummingtonite, biotite, magnetite, apatite	Quartz, carbonate, fluorite, feldspars, spessartine, rhodonite, bustamite, apatite, gahnite
Palaeogeographic setting	Shelf-type deposition, horst or shallow rift	Deep rift
Similar deposits	Gamsberg, Aggeneys (South Africa); Pinnacles, (N.S.W.)	Bergslagen area (Sweden)

Fig. 4. Outcrop of tourmalinite unit, Anitra, Qld.

Willyama Supergroup

The Lower-Middle Proterozoic Willyama Supergroup crops out in western N.S.W. (Broken Hill and Euriowie Blocks) and South Australia (Olary, Mount Painter, and Peake-Denison Blocks) with correlatives in southern South Australia on Yorke and Eyre Peninsulae. Detailed stratigraphic studies have been undertaken only in the Broken Hill and Euriowie Blocks (Willis et al. 1983); however, mineral exploration and some mapping in the other Blocks has enabled a broad stratigraphic correlation (Fig. 5).

The Willyama Supergroup is partly overlain by Late Proterozoic cover sediments and volcanics of the Adelaide Fold Belt and is extensively covered by Paleozoic, Mezozoic, Tertiary, and Quaternary sediments and sedimentary rocks. The Supergroup is composed of rocks dated at a minimum of 1820 m.y. which have underdone a number of episodes of regional low pressure-high temperature metamorphism at about 1660 and 1570-1590 Ma and a number of events of medium pressure-low temperature retrogression, the last of which (Delamerian Orogeny) is dated at 500 m.y. (Pidgeon 1967; Gulson 1984). Prograde facies of metamorphism are zoned from greenschist to granulite.

Although primary stratigraphic relationships are obscured by a complex history of deformation, three main phases of deformation have been recognized (Glen et al. 1977). In high metamorphic grade rocks, D1 gave rise to a layer-parallel schistosity, whereas D2 is characterized by tight folds with a high metamorphic grade axial plane

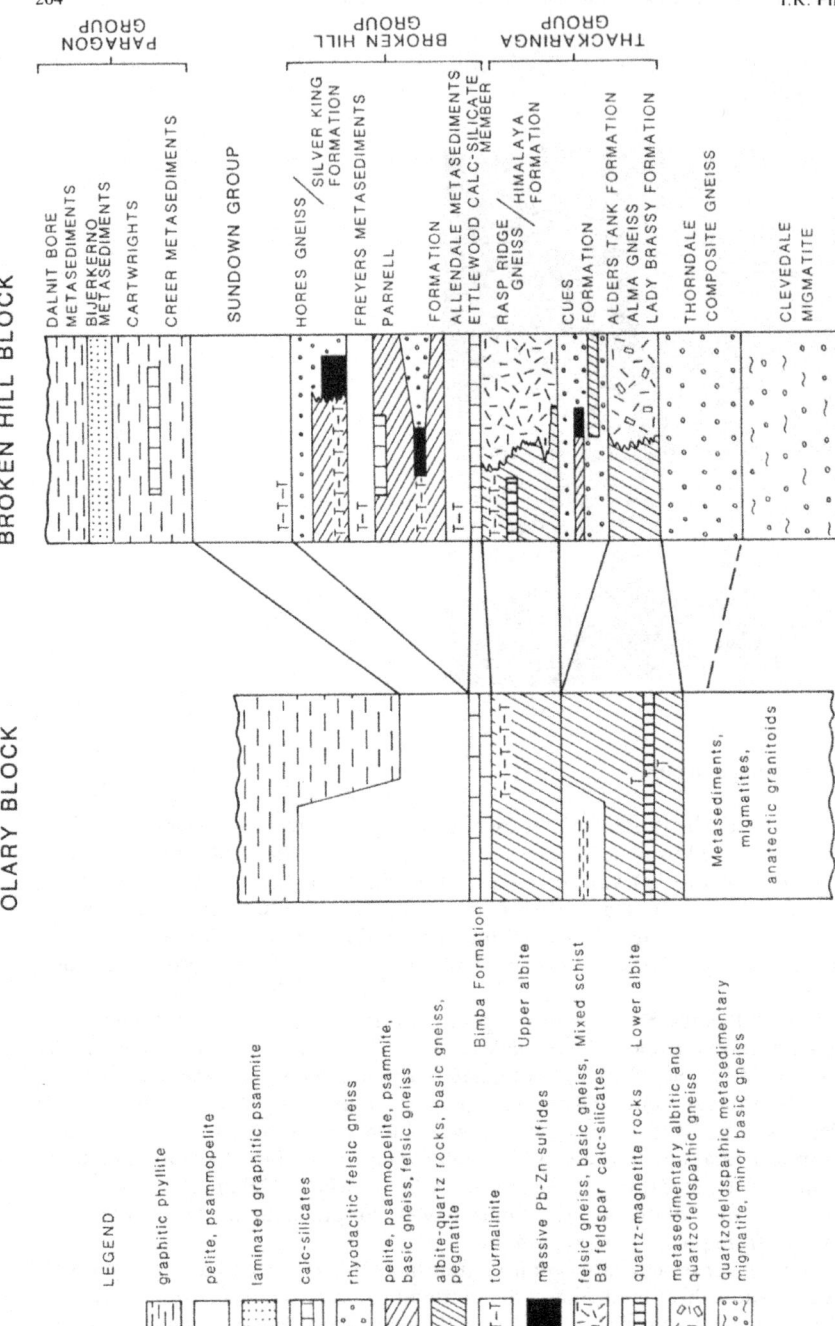

Fig. 5. Stratigraphic correlation of the Willyama Supergroup in the Olary and Broken Hill Blocks. (After Willis et al. 1983)

schistosity. Both D1 and D2 are coeval with the two events of low pressure-high temperature regional metamorphism whereas D3 (which produced relatively open folds) was associated with retrogression.

Olary Block, South Australia

The principal differences between the Olary and the Broken Hill Blocks is the presence of sequences within the Broken Hill Block which contain deep water turbidites, abundant mafic gneiss (tholeiitic basalt precursor), quartz + feldspar + garnet + biotite gneiss (rhyodacitic precursor), numerous facies of exhalites (e.g., quartz-gahnite, quartz + magnetite + F-apatite + spessartine, coticule rocks, spessartine + quartz, scheelite-bearing calc-silicates, tourmaline + quartz + gahnite + spessartine), and numerous sulfide deposits (e.g., Broken Hill in Hores Gneiss; Southern Cross, Parnell, and Nine Mile within Parnell Formation; Pinnacles with Cues Formation). These sequences are interpreted as having been formed in deep grabens or the axial valley of a rift (Plimer 1985) in contrast to the sequences of the Olary Block, which are more akin to platform or horst-type environments of a rift. Nevertheless, although the Olary Block contains no known large massive sulfide deposits like Broken Hill, the "Upper Albite" and "Lower Albite" correlatives of the Thackaringa Group contain minor exhalative mineralization and associated exhalites as does the Thackaringa Group. Exhalites recognized in the Thackaringa Group correlatives of the Olary Block are quartz + magnetite, quartz + magnetite + chalcopyrite + pyrite + Co minerals, barite + magnetite + quartz, barite, quartz + magnetite + Fe-silicates + pyrite + base metal sulfides, coticule rocks, hematite + magnetite + quartz rocks, and Mn-silicate rocks (e.g., piemontite + garnet + quartz + albite + tremolite + phlogopite + hematite, Ashley 1984).

Tourmalinites of the Olary Block have not previously been recorded. They occur principally within the "Upper Albite" (e.g., Alconie Hill, Old Boolcoomatta, Mt. Howden) although minor tourmaline within the "Lower Albite" is present in footwall and hanging wall assemblages to the Dome Rock Cu-Co deposit and tourmaline occurs in transgressive grunerite-garnet alteration assemblages and exhalites at Weekeroo (Ashley 1984).

In the "Lower Albite" sequence, the Dome Rock Cu-Co deposit, tourmaline and other ferromagnesian silicates vary from Mg>Fe species in the footwall pelitic metasediments and ore zone to Fe>Mg species in the hanging wall pelitic metasediments, iron formations, and albite-rich rocks.

Tourmalinites from the "Upper Albite" crop out as very fine-grained laminated tourmaline-quartz-K feldspar-albite rocks concordant with stratigraphy and structure as narrow (<3 m wide) discontinuous (<2 km strike length) horizons which can be traced around fold hinges. In the hinges of folds, and adjacent to transgressive structures (e.g., folds, shears, quartz veins, pegmatite dykes), the tourmaline has undergone recrystallization from grains less than 1 mm in size to coarse euhedral grains up to 3 cm in size. Bedding defined by modal variations in feldspars and quartz was the only sedimentary structure observed. Individual laminae vary from 0.5 to 10 mm in thickness and are stratigraphically continuous for tens of meters.

At one locality (Old Boolcoomatta), very fine-grained tourmalinite within amphibolite facies quartz-albite rocks of the "Upper Albite" coarsens in grain size along strike over a distance of 500 m. At its coarsest (10–20 cm grainsize), the rock is a strata-bound pegmatite partially zoned from a quartz core to a tourmaline + albite + K-feldspar blocky zone partially altered to a muscovite + quartz.

Broken Hill Block, N.S.W.

Tourmalinites occur in the Thackaringa, Broken Hill and Sundown Groups (Fig. 5) with the greatest abundance of tourmaline-bearing rocks intimately associated with submarine exhalative Pb-Zn and W mineralization in the Parnell Formation of the Broken Hill Group (Plimer 1980, 1983).

Minor plagioclase-quartz-tourmaline rocks occur within the quartz-albite rocks associated with quartz-magnetite exhalites of the Himalaya Formation of the Thackaringa Group (Fig. 5). These are correlated with the tourmalinites of the "Upper Albite" of the adjacent Olary Block. Tourmalinite and tourmaline-rich pelitic metasediments occur within the Allendale Metasediments immediately above and below the segments of the Ettlewood Calc-Silicate Member enriched in scheelite, sphalerite, galena, and pyrrhotite which are considered to be of exhalative origin. Numerous small submarine exhalative Pb-Zn and scheelite deposits occur within the Parnell Formation (e.g., Southern Cross, Parnell, Nine Mile, Champion, Great Western, Corruga). These deposits are hosted by pelitic, psammopelitic and psammitic metasediments intimately associated with amphibolite (which has undergone premetamorphic hydrothermal alteration), quartz + feldspar + garnet + biotite gneiss (rhyodacitic volcanic which has undergone premetamorphic alteration), and numerous facies of exhalites. The most common exhalites observed are quartz-gahnite rocks (which contain small and variable contents of plumbian K-feldspar, zincian biotite, F-apatite, spessartine, feldspars and sulfides), tourmalinite, quartz + gahnite + tourmaline + biotite rocks, quartz + tourmaline + $CaWO_4$ rocks, garnet + quartz + scheelite + epidote + amphibole rocks, and plumbian K-feldspar + quartz + zincian biotite rocks. These exhalites are enclosed by or merge into garnet-, tourmaline-, biotite-, gahnite-, or plumbian K-feldspar-rich metasediments. Although discontinuous horizons of quartz-gahnite rocks and tourmalinite outline the ore horizon for tens of kilometers, centers of mineralization are characterized by a great diversity and stacking of small discontinuous exhalite horizons.

Tourmalinites, tourmaline-bearing quartz-gahnite rocks and associated tourmaline-rich pelitic metasediments from the Parnell Formation crop out as narrow (< 1m) black discontinuous (< 50 m strike length) horizons. These horizons commonly display bedding defined by quartz and tourmaline laminae less than 1 mm in thickness. Although the Supergroup has undergone high grade metamorphism and multiphase deformation, the tourmalinites are extremely fine-grained, have no lineation or axial plane schistosity and bedding coplanar with bedding in the enclosed metasediments. Other rare sedimentary structures are crossbedding, graded bedding (Figs. 6 and 7), intraformational slump structures, and small-scale

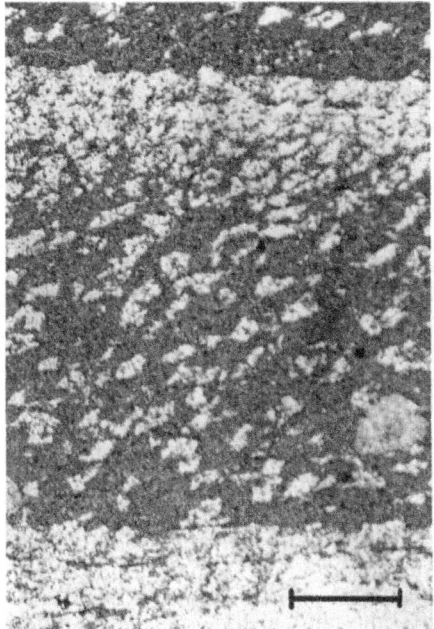

Fig. 6. Graded bedding in quartz-tourmaline rock from the Broken Hill Group, Broken Hill Block showing truncated tourmaline-rich base (*darker*) and gradation upwards to a truncated tourmaline-rich top (*lighter*). Bar = 1 mm

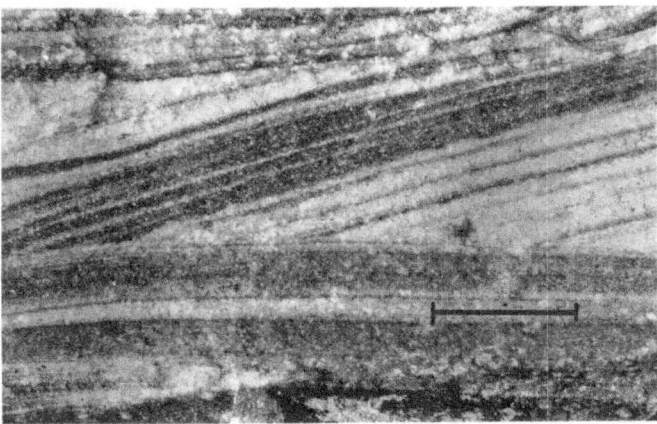

Fig. 7. Crossbedding in quartz-tourmaline rock. Hores Gneiss, Broken Hill Block. Bar = 1 cm

intraformational breccias (Slack et al. 1984). Minor tourmalinites are also associated with quartz-gahnite exhalites in the Freyers Metasediments.

The main Broken Hill Pb-Zn-Ag deposit occurs in the Hores Gneiss (Fig. 5). Tourmaline is an extremely rare species in an orebody which contains over 350 mineral species. Along strike 3 km from the orebody, tourmaline is a common mineral in a package of stacked exhalite horizons at the Globe Mine comprising quartz + sulfide, quartz + gahnite + sulfide, spessartine + quartz ± sulfides ± apatite, tourmaline + gahnite + spessartine, plumbian K-feldspar + quartz, quartz + gahnite + tourmaline and biotite + gahnite + tourmaline ± minor spessartine + quartz + sulfides. The strike-equivalent Silver King Formation contains a number of tourmalinite horizons up to 2 m thick and tourmaline-rich pelitic metasediments which have continuity of outcrop for more than 5 km of strike. Tension gash fillings of quartz and coarse-grained tourmaline contain scheelite and wolframite, tourmalinite and quartz + feldspar + garnet + biotite gneiss contain gahnite and minor wolframite and pelitic intercalations within tourmalinite and tourmaline-rich metasediments contain minor cassiterite. Tourmalinites of the Silver King Formation display extremely delicate bedding defined by tourmaline and quartz laminae less than 0.5 mm thick and rare cross bedding (Fig. 7). Minor quartz-tourmaline rocks (with rare gahnite) are present within the Sundown Group stratigraphically above the areas of the Silver King Formation which contain abundant persistent thick tourmalinite and tourmaline-rich metasediments.

All quartz-tourmaline rocks and tourmalinites of the Broken Hill Block have been transgressed by quartz veins (up to 10 cm in width). Bedding and cross bedding can be rejoined across these quartz veins which can contain a minor quantity of coarser grained tourmaline. In one place, tourmalinites associated with the Ettlewood Calc-Silicate Member have been transgressed by a post-tectonic granite (Mundi Mundi Granite, 1540 m.y. Pidgeon 1967). The granite contains xenoliths and schlieren of tourmalinite and is enriched in tourmaline, whereas tourmaline is an extremely rare phase elsewhere in the granite.

Petrography of Tourmaline-Rich Rocks

Tourmaline-rich rocks from the Golden Dyke Dome vary from tourmalinite to tourmaline-bearing pelitic metasediments. In thin section, tourmaline is associated with very fine grained quartz, albite, chlorite, biotite, graphite, sulfides, stilpnomelane, muscovite and rutile. The tourmaline occurs with quartz (Fig. 8) and rarely with graphite in thin lamellae, as layered aggregates or clots defining bedding and, in the carbonaceous metapelites, defining axial plane cleavage. The tourmalinites contain 20–50% tourmaline although some laminae contain > 80% tourmaline. Tourmaline-bearing rocks with < 20% tourmaline grade into the enclosing metasediments. Tourmaline is greenish, zoned from a deeper green core to a lighter green rim, and rarely contains inclusions of quartz, graphite, and chlorite.

In the Rum Jungle area, tourmalinites and quartz-tourmaline rocks contain 10–85% tourmaline. Tourmaline is greenish, optically zoned from a darker core to a lighter green rim, contains inclusions of quartz and phyllosilicates, especially in the core, and in many places is fractured or oriented parallel to schistosity.

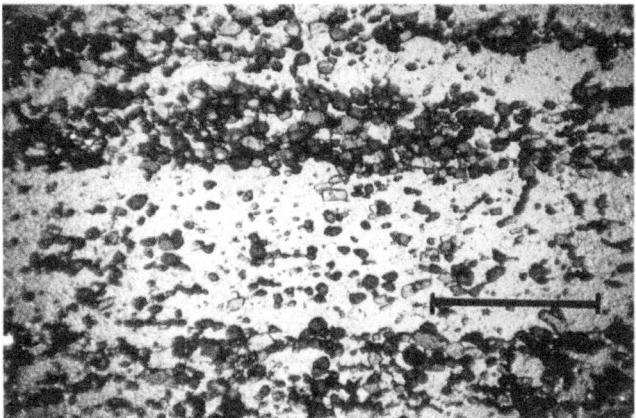

Fig. 8. Tourmaline laminae defining bedding in quartz-tourmaline rock, Golden Dyke, N.T. Bar = 0.5 mm

The quartz-tourmaline rocks of the Eastern Creek Volcanics in the Mt. Isa Orogen contain subhedral granoblastic unzoned green tourmaline grains of 0.5–2 mm grain size aggregated in laminae which define bedding. Tourmaline is very rarely zoned from a bluish green core to a deeper green rim, contains numerous inclusions of quartz (especially in the core) and is associated with coarse-grained muscovite and quartz. The enclosing tourmaline-bearing pelites contain chlorite, biotite, quartz, sericitized feldspars, tourmaline, and sericite-chlorite pseudomorphing cordierite or andalusite.

At Anitra and Pegmont, tourmalinite contains euhedral color zoned granoblastic poikiloblastic lamellar twinned tourmaline with interstitial quartz, magnetite, feldspars and muscovite. Tourmaline is pleochroic from pink to blue-brown-green and is prominently zoned from a bluish core to a brown outer zone with a greenish brown rim (Fig. 9). Quartz-tourmaline rocks from Anitra and Pegmont contain alternating layers of coarse-grained or fine-grained granoblastic quartz with minor quantities of interstitial tourmaline, thinner layers of fine-grained poikioblastic tourmaline with interstitial quartz and layers with granoblastic quartz and interstitial greenish-brown tourmaline.

Tourmalinites form the Olary Block contain brown poikioblastic granoblastic tourmaline (Fig. 10) associated with quartz, albite, muscovite, apatite, chlorite, and biotite. The core of tourmaline grains is generally slightly more bluish than the rim and the core generally contains inclusions of quartz and albite.

In the granulite facies Broken Hill Block, tourmalinites contain very fine grained tourmaline (0.02–1 mm) associated with quartz, apatite, muscovite, K-feldspar, biotite and spessartine (Fig. 11). Tourmaline from tourmalinite is brown to greenish-brown and is rarely color zoned from a blue core to a green-brown rim. Bedding is defined by changes in the modal content of the rock and, associated with the Ettlewood Calc-Silicate Member (Fig. 5), laminae less than 0.5 mm in thickness of brownish-green tourmaline are coplanar with other thin laminae of blue tour-

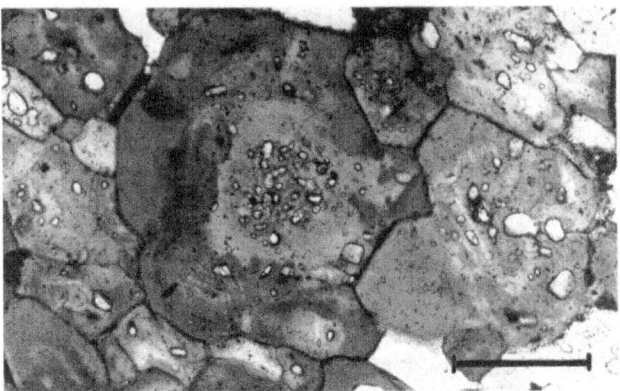

Fig. 9. Photomicrographic showing tourmalinite comprising zoned poikioblastic tourmaline euhedra with quartz inclusions and interstitial quartz, Anita, Mt. Isa Orogen, Qld. Bar = 0.5 mm, plane polarized light

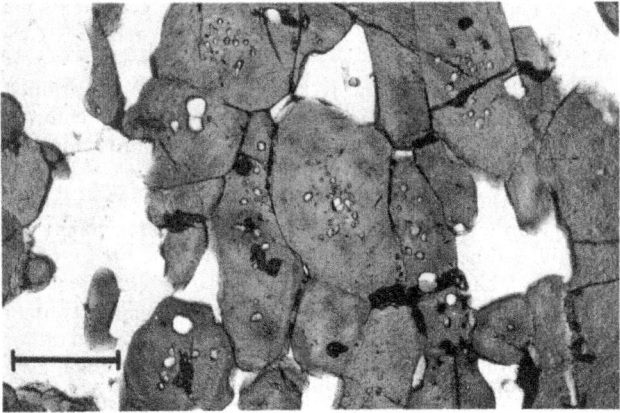

Fig. 10. Photomicrograph of tourmalinite from the Olary Block showing poikioblastic tourmaline with quartz and albite inclusions and quartz. Bar = 0.5 mm, plane polarized light

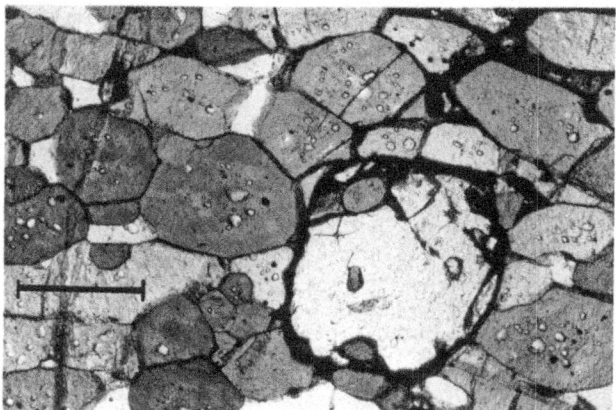

Fig. 11. Photomicrograph of granoblastic tourmalinite from the Globe Mine, Broken Hill showing poikioblastic tourmalinite with quartz inclusions, spessartine and quartz. Bar = 0.5 mm, plane polarized light

maline and brown tourmaline with a blue core. Poikioblastic granoblastic tourmaline contains inclusions of quartz, muscovite, feldspars, and garnet and, in metapelites, tourmaline is more commonly present in sillimanite and muscovite-rich foliae; sillimanite foliae wrap around tourmaline grains and the edge of brown tourmaline grains in contact with biotite are blue.

Geochemistry of Tourmalinite

Rock Chemistry

The tourmaline-rich rocks are characterized by a high SiO_2 content except where the rock is rich in sulfides, a small but detectable TiO_2 content, an Al_2O_3 content of 6.6 to 17%, very variable Fe contents (depending upon the amount of sulfides, chlorite, biotite, and tourmaline present), a low content of alkalis present with tourmaline being the principal alkali-bearing mineral, and a variable content of S, As, Au, base metals, rare earth elements, chlorine, and fluorine (Table 2). The analyzed tourmaline-bearing rocks are very variable in composition (especially the minor elements); however, tourmalinites and quartz-tourmaline rocks are considered to be siliceous sediments analagous to boron-bearing siliceous iron formations.

Mineral Chemistry

Tourmaline was analyzed by a JEOL JXA840 SEM-electron microprobe and crosschecked with a wavelength dispersive JEOL JX5A electron microprobe using synthetic and natural mineral standards. Analyses of iron minerals containing light elements (boron, lithium, fluorine, oxygen, hydrogen) and complex coupled vari-

Table 2. Chemical composition of tourmalinites and tourmaline-rich rocks from Lower-Middle Proterozoic areas, Australia (in wt.%)

	Golden Dyke, Northern Territory			Mt. Isa Orogen		Willyama Supergroup		
	1	2	3	4	5	6	7	8
SiO_2	75.43	71.76	46.13	81.02	68.59	67.38	72.16	85.56
TiO_2	0.48	0.86	0.20	0.29	1.23	0.69	0.46	0.37
Al_2O_3	12.72	12.00	8.43	9.84	16.53	16.95	14.77	6.57
B_2O_3	1.57	1.86	1.14	1.58	1.97	1.51	2.14	0.69
Cr_2O_3	0.02	0.03	0.04	0.02	0.02	0.04	0.03	0.05
Fe_2O_3	0.74	0.54	2.00	1.76	1.23	1.69	0.76	0.17
FeO	3.19	2.98	25.35	1.58	5.41	1.49	3.62	1.05
NiO	0.00	0.00	0.04	0.00	0.00	0.02	0.01	–
MnO	0.02	0.02	0.01	0.02	0.03	0.08	0.07	0.04
MgO	3.30	1.90	1.03	2.11	1.91	1.66	1.44	0.25
CaO	0.27	0.33	0.17	0.27	0.54	1.08	0.39	0.72
Na_2O	0.71	0.52	0.33	0.66	0.53	1.84	0.95	0.29
K_2O	0.10	0.58	0.07	0.13	0.36	4.76	0.99	0.70
P_2O_5	0.00	0.00	0.02	0.02	0.00	0.00	0.20	1.47
H_2O	1.48	2.07	4.32	1.33	2.91	1.31	1.97	0.55
TOTAL	100.04	95.20[a]	89.15[b]	100.59	101.26	100.49	99.96	99.29[c]

1. Tourmalinite, Golden Dyke Dome (average of 5 analyses from Plimer 1986)
2. Quartz-tourmaline-pyrite rock, Cosmopolitan Howley Mine, Golden Dyke area (from Plimer 1986)
3. Pyrite-arsenopyrite-tourmaline rock, Heatleys Prospect, Golden Dyke Dome (from Plimer 1986)
4. Quartz-tourmaline rock, Eastern Creek Volcanics, Mt. Isa (average of 3 analyses)
5. Quartz-tourmaline rock, Anitra Prospect, Mary Kathleen Fold Belt
6. Tourmaline-K-feldspar-albite rock, Alconnie Hill, Olary Block
7. Tourmalinite, Mt. Howden, Olary Block
8. Quartz-tourmaline-muscovite rock, Hores Mine, Broken Hill Block

[a]Excludes S,
[b]Excludes S and As,
[c]Excludes F, includes 0.81% Ce_2O_3, La_2O_3, ZnO

able valency cationic substitutions by energy-dispersive electron microprobe methods are difficult. The large beam diameter used (to avoid volatilization of light elements), assumption of the B_2O_3 substitution factor, assumption of 3 atoms of B per formula unit, ignoring of Li_2O (unlikely in the analyzed schorl-dravite) and assumption that all iron is Fe^{2+} (because the Fe^{3+} content of tourmalines is generally low and analyzed tourmaline rocks have a low ferric content) renders the tourmaline electron microprobe data not nearly as accurate as that for other common minerals (Table 3). Data were reduced using a modified ZAF data reduction program, results were internally consistent and data herein are analyzed using oxide ratios in the W and X sites to minimize analytical and structural formulae calculations (made on the basis of 31 (O, OH)).

The tourmaline compositions plot in the schorl-dravite series (Fig. 12). At Golden Dyke, tourmaline is zoned with cores containing higher Na and Fe and lower Ca and Mg contents than the rims. There is no systematic change in tourmaline chemistry from the Phyllite Member to the Kapalga Formation (Fig. 1), although the wide range of compositions indicates that tourmaline compositions reflect the bulk

Table 3. Representative wavelength and energy-dispersive electron microprobe data of tourmaline from tourmalinites (in wt.%)

	1	2	3	4	5	6	7	8
SiO$_2$	36.07	37.40	35.07	35.71	35.88	36.37	38.59	31.87
TiO$_2$	0.68	0.29	0.29	0.29	0.25	0.93	0.95	1.80
Al$_2$O$_3$	32.78	33.08	31.41	32.02	31.42	32.47	32.23	31.73
FeO	7.81	6.33	12.35	13.22	8.91	6.70	11.18	11.47
MnO	0.00	0.00	0.27	0.13	0.07	0.00	0.05	0.04
MgO	6.00	8.07	4.65	3.46	6.64	6.71	2.37	3.01
CaO	0.90	0.46	0.21	0.12	1.02	0.27	1.36	1.31
Na$_2$O	1.78	2.34	2.50	2.33	1.90	2.43	0.22	1.05
K$_2$O	0.02	0.01	0.04	0.07	0.07	0.02	0.07	0.13
	86.36	88.08	86.63	87.36	86.34	85.79	87.57	85.36

1. Golden Dyke Dome, N.T. (includes 0.11 Cr$_2$O$_3$, 0.04% NiO, 0.10% BaO)
2. Eastern Creek Volcanics, Qld. (includes 0.01% Cr$_2$O$_3$, 0.07% NiO, 0.02% BaO)
3. Pegmont, Qld. (includes 0.02% BaO)
4. Anitra, Qld.
5. Olary Block, S.A. (includes 0.05% Cr$_2$O$_3$, 0.12% NiO)
6. Thackaringa Group, N.S.W.
7. Broken Hill Group, N.S.W. (includes 0.52% F, 0.02% Cr, 0.01% P$_2$O$_5$)
8. Sundown Group, N.S.W.

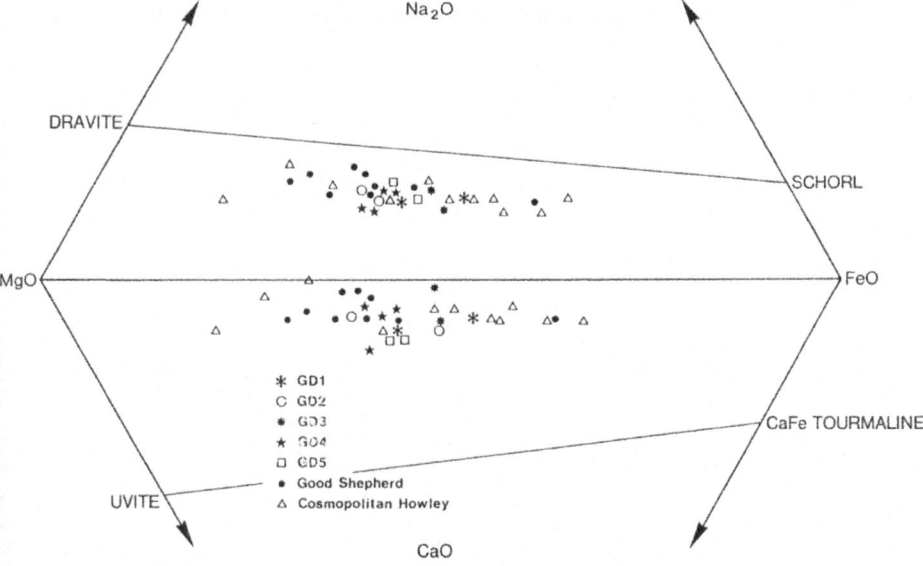

Fig. 12. Na$_2$O-MgO-FeO-CaO plot of tourmaline composition from tourmalinites and tourmaline-bearing iron formations of the Golden Dyke Dome, (GD1-GD5) and major mines, N.T.

rock chemistry. This is supported by the presence of the most Fe-rich tourmalines in iron formations, whereas the most Mg-rich tourmalines are in chlorite-albite bearing pelitic metasediments.

Tourmaline from Rum Jungle varies in composition from dravite$_{93-18}$ to uvite$_{2-10}$ with dravite≫schorl. The Ti, Fe, and Ca content decrease and the Mg content increases from core to rim of the grain (Bone pers. commun. 1986). In contrast to tourmalines from the Golden Dyke dome, those from Rum Jungle are Mg-rich, which reflects the presence of abundant Mg-rich rocks (e.g., magnesite) which formed in a shallow alkaline lacustrine environment in a continental rift.

In the Mount Isa Orogen, tourmaline compositions again reflect the bulk rock chemistry. Tourmaline from the basaltic Eastern Creek Volcanics is far richer in Mg than that associated with iron formations from the Mary Kathleen Fold Belt, although both plot in the schorl-dravite solid solution series (Figs. 13 and 14). Tourmaline from the Eastern Creek Volcanics is zoned from a core slightly depleted in Fe and Ca to a rim slightly enriched in Mg and Na, although reverse chemical zoning is common.

The general paucity of Mg-rich rocks is the Willyama Supergroup is also reflected in the tourmaline composition. Tourmaline from the Olary Block plots on the schorl-dravite solid solution series (Fig. 15) with most analyses towards the schorl apex. Minor quantities of Cr are commonly recorded from tourmaline from the Olary Block. Zoning is from a Fe- and Ca-rich core to a Mg- and Na-rich rim. At the Dome Rock mine, tourmaline from the footwall metasediments and banded albite rocks are more magnesian (dravitic) and more calcic than those from the hanging

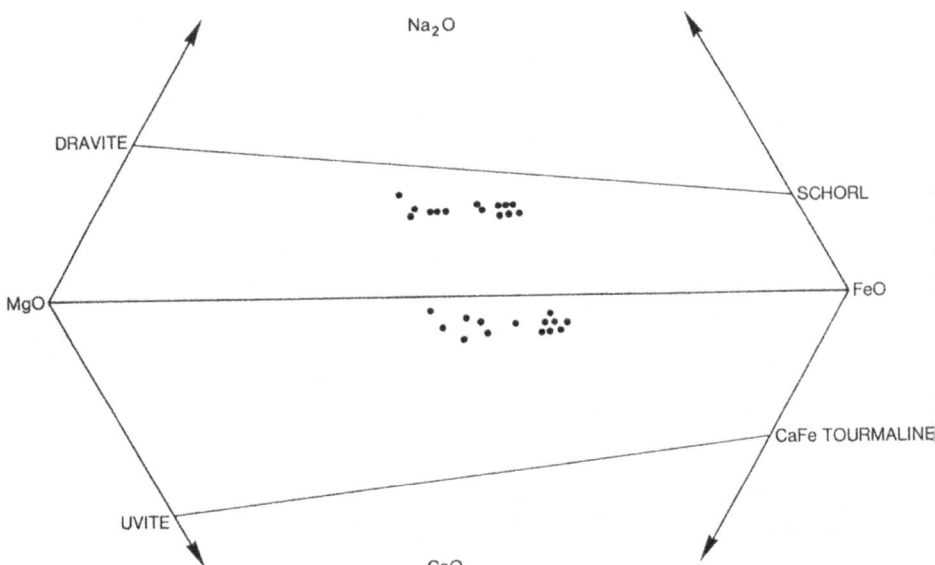

Fig. 13. Na$_2$O-MgO-FeO-CaO plot of tourmaline composition from quartz-tourmaline rocks of the Eastern Creek Volcanics, Mt. Isa Orogen, Qld.

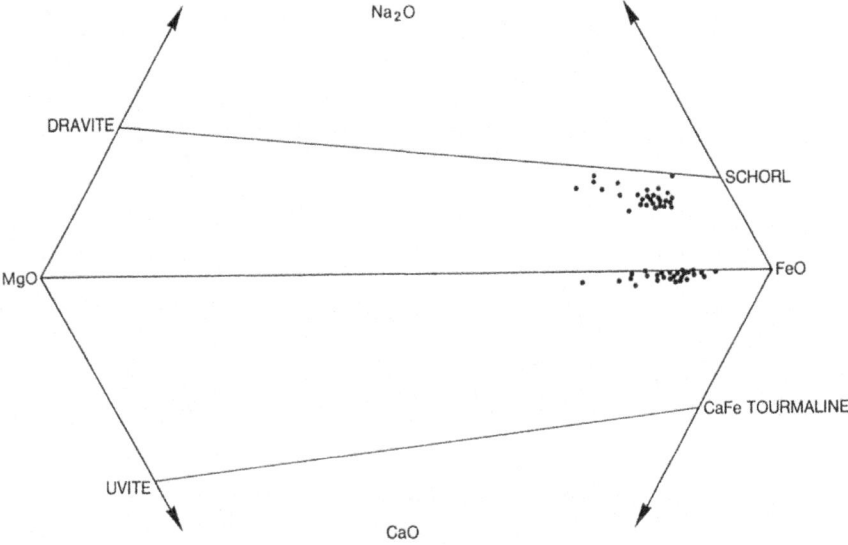

Fig. 14. Na₂O-MgO-FeO-CaO plot of tourmaline from tourmalinites at Pegmont and Anita, Mt. Isa Orogen. Qld.

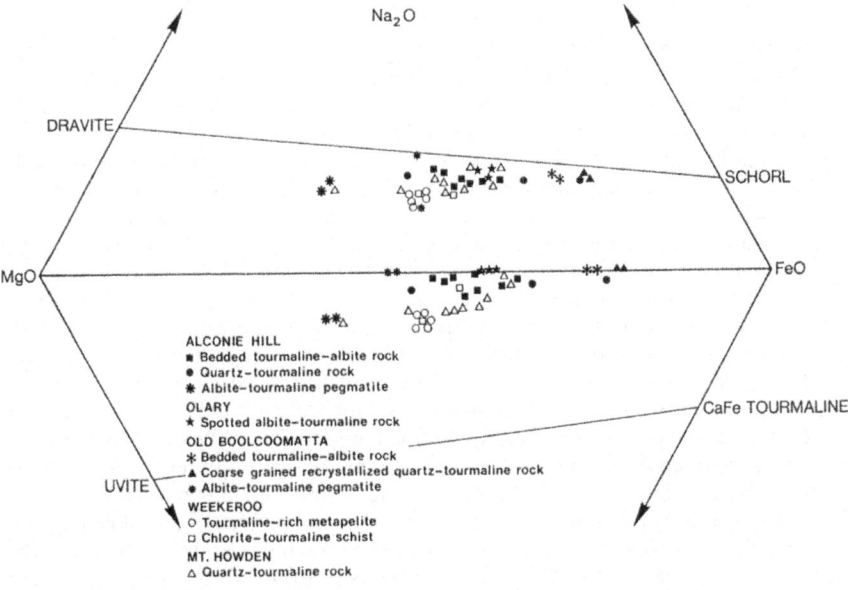

Fig. 15. Na₂O-MgO-FeO-CaO plot of tourmaline from tourmalinites of the Thackaringa Group equivalent rocks, Willyama Supergroup, Olary Block, S.A.

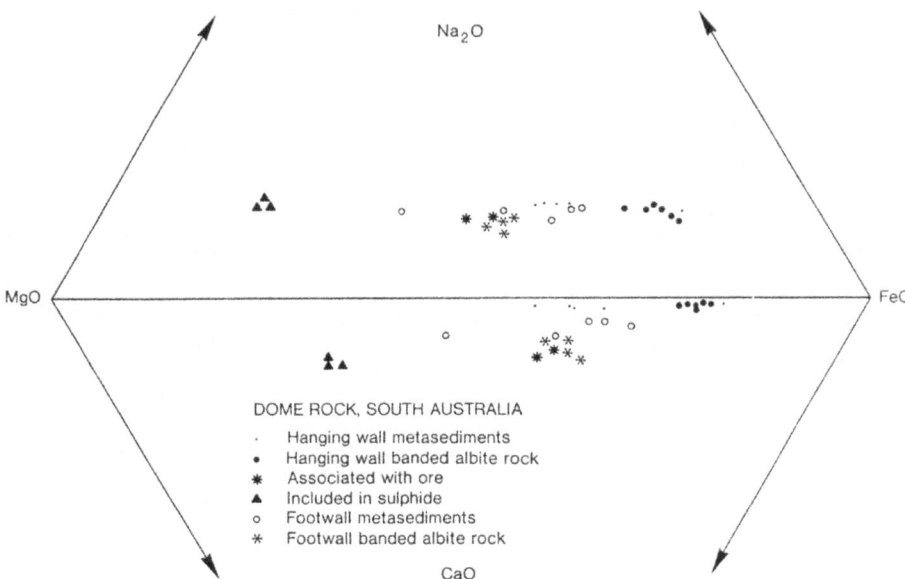

Fig. 16. Na$_2$O-MgO-FeO-CaO plot of tourmaline from metasediments of the Dome Rock Mine. Thackaringa Group equivalent, Olary Rock, S.A.

wall metasediments and banded albite rocks (Fig. 16). Because the ore zone contains abundant iron minerals such as pyrite, pyrrhotite, chalcopyrite, and biotite and tourmaline are the only Mg-bearing phases, tourmaline from the ore zone is dravitic with the most magnesian tourmaline being inclusions in sulfides. Furthermore, tourmaline from the ore zone contains Cr and Ti contents considerably higher than that from wall rock tourmaline. Tourmaline with the highest Na and lowest Ca contents occur in the hanging wall albite rocks. The change in chemistry from footwall to hanging wall from magnesian tourmaline (schorl>dravite) to ferroan tourmaline (schorl>dravite) is in accord with the petrographic data.

The detailed chemistry of tourmalines from the Broken Hill Block has been reported elsewhere (Plimer 1983) and new analyses have confirmed original data. Tourmaline plots in the schorl-dravite solid solution series although a very variable Fe and Mg content was recorded. The most Mg-rich tourmalines occur in the Thackaringa Group: the lowermost tourmalinite-bearing section of the stratigraphy (Fig. 5). Tourmalinite is commonly within banded albite rocks and, in both prograde and retrograde units, the tourmaline contains the maximum permissable Na content, thereby demonstrating that tourmaline composition is in response to bulk rock composition. The composition of tourmalines from the overlying Broken Hill Group (Fig. 5) is more Fe-rich than those from the Thackaringa Group and is characterized by the presence of >1 wt.% F, reflecting the association of F-bearing phases such as F-apatite, fluorite, F-vesuvianite and spessartine. Tourmaline from the Broken Hill Group plots partially in the field of granite field (Fig. 17). There is no compositional difference between tourmaline from the top and bottom of graded beds and tourmaline from the associated tourmalinites. Tourmaline from the Sundown

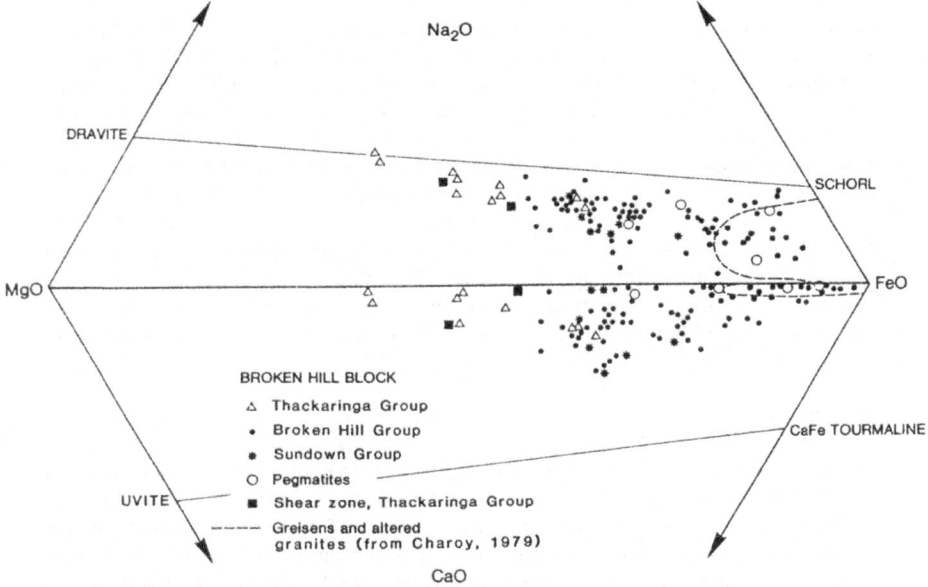

Fig. 17. Na₂O-MgO-FeO-CaO plot of tourmaline from tourmalinites in three major stratigraphic units of the Willyama Supergroup, Broken Hill Block, N.S.W.

Group (Fig. 5) is compositionally very similar to that from the Broken Hill Group (Fig. 17).

Most tourmalines from the Broken Hill Block show compositional zoning. In the Thackaringa Group, the Na, Mg, Ca, F, and Cl content decrease and Fe and Ti increase from the cores to the rims of grains although rare unzoned grains were observed. Tourmaline from the Broken Hill Group is strongly zoned with an increase in Ca, Fe, F, and Ti and decrease in Mg, Na, and Cl from core to rim. Rare reverse zoning was observed and, in some sections, layers of reversely zoned grains are adjacent and coplanar with layers of normally zoned grains. Very weak zoning is present in tourmaline from the Sundown Group with decreases in Na and Mg and increases in Fe, Ca, and Mg from core to rim. In general, the cores of tourmaline from the Broken Hill Block are more dravitic, whereas the rim is richer in the schorl and uvite end members.

Discussion

Timing of Formation of Tourmaline-Rich Rocks

Tourmaline-rich rocks of Lower-Middle Proterozoic age described herein occur stratigraphically below (e.g., Broken Hill stratiform scheelite, Dome Rock Cu-Co), within (e.g., Golden Dyke Au), stratigraphically equivalent to (e.g., Golden Dyke, Broken Hill Pb-Zn) and above (e.g., Pegmont) submarine exhalative deposits of Pb-Zn-Ag, Au, W, Sn, Cu, and Cu-Co. Furthermore, these tourmaline-rich rocks need not necessarily be associated with ore deposits (e.g., Olary Block, Rum Jungle). The tourmaline-rich rocks display bedding, graded bedding, crossbedding, intraformational slumps, and pull-apart structures recognizable in greenschist (Golden Dyke, Rum Jungle), amphibolite (Mt. Isa Orogen, Olary Block) and granulite facies terrains (Broken Hill Block). The stratigraphic continuity, stratigraphically associated submarine exhalative mineralization, sedimentary structures, general lack of associated granitic or metasomatized rocks, metamorphic textures, and deformation structures all indicate that the tourmaline-rich rocks enjoyed the same metamorphism and deformation as the enclosing rocks and are an integral part of the stratigraphic package. Therefore, the tourmaline-rich rocks formed as a primary sedimentary synsedimentary unit and are interpreted as a boron-rich exhalite intimately associated with submarine hot spring precipitation.

Source of Boron

Present-day terrestrial hot springs can contain an appreciable boron content (e.g., up to 1020 ppm, Henley et al. 1984), black smokers of the East Pacific Rise at 13° and 21°N contain 4.5 ppm B, whereas the black smokers of the Guaymas Basin contain 13–24 ppm B (Spivack pers. commun. 1985). If all submarine exhalative ores formed from an ore fluid which was predominantly modified seawater, then boron-rich exhalites should be associated with all types of submarine exhalative ore deposits. This is clearly not the case.

Boron-rich hot springs are not associated with the cooling of pristine oceanic crust (e.g., East Pacific Rise) whereas minor quantities of boron have been added from circulating seawater into the uppermost 300 m of Recent seafloor basalts and the basalts of the Troodos Complex (Habermehl and Erzinger 1986). Clearly boron has been extracted from the circulating seawater into the oxidative alteration facies of the seafloor basalt above the zone of reduction and metal leaching. Therefore, boron-rich exhalites are not expected to be associated with Cyprus-type deposits or umbers; however, if the circulating seawater passes through both seafloor basalt and overlying argillaceous sediments before debouchment as smokers (e.g., Guaymas Basin), then large quantities of hydrothermal boron can be derived from the leaching of clays. Clays can contain up to 2000 ppm boron substitution in the tetrahedral site and 1000 ppm adsorbed boron (Harder 1959; Stubican and Roy 1962). This is commensurate with boron-rich exhalites associated with back-arc basin type deposits (e.g., tourmaline-bearing iron formations associated with Norwegian Caledonide deposits; Sazare Besshi-type deposits of Japan).

Similarly, circulating seawater in acid volcanic piles in island arcs encounters small volumes of argillaceous sediments and hence ore fluids would have little boron derived from leaching. This is in accord with the general paucity of tourmaline and tourmalinites associated with submarine exhalative ores in acid volcanic settings (e.g., Iberian Pyrite Belt, Kuroko deposits of Japan) except where large quantities of argillaceous sediments or evaporites are within the sequence or the basement (e.g., Elizabeth, U.S.A.; Rosebery, Australia).

If heated seawater circulates through a thick argillaceous sediment pile (e.g., Golden Dyke) or a thick argillaceous sediment-evaporite pile (e.g., Mt. Isa, Rum

Jungle), then large quantities of boron are available. This modified and possibly supersaline seawater can therefore have a greatly enhanced boron content which is commensurate with the observation that boron-rich exhalites are most commonly associated with sediment-hosted submarine exhalative ores (Slack 1982).

Precursor to Tourmaline-Rich Exhalites

The most likely precursors to tourmalinite and quartz-tourmaline rocks are evaporites containing borates and/or borosilicates, a sediment containing silica and boron-rich clay, or a silica-tourmaline sediment.

The paleogeographic setting of the tourmalinies of the Golden Dyke area, Mt. Isa Orogen, and Willyama Supergroup preclude derivation of tourmalinite or quartz-tourmaline rocks by metamorphism of an evaporite. However, evaporites within the Lower-Middle Proterozoic sequences of Northern Territory (e.g., Rum Jungle) and northern Queensland indicate that circulating seawater could have come into contact with evaporites or have derived boron-rich fluids by dewatering. It is expected that during the onset of metamorphism, the highly soluble borate minerals characteristic of evaporites would be quickly removed during the initial dewatering.

If clays can contain a maximum of 2000 ppm adsorbed and substituted boron, then metamorphism of a clay rock could only produce an Al silicate/feldspar-quartz-tourmaline rock with 4 vol.% tourmaline. This may explain the tourmaline-bearing felsic rocks of the Thackaringa Group in the Broken Hill Block; however, it is clear that tourmalinites with 20–80 vol.% tourmaline, quartz-tourmaline rocks, and tourmaline-rich sediments with more than 4 vol.% tourmaline could not have formed by this mechanism.

Because tourmaline has such a large stability range (Werding and Schreyer 1984) and can form as overgrowths in sediments at very low temperatures and pressures, it is suggested that the precursor to tourmaline in tourmalinite was tourmaline. This is strongly supported by the sedimentary structures and the presence of extremely thin laminae of tourmaline (<0.5 mm) which are zoned and compositionally different from adjacent coplanar laminae.

Facies Relationships to Mineralization

The uptake of boron in the uppermost more oxidized portion of altered Recent seafloor basalts and at Troodos (Habermehl and Erzinger 1986) indicates that boron fixation takes place at somewhat lower temperature under more oxidized conditions. However, the solubility and speciation of boron complexes at high temperature are unknown.

Field relationships show that tourmaline is rarely found within footwall alteration zones or exhalative ore and more commonly occurs laterally equivalent to or stratigraphically above exhalative sulfide mineralization, suggesting that precipitation was in somewhat lower temperature, higher pH and more oxidizing conditions. This is supported by the occurrence of tourmalinite and quartz-tourmaline rocks over very large areas which are unrelated to economic sulfide mineralization and the elevated boron (250–800 ppm) content of seafloor hydrothermal precipitates (Habermehl and Erzinger 1986). It is suggested that tourmalinites and quartz-tourmaline rocks are considered as a boron-rich cherty iron formation which commonly occurs in response to a slight increase in geothermal gradient, commensurate with its most common occurrence as a distal exhalite facies.

Implications

Tourmalinites and tourmaline-rich exhalites are far more common than previously thought and can be regarded as a boron-rich siliceous iron formation. They have formed in response to an elevated geothermal gradient and leaching principally of areas dominated by pelitic and evaporitic sediments. Although tourmalinites can be the host for or are intimately associated with exhalative mineralization, they need not be associated with mineralization. Most commonly, the tourmalinites in Australia and elsewhere are in Lower-Middle Proterozoic terrains and are immediately above or stratigraphically equivalent to exhalative mineralization. However, there are a number of deposits hosted by thick sequences of pelitic sediments and evaporites (e.g., Mt. Isa) which have no associated tourmalinites, but it is not known whether the ore-associated shales are enriched in boron. Although no rigid facies relationship between tourmaline-rich rocks and mineralization can be established, their presence indicates that exhalative processes took place, thereby raising the exploration interest in an area. Furthermore, they can be the host for invisible mineralization such as scheelite, gold, and rare earth elements.

Although tourmaline is present in some siliceous iron formations, it is most commonly absent from submarine hydrothermal iron formations (e.g., Lahn-Dill ores). Clearly these deposits result from the convective fluid circulation associated with mafic volcanism (Quade 1970); however, the lack of tourmaline may be related to the higher f_{o2} conditions of a saline fluid which would allow only nonspecific complexing.

The complexing and speciation of boron at elevated temperatures in unknown. However, the uncanny association of tourmalinite with calc-silicate, altered amphibolite, and tourmalinite-hosted stratiform scheelite deposits of the Willyama Supergroup, the Namaqua Metamorphic Sequence of southern Africa, the Bergslagen Supergroup of Fennoscandia, the Archean Malene Supracrustal rocks of Greenland, and the Bohemian Massif very strongly suggests that boron was not only a component of the ore fluid but may well have transported tungsten. Similar arguments can be used for the selective transport of gold (e.g., Golden Dyke, Passagem de Mariana) however, the abundance of sulfides and arsenopyrite indicates that transport may more probably have been by thio or arseno complexes.

The presence of tourmaline associated with exhalative deposits of Sn and W (Plimer 1980) and the well known presence of tourmaline associated with granite-associated Sn-W deposits indicates that B is commonly a component of ore fluids which form Sn and W deposits. In some areas, it can be demonstrated that plutons associated with Sn-W deposits have tourmaline alteration where they intrude pelitic rocks, and sericitic alteration where they intrude acid volcanics thereby suggesting that some components of alteration derived from the intruded rocks (Plimer and Kleeman 1985). Similarly, anatexis of thick pelitic metasediment sequences of Lower-Middle Proterozoic sources containing tourmalinites may be able to produce granitic rocks with elevated tourmaline contents.

Conclusions

Tourmalinites are present in Lower-Middle Proterozoic terrains of Australia associated with exhalative mineralization although in some places there is no known association with mineralization. They occur in greenschist, amphibolite, and granulite facies metamorphic terrains and have undergone minor to intense deformation. Minor remobilization of quartz and tourmaline has occurred during metamorphism and deformation. Tourmaline-rich rocks associated with mineralization occur in footwall alteration zones, ore zones, strike-equivalent exhalites, hanging wall alteration zones and overlying exhalites. Chemical analyses of tourmaline-rich rocks show that they are boron-rich siliceous exhalites which contain tourmaline of the schorl-dravite solid solution series. The composition of the tourmaline reflects the bulk rock composition and cannot be used as a facies guide to mineralization. Tourmaline is commonly zoned from a more dravitic core to a schorl-rich rim.

It is suggested that tourmalinites were precipitated as a primary silica-tourmaline chemical sediment from exhalative fluids which passed though a thick column of clay-rich sediments and/or evaporites.

Acknowledgments. The assistance in the field by Esso Australia Ltd., The Geological Survey of N.S.W., Geopeko Ltd., Mt. Isa Mines Ltd., North Broken Hill Ltd., Shell Company of Australia, and the University of New England is gratefully acknowledged. The author has benefited from discussions with P. Appel, P. Ashley, R. Barnes, R. Beeson, Y. Bone, J. Erzinger, J. Lawton, T. Lees, P. Nicholson, R. Newport, D. Patterson and J. Slack .

References

Ahmed R, Wilson CJL (1981) Uranium and boron distributions related to metamorphic fluid activity. Contrib Miner Pet 76:24–32
Allman-Ward P, Halls C, Rankin A, Bristow CM (1982) An intrusive hydrothermal breccia body at Wheal Remfry in the western part of the St. Austell granite pluton, Cornwall, England. In: Evans A M (ed) Metallization associated with acid magmatism. Wiley, New York, pp 1–28
Appel PWU (1983) Tourmaline in the Early Archean Isua supracrustal belt, West Greenland. J Geol 92:599–605
Appel PWU (1985) Strata-bound tourmaline in the Archean Malene supracrustals, West Greenland. Can J Earth Sci 22:1485–1491
Ashley PM (1984) Piemontite-bearing rocks from the Olary district, South Australia. Aust J Earth Sci 31:203–216
Augustithis SS (1967) On the texture and paragenesis of the gold-quartz-tourmaline veins of Oudonoc, W. Ethiopia. Miner Dep 3:48–55
Barnes RG (1983) Stratiform and stratabound tungsten mineralization in the Broken Hill Block, New South Wales. J Geol Soc Aust 30:225–235

Behr H-J, Ahrendt H, Martin H, Porada H, Rohrs J, Weber K (1983) Sedimentology and mineralogy of Upper Proterozoic playa lake deposits in the Damara Orogen. In: Martin H, Eder F W (eds) Intra continental fold belts. Springer, Berlin Heidelberg New York, pp 577–610

Benard F, Monton P, Pichavant M (1985) Phase relations of tourmaline leucogranites and the significance of tourmaline in silicic magmas. J Geol 93:271–291

Carstens CW (1927) Über das Auftreten von Tourmalin in Norwegischen Kiesvorkommen. Nor Geol Tidsskr 9:331–336

Charoy B (1979) Définition et importance de phenomènes dentériques et des fluides associés dans les granites. Conséquences metallogéniques. Sci Terre Mem 37:305–314

Chorlton LB, Martin R F (1978) The effect of boron on the granite solidus. Can Miner 16:239–244

Clarke GW, Paterson R G, Taylor R G (1986) The nature and origin of brecciation and mineralization at the White Crystal ore deposit, Ardlethan tin mine, New South Wales. Aust J Earth Sci 32:343–349

Cunningham WB, Holl R, Taupitz K C (1973) Two new tungsten-bearing horizons in the older Precambrian of Rhodesia. Miner Dep 8:200–203

Ethier VG, Campbell FA (1977) Tourmaline concentrations in Proterozoic sediments of the southern Cordillera of Canada and their economic significance. Can J Earth Sci 14:2348–2363

Exley CS (1959) Magmatic differentiation and alteration in the St. Ausell granite. J Geol Soc (Lond) 114:197–230

Fleischer R, Routhier P (1973) The "consanguinous" origin of a tourmaline-bearing gold deposit: Passagem de Mariana (Brazil). Econ Geol 68:11–22

Glen RA, Laing WP, Parker AJ, Rutland RWR (1977) Tectonic relationships between the Proterozoic Gawler and Willyama orogenic domains, Australia. J Geol Soc Aust 24:125–150

Grant JN, Halls C, Sheppard SMF, Avila W (1980) Evolution of porphyry tin deposits of Bolivia. Soc Mining Geol Jpn, Spec Issue 8:151–173

Gulson BL (1984) Uranium-lead and lead-lead investigation of minerals from the Broken Hill lodes and mine sequence rocks and their implications for ore genesis. CSIRO Res Rev 1983:21–23

Habermehl K, Erzinger J (1986) Zur Geochemie von Bor in ozeanischer Umgebung. Abstr 76 Jahrest Geol Verein

Harder H (1959) Beitrag zur Geochemie des Bors, III Bor in Sedimenten. Nachrichten Akad Wissensch Gottingen II, Math-Phys Klasse 6:123–175

Henley RW, Barton PB, Truesdell AH, Whitney JA (1984) Fluid-mineral equilibria in hydrothermal systems. Rev Econ Geol 1:15

Henry DJ, Guidotti CV (1985) Tourmaline as a petrogenetic indicator mineral: an example from the staurolite-grade metapelites of NW Maine. Am Miner 70:1–15

Howard PF (1959) The geology of the Elizabeth Mine, Vermont. Ver Geol Surv 5

Joubert P, Moore JM (1986) Mineralogical anomalies in the Namaqualand Metamorphic Complex. Rpt Nat Geosci Prog, Pretoria

Kents P (1964) Special breccias associated with hydrothermal developments in the Andes. Econ Geol 59:1551–1613

Manning DAC, Pichevant M (1983) The role of flourine and boron in the generation of granitic melts. In: Atherton MP, Gribble CD (eds) Migmatites, melting and metamorphism. Shiva, pp 94–109

Mendelsohn F (1957) The structure and metamorphism of the Roan Antelope deposit. Thesis, University of the Witwatersrand, Johannesburg

Milton C (1971) Authigenic minerals of the Green River Formation. Wyo Univ Contrib Geol 10:57–63

Nicholson PM (1980) The geology and economic significance of the Golden Dyke Dome, Northern Territory. In: Ferguson J, Goleby AB (eds) Uranium in the Pine Creek Geosyncline. IAEA Vienna, pp 319–334

Page RW, Compston W, Needham RS (1980) Geochronology and evolution of the late-Archean basement and Proterozoic rocks in the Alligator Rivers Uranium Field, Northern Territory, Australia. In: Ferguson J, Goleby AB (eds) Uranium in the Pine Creek Geosyncline. IAEA Vienna, pp 39–68

Parr JM, Rickard DT (1986) Early Proterozoic subaerial volcanism and its relationship to Broken Hill — type mineralization in Central Sweden. Geol Soc Lond IGCP 217 Keyworth 1986, p 12

Phillips GN, Groves DI (1983) The nature of Archean gold-bearing fluids as deduced from gold deposits of Western Australia. J Geol Soc Aust 30:25–39

Pidgeon RT (1967) A rubidium-stontium geochronological study of the Willyama Complex, Broken Hill, Australia. J Pet 8:283–324

Plimer IR (1980) Exhalative Sn and W deposits associated with mafic volcanism as precursors to Sn and W deposits associated with granites. Miner Dep 15:275–289

Plimer IR (1983) The association of tourmaline-bearing rocks with mineralization at Broken Hill, N.S.W. Proc Ann Aust Min Metall Conf Broken Hill, pp 157–176

Plimer IR (1985) Broken Hill Pb-Zn-Ag deposit — a product of mantle metasomatism. Miner Dep 20:147–153

Plimer IR (1986) Tourmalinites from the Golden Dyke Dome, Northern Australia. Miner Dep 21:263–270

Plimer IR, Kleeman JD (1985) Mineralization associated with the Mole Granite, Australia. In: Halls C High heat production (HHP) granites, hydrothermal circulation and ore genesis. Inst Mining Metall, London, pp 562–570

Plumb KA, Derrick GM, Wilson IH (1980) Precambrian geology of the McArthur River-Mount Isa region, northern Australia. In: Henderson RA, Stephenson PJ (eds) The geology and geophysics of northeastern Australia. Geol Soc Aust Brisbane, pp 71–88

Quade H (1970) Der Bildungsraum und die genetische Problematik der vulkansedimentaren Eisenerze. Clausth Hefte 9:27–65

Rouhunkoski P (1968) On the geology and geochemistry of the Vihanti zinc ore deposit, Finland. Comm Geol Fin Bull 236:15–77

Sillitoe RH, Sawkins FJ (1971) Geologic, mineralogic and fluid inclusion studies relating to the origin of copper-bearing tourmaline breccia pipes, Chile. Econ Geol 66:1028–1041

Slack JF (1982) Tourmaline in Appalachian-Caledonian massive sulphide deposits and its exploration significance. Trans Inst Miner Metall B91:B81–B89

Slack JF, Herriman N, Barnes RG, Plimer IR (1984) Stratiform tourmalinites in metamorphic terranes and their geologic significance. Geology (Boulde) 12:713–716

Smith GI (1979) Subsurface stratigraphy and geochemistry of Late Quaternary evaporites, Searles Lake, California. US Geol Surv Prof Pap 1043

Stillwell FL (1934) Observations on the lead-zinc lode at Rosebery, Tasmania. Proc Aust Inst Miner Metall 94:43–67

Stubican V, Roy R (1962) Boron substitution in synthetic micas and clays. Am Miner 47:166–173

Vaughan JP, Stanton RL (1984) Stratiform lead-zinc mineralization in the Kuridala Formation and Soldiers Cap Group, Mount Isa Block, NW Queensland. Proc Ann Inst Miner Metall Conf Darwin, pp 1–11

Werding G, Schreyer W (1984) Alkali-free tourmaline in the system $MgO-Al_2O_3-B_2O_3-SiO_2-H_2O$. Geochim Cosmochim Acta 45:1331–1344

Willis IL, Brown RE, Stroud WJ, Stevens BPJ (1983) The Early Proterozoic Willyama Supergroup: stratigraphic subdivision and interpretation of high to low grade metamorphic rocks in the Broken Hill Block, New South Wales. J Geol Soc Aust 30:95–124

Subject Index

Made in United States
Orlando, FL
22 March 2026

79555861R00168